Energy for Sustainable Future

Energy for Sustainable Future

Editors

T M Indra Mahlia
Islam Md Rizwanul Fattah

MDPI • Basel • Beijing • Wuhan • Barcelona • Belgrade • Manchester • Tokyo • Cluj • Tianjin

Editors
T M Indra Mahlia
University of Technology
Sydney
Australia

Islam Md Rizwanul Fattah
University of Technology
Sydney
Australia

Editorial Office
MDPI
St. Alban-Anlage 66
4052 Basel, Switzerland

This is a reprint of articles from the Special Issue published online in the open access journal *Energies* (ISSN 1996-1073) (available at: https://www.mdpi.com/journal/energies/special_issues/ energy_sustainable_future).

For citation purposes, cite each article independently as indicated on the article page online and as indicated below:

LastName, A.A.; LastName, B.B.; LastName, C.C. Article Title. *Journal Name* **Year**, *Volume Number*, Page Range.

ISBN 978-3-0365-2614-0 (Hbk)
ISBN 978-3-0365-2615-7 (PDF)

Contents

About the Editors

T M Indra Mahlia (Distinguished Professor)

T M Indra Mahlia is a Distinguished Professor in the School of Information, Systems, and Modelling at UTS, and a highly cited researcher in the field of engineering. Professor Mahlia is a core member of the University's Centre for Green Technology, which aims to develop novel technology for future clean fuels, vehicle emissions measurement, and environmental decontamination. His research interests include sustainable energy, renewable energy conversion and storage, forestry biomass and bioproducts, automotive combustion, and fuel engineering. He is involved as a researcher in a $5 million project funded by various sources. He was listed as a Highly Cited Researcher by Clarivate Analytics in 2017, 2018, 2019, 2020, and 2021, and a field leader in Sustainable Energy (Engineering & Computer Science) in a special report published by The Australian in 2019. He was appointed as a professor at the Department of Mechanical Engineering at the University of Malaya, Kuala Lumpur in 2009. His publications include more than 350 journal articles, proceedings, and research reports, of which more than 300 are in WOS and Scopus indexed. A polyglot, Professor Mahlia is fluent in and can peer review in English, Indonesian, Malay, and Achinese.

Islam Md Rizwanul Fattah (Dr)

Dr IMR Fattah is a Postdoctoral Research Fellow in the School of ISM, Faculty of Engineering and IT at the University of Technology Sydney (UTS), researching the effective use of waste for sustainable energy applications. He has accomplished his PhD in reducing PM and soot emissions from diesel combustion from the University of New South Wales (UNSW, Sydney) in 2019. Previously, he completed his Master of Engineering Science from the University of Malaya (UM) in 2014 and Bachelor of Science in Mechanical Engineering from Bangladesh University of Engineering and Technology (BUET) in 2011. He has been actively engaged in the field since 2012 by publishing over 70 articles and gaining over 5000 citations of his works. He serves as an Editorial Board Member at Energies (MDPI). He also has peer-reviewed over 250 WOS-indexed journals throughout his career.

Editorial

Energy for Sustainable Future

T. M. Indra Mahlia * and **I. M. Rizwanul Fattah ***

Centre for Green Technology, Faculty of Engineering and I.T., University of Technology Sydney, Ultimo, NSW 2007, Australia
* Correspondence: TMIndra.Mahlia@uts.edu.au (T.M.I.M.); IslamMdRizwanul.Fattah@uts.edu.au (I.M.R.F.)

Energy and the environment are interrelated, and they are critical factors that influence the development of societies. The pollution of the environment, without considering various consequences, has become one of the most important global issues today. This environmental pollution is mainly the result of increases in economic activities, population, transportation, electricity generation, agriculture, forestry, and land use. The exigency of energy for these activities, the rapidly rising price of petroleum oil, the harmful effect of greenhouse gases, and the quest for energy security have steered our attention towards sustainable sources of energy. It is fundamental to find innovative solutions that are sustainable from the perspective of energy management and environmental protection. These solutions will provide a promising future in terms of energy sources meeting energy demand, together with maintaining the environment.

This book includes three review articles, which review the state-of-the-art of different sustainable energy resources. These articles include ammonia as a renewable energy carrier [1], integration of solar photovoltaic [2], and bio-oil from waste tires for automotive engine applications [3]. In addition, eight research studies reveal new knowledge about energy for a sustainable future. The topics covered span many diverse areas associated with sustainable energy, including various biofuels [4,5], photovoltaic [6], and other aspects of sustainability [7–11]. These complementary contributions provide a substantial body of knowledge in the field of renewable and sustainable energy.

We would like to thank the academic and managing editors, as well as the reviewers, for their efforts and input. We enjoyed editing the papers for this collection.

Author Contributions: Writing—original draft preparation, I.M.R.F.; writing—review and editing, T.M.I.M.; supervision, T.M.I.M. All authors have read and agreed to the published version of the manuscript.

Funding: This research received no external funding.

Institutional Review Board Statement: Not applicable.

Informed Consent Statement: Not applicable.

Conflicts of Interest: The author declares no conflict of interest.

Citation: Mahlia, T.M.I.; Fattah, I.M.R. Energy for Sustainable Future. *Energies* **2021**, *14*, 7962. https://doi.org/10.3390/en14237962

Received: 22 November 2021
Accepted: 23 November 2021
Published: 29 November 2021

References

1. Hasan, M.H.; Mahlia, T.M.I.; Mofijur, M.; Rizwanul Fattah, I.M.; Handayani, F.; Ong, H.C.; Silitonga, A.S. A Comprehensive Review on the Recent Development of Ammonia as a Renewable Energy Carrier. *Energies* **2021**, *14*, 3732. [CrossRef]
2. Samoita, D.; Nzila, C.; Østergaard, P.A.; Remmen, A. Barriers and Solutions for Increasing the Integration of Solar Photovoltaic in Kenya's Electricity Mix. *Energies* **2020**, *13*, 5502. [CrossRef]
3. Jahirul, M.I.; Hossain, F.M.; Rasul, M.G.; Chowdhury, A.A. A Review on the Thermochemical Recycling of Waste Tyres to Oil for Automobile Engine Application. *Energies* **2021**, *14*, 3837. [CrossRef]
4. Pourzolfaghar, H.; Abnisa, F.; Wan Daud, W.M.A.; Aroua, M.K.; Mahlia, T.M.I. Catalyst Characteristics and Performance of Silica-Supported Zinc for Hydrodeoxygenation of Phenol. *Energies* **2020**, *13*, 2802. [CrossRef]
5. Ong, M.Y.; Nomanbhay, S.; Kusumo, F.; Raja Shahruzzaman, R.M.H.; Shamsuddin, A.H. Modeling and Optimization of Microwave-Based Bio-Jet Fuel from Coconut Oil: Investigation of Response Surface Methodology (RSM) and Artificial Neural Network Methodology (ANN). *Energies* **2021**, *14*, 295. [CrossRef]
6. Dumlao, S.M.G.; Ishihara, K.N. Weather-Driven Scenario Analysis for Decommissioning Coal Power Plants in High PV Penetration Grids. *Energies* **2021**, *14*, 2389. [CrossRef]
7. Hassan, O.; Morse, S.; Leach, M. The Energy Lock-In Effect of Solar Home Systems: A Case Study in Rural Nigeria. *Energies* **2020**, *13*, 6682. [CrossRef]
8. Jadim, R.; Kans, M.; Schulte, J.; Alhattab, M.; Alhendi, M.; Bushehry, A. On Approaching Relevant Cost-Effective Sustainable Maintenance of Mineral Oil-Filled Electrical Transformers. *Energies* **2021**, *14*, 3670. [CrossRef]
9. Jasiński, R.; Galant-Gołębiewska, M.; Nowak, M.; Ginter, M.; Kurzawska, P.; Kurtyka, K.; Maciejewska, M. Case Study of Pollution with Particulate Matter in Selected Locations of Polish Cities. *Energies* **2021**, *14*, 2529. [CrossRef]
10. Sofyan, S.E.; Hu, E.; Kotousov, A.; Riayatsyah, T.M.I.; Thaib, R. Mathematical Modelling and Operational Analysis of Combined Vertical–Horizontal Heat Exchanger for Shallow Geothermal Energy Application in Cooling Mode. *Energies* **2020**, *13*, 6598. [CrossRef]
11. Yudha, S.W.; Tjahjono, B.; Longhurst, P. Stakeholders' Recount on the Dynamics of Indonesia's Renewable Energy Sector. *Energies* **2021**, *14*, 2762. [CrossRef]

Review

A Comprehensive Review on the Recent Development of Ammonia as a Renewable Energy Carrier

Muhammad Heikal Hasan [1,*], **Teuku Meurah Indra Mahlia** [1,*], **M. Mofijur** [1], **I.M. Rizwanul Fattah** [1], **Fitri Handayani** [2], **Hwai Chyuan Ong** [1] and **A. S. Silitonga** [3]

[1] Centre for Green Technology, Faculty of Engineering and IT, University of Technology Sydney, Sydney, NSW 2007, Australia; MdMofijur.Rahman@uts.edu.au (M.M.); IslamMdRizwanul.Fattah@uts.edu.au (I.M.R.F.); HwaiChyuan.Ong@uts.edu.au (H.C.O.)

[2] Department of Mechanical Engineering, Syiah Kuala University, Banda Aceh 23111, Indonesia; fitri.mech@gmail.com

[3] Department of Mechanical Engineering, Politeknik Negeri Medan, Medan 20155, Indonesia; ardinsu@yahoo.co.id

* Correspondence: muhammadheikal.hasan@student.uts.edu.au (M.H.H.); TMIndra.Mahlia@uts.edu.au (T.M.I.M.); Tel.: +61-4123-08541 (M.H.H.); +61-2951-490571 (T.M.I.M.)

Abstract: Global energy sources are being transformed from hydrocarbon-based energy sources to renewable and carbon-free energy sources such as wind, solar and hydrogen. The biggest challenge with hydrogen as a renewable energy carrier is the storage and delivery system's complexity. Therefore, other media such as ammonia for indirect storage are now being considered. Research has shown that at reasonable pressures, ammonia is easily contained as a liquid. In this form, energy density is approximately half of that of gasoline and ten times more than batteries. Ammonia can provide effective storage of renewable energy through its existing storage and distribution network. In this article, we aimed to analyse the previous studies and the current research on the preparation of ammonia as a next-generation renewable energy carrier. The study focuses on technical advances emerging in ammonia synthesis technologies, such as photocatalysis, electrocatalysis and plasmacatalysis. Ammonia is now also strongly regarded as fuel in the transport, industrial and power sectors and is relatively more versatile in reducing CO_2 emissions. Therefore, the utilisation of ammonia as a renewable energy carrier plays a significant role in reducing GHG emissions. Finally, the simplicity of ammonia processing, transport and use makes it an appealing choice for the link between the development of renewable energy and demand.

Keywords: ammonia; renewable energy storage; hydrogen storage

1. Introduction

Industrial activity in parts of the world often has several direct and indirect adverse environmental consequences. Carbon dioxide (CO_2), a natural greenhouse gas (GHG) that helps keep the globe warm, is out of control and triggering a climate crisis. This anthropogenic emission endangers human health, agriculture, natural ecosystems and atmospheric stability [1]. As reported in the Intergovernmental Panel on Climate Change (IPCC) Climate Change Mitigation Report (2014), CO_2 is the main contributor to GHG emissions, which have 76% (including 11% of forests and land use) of the overall share, while CH_4 and N_2O accounted for 16% and 6%, respectively [2]. According to National Oceanic and Atmospheric Administration (NOAA), global ambient CO_2 concentrations rose from 280 ppm to 407.4 ppm in 2018, setting a new high for the last 800,000 years [3,4].

A growing number of international reports illustrates the health and environmental effects of GHG emissions [5–7]. The ultimate goal is to maintain GHG concentrations at a point where the dangerous effect of climate change can be prevented. In 1997, the convention was supplemented and updated by the Kyoto Protocol [8]. Unlike the United

Nations Framework Convention on Climate Change (UNFCCC), the Kyoto Protocol binds parties from a developed country to reduce GHG emissions [9]. Recently, under the terms of the Paris Agreement of 2015, the United Nations (UN) has committed itself to a long-term target to keep temperatures below 2 °C, compared to pre-industrial levels, to avoid the worst consequences of global warming [10,11]. Figure 1 shows the top 10 countries on CO_2 emissions worldwide and their goal in the Paris Agreement.

In the last decades, the world has put great effort and investment into developing renewable energy to decarbonise the economy. The global pandemic disruption caused by COVID-19 teaches humanity that reducing the proportion of fossil fuels in human activity will substantially increase the regeneration of the Earth's atmosphere due to a reduction in air pollution. For example, in China, restricting the activity of people caused by the COVID-19 pandemic resulted in a 25% and 50% reduction in CO_2 and NO_X emissions, respectively [12,13]. However, the problem with these renewable sources of energy is that they are highly weather-dependent. Therefore, it is necessary to have an affordable and efficient method for storing energy. Several prevailing storage technologies, such as chemical, mechanical, electrical and thermal energy storages, can be applied from small to large-scale applications. Prominent storage solutions, such as batteries, provide durability and technological sophistication but are far too expensive and cannot provide the power needed for seasonal and grid-scale demand. Consequently, hydrogen has been the best candidate since the 1970s, positioning this as both energy storage and a clean energy source [14]. However, storage and delivery issues associated with H_2 remain an obstacle to its implementation [15]. To store and distribute H_2 efficiently, other indirect storage media such as ammonia (NH_3), with its proven transportability and high flexibility, is required [16].

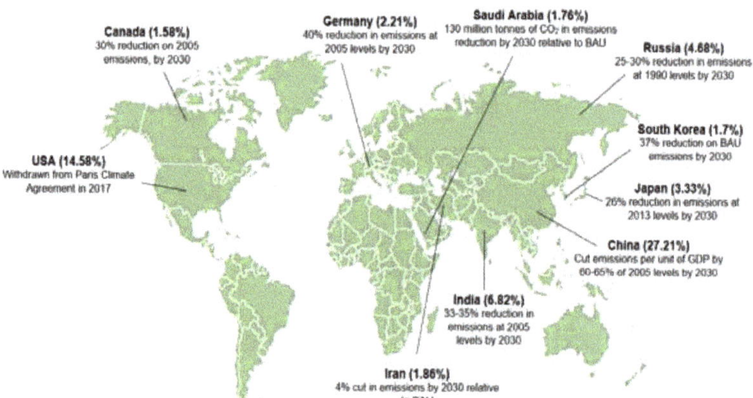

Figure 1. Top 10 global CO_2 emitters and the target in the Paris Agreement [17].

In some countries, ammonia has become a part of their energy roadmap. For example, Japan's Cross-Ministerial Strategic Innovation Program (SIP) attempts to show ammonia, hydrides and hydrogen as essential elements of the nation's hydrogen energy system [18]. The Japanese Ministry of Economy, Trade and Industry (METI) described ammonia in conjunction with "the concept of importing renewable energy produced in other countries" [19]. In the United States, The Advanced Research Projects Agency-Energy (ARPA-E) announced a cumulative grant of $32.7 million for 16 projects, 13 of which focus on ammonia [16]. Governments in New Zealand and Australia have announced federal grants to support the development of ammonia plants driven by renewable energy. More recently, Australia has awarded Yara and ENGIE AU$995,000 for their solar ammonia project, Yara Pilbara, from the Australian Renewable Energy Agency (ARENA). As for New Zealand, Ballance-Agri Nutrients and Hiringa Energy received NZ$19.9 million from the Provincial Growth Fund for their wind-fed ammonia plant in Kapuni [20]. Similarly to Japan, the country is also looking into the possibility of exporting renewable energy in the form of ammonia. In addition, Toyota and Commonwealth Scientific and Industrial Research Organization (CSIRO) designed the first ammonia fuel car tested in Australia in 2018 [21].

Today, with a global production rate of more than 176 million metric tonnes of ammonia, the chemical is being used as fertiliser and a building block in the manufacture of many products [22,23]. With 28.5% of global production, China is known as the main producer of ammonia [24]. The uses of ammonia as an intermediate for the production of fertilisers account for over 80% of the total production of ammonia [25]. Other applications include fibre and plastics, pharmaceuticals, mining and metallurgy, pulp and paper, refrigeration and explosives [26]. Other than that, ammonia has also being recently proposed to be used in automotive applications for NOx emission control (DeNOx) technologies [27]. Furthermore, ammonia has also been researched as a source of energy for fuel cells, transport, industry and power generation [28].

Unfortunately, the current industrial ammonia synthesis method is complicated, energy-intensive and heavily dependent on hydrocarbon. The Haber–Bosch process that is currently used to synthesise ammonia is responsible for almost 11% of global industrial CO_2 emissions [29,30]. In addition, the nature of renewable energy sources, which are irregular, requires turnkey systems that can be instantly switched on and off [31]. Thus, the challenge for the global deployment of ammonia as energy storage is, therefore, the simpler and more efficient production of ammonia from abundant sources, such as ambient air and water, with a ready to go system, which can be driven by intermittent energy sources.

The number of research studies on renewable ammonia, its innovation on the production route and its performances as a fuel has increased substantially in recent years, as shown by a growing number of scientific papers and review papers (Figure 2a–c). There has been a significant increase in the numbers of scientific articles related to ammonia, renewable energy, energy storage or energy carriers in recent years. For innovative approaches of ammonia synthesis, electrochemistry gained great attention in recent years due to the direct conversion of electricity into ammonia. Of all the devices capable of converting ammonia into energy, the Internal Combustion (IC) engine appears to receive the most development effort, noting that use for other applications is less explored.

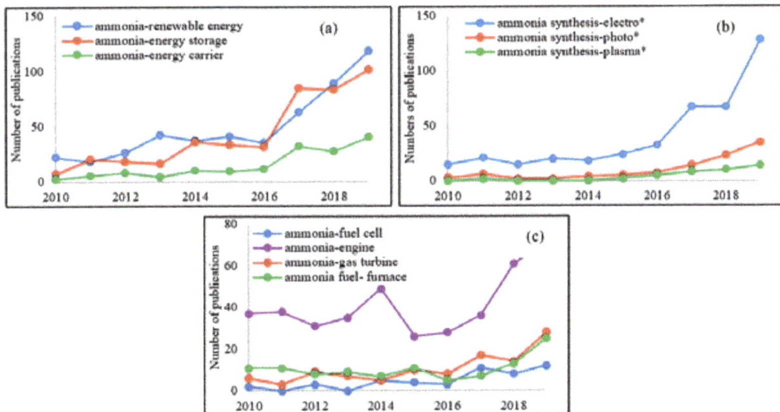

Figure 2. Recent publications of (**a**) ammonia as a renewable energy carrier, (**b**) innovations in ammonia production, (**c**) ammonia as a fuel in Scopus databases.

This study comprehensively reviews all aspects of ammonia as energy storage, including innovative approaches for converting renewable energy into ammonia and devices that convert ammonia into energy. The comparison of the recent review article of ammonia as renewable energy and this work is given in Table 1.

Table 1. The comparison of recent review articles on ammonia as a renewable energy carrier and this work.

Category	[16]	[32]	[33]	[34]	[35]	[36]	[37]	This Work
Ammonia economy	√	-	-	-	√	-	√	√
Photocatalysis	-	-	-	-	-	-	-	√
Electrocatalysis	-	√	√	√	-	√	-	√
Plasmacatalysis	-	-	√	-	-	√	-	√
Fuel cell	√	-	-	-	√	-	√	√
IC engine	√	-	-	-	√	-	-	√
Gas turbine	√	-	-	-	√	-	-	√

2. Key-Driver of Ammonia Economy

The issue of the storage and distribution of hydrogen creates opportunities for ammonia to be seen as alternative storage of renewable energy. The previous study on the potential hydrogen storage material by Kojima [38] has revealed that ammonia has the highest gravimetric densities with the highest volumetric densities, as shown in Figure 3 [16,38]. Consequently, attempts are being made to leverage ammonia over others to replace hydrogen as a central energy distribution block [16]. In addition, ammonia can be used to replenish soil nutrients, boosting the growth of crops and accelerating afforestation that will indirectly help to balance CO_2 gas in the atmosphere through photosynthesis by plants.

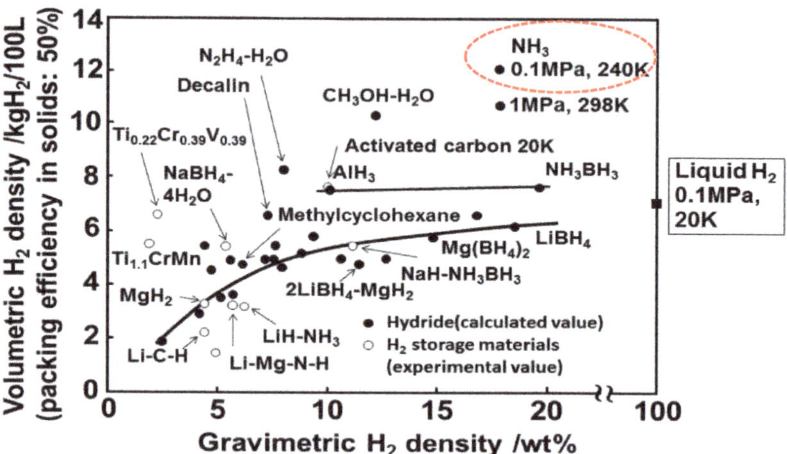

Figure 3. The density of hydrogen in hydrogen carriers [38].

The sustainable future of renewable energy-driven ammonia production is not a novel idea but has never been widely embraced over methane or coal-based ammonia. By transforming into ammonia that can be liquefied under moderate pressure, renewable energy can be transported from places where renewable energy is cheap or excessive to where it is limited or expensive. This synergy could open up opportunities for exports and imports of renewable energy, similar to today's hydrocarbon fuel. In addition to being used directly in the form of ammonia, this may also be dissociated into its component for use as hydrogen fuel at a relatively low cost. Cost comparison of the hydrogen produced from a variety of feedstock is presented in Table 2.

Table 2. Cost comparison of hydrogen production via various feedstock [39–43].

Method	Electricity Source	Hydrogen Production (kg/day)	Hydrogen Cost ($/kg)
Water electrolysis	Wind	1400–62,950	5.10–23.37
	Solar PV	1356	10.49
	Solar Thermal	1000	7.00
	Nuclear	1000	4.15
Thermochemical water splitting	Solar	6000	7.98–8.40
	Nuclear	7000–800,000	2.17–2.63
Natural Gas steam reforming	With carbon capture storage	-	2.27
	Without carbon capture storage	-	2.08
Coal gasification	With carbon capture storage	-	1.63
	Without carbon capture storage	-	1.34
Biomass gasification	-	-	1.77–2.05

In addition to the aforementioned factors, ammonia transportation and storage facilities already exist today, where around 18 million tonnes of ammonia are exchanged annually. Unlike hydrogen that cannot be liquefied under a pressurised tank, ammonia may be kept in liquid form when at least 8.58 bar is maintained. Under this load, carbon steel tanks are sufficient to be used. Ammonia may also be contained as a liquid in lower temperature storage if the temperature of $-33\ °C$ is preserved while hydrogen only can be liquefied under cryogenic cooling at $-252.87\ °C$. The energy density in this form is roughly half that of petrol and more than ten times that of batteries. Ammonia also offers a much lower storage cost compared to hydrogen. Commonly, long-distance ammonia transport is accomplished by using carbon steel pipelines that are opposed to hydrogen, which now still has material issues with the pipeline. For transporting ammonia via pipeline over 1610 km, it requires 1119 kJ/kg-H_2, which is much lower than that of hydrogen transport of 14,814 kJ/kg-H_2. This disparity can be described by the state of the two fluids where hydrogen gas is distributed with the aid of compressors while the ammonia is carried as a liquid using a pump. The comparison between ammonia and hydrogen is given in Table 3.

Table 3. Difference between NH_3 and H_2 in terms of storage and distribution [39,44,45].

Parameter	Ammonia	Hydrogen	Hydrogen
Storage method	Liquid	Compressed	Liquid
Storage pressure (MPa)	1.1	70	-
Liquefy temperature (°C)	−33	-	−252.87
Fuel density (kg m^{-3})	600	39.1	70.99
Storage cost over 182 days ($/kg-$H_2$)	0.54	-	14.95
Transporting over 1610 km pipeline (kJ/kg-H_2)	1119	14,814	-

Ammonia also can be transported as a pressurised liquid via truck and rail. Trailers can transport 43,530 L ammonia at 2.07 MPa. Such a tank could hold up to 26 tonnes or 600 GJ of energy of ammonia [46]. In contrast, a hydrogen lorry only can be used to transport around 340 kg of hydrogen gas at 17.91 MPa, equivalent to 48 GJ of hydrogen energy content, while transporting in liquid form in a hydrogen trailer could hold around 3.9 tonnes of hydrogen, equal to 553 GJ of energy [47]. Rail transport uses a pressurised tank with a capacity of 126,810 L at 1.55 MPa, which indicates the capability to carry 77.5 tonnes of NH_3 [46]. Ammonia can be transported by ship or barge using pressurised storage vessels. By using these vessels, existing ocean-going ships can transport 55,000 tonnes of ammonia [48]. The NH_3 and H_2 transport methods are summarised in Table 4.

Table 4. Differences in NH_3 and H_2 transport methods [44].

Parameter	Ammonia			Hydrogen	
	Truck	Rail	Ship	Truck	Truck
Storage method	Liquid	Liquid	Liquid	Compressed	Liquid
Storage pressure (MPa)	2.07	1.55	-	17.91	-
Capacity (tonnes)	26	77.5	55,000	0.34	3.9
Energy capacity (GJ-HHV)	600	1746	1,240,000	48	553

Note: HHV (higher heating value).

In the recent development, storing ammonia as metal ammine complexes, i.e., hexaamminemagnesium chloride, $Mg(NH_3)_6Cl_2$ (Figure 4), also gives a beneficial advantage for storing and transporting since it has very low vapour pressure (0.002 bar at ambient temperature) [49,50]. Hexaamminemagnesium chloride is formed simply bypassing ammonia at room temperature over anhydrous magnesium chloride. $Mg(NH_3)_6Cl_2$ can be formed into a small and dense solid without any void space. The amine has volumetric hydrogen content between 105 and 110 kg H_2 m^{-3} and gravimetric hydrogen content over 9 wt.% [49].

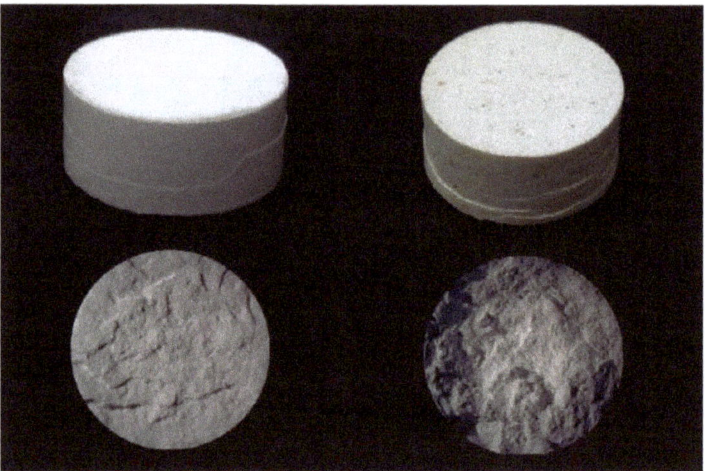

Figure 4. $Mg(NH_3)_6Cl_{12}$ for ammonia storage [50].

3. Ammonia-Based Energy Storage

3.1. Characteristics

Ammonia is comprised of N_2 and H_2 with the chemical formula of NH_3. A highly irritating, colourless gas with a distinctive pungent scent characterises the chemical. With a density of 0.769 kg m^{-3}, ammonia is lighter than air. It also liquefies easily, which is caused by a resilient hydrogen bond among molecules. The boiling point and the freezing point at standard temperature and pressure are -33.35 °C and -77.65 °C, respectively. Ammonia in the air has a flammability limit between 15 and 25%. Ammonia burns with a light yellowish-green flame. Where a suitable catalyst and high temperature is present, ammonia is decomposed into the constituent elements.

The ammonia molecule has a pyramidal trigonal form with an angle of the bond of 106.7 °C. The atom of nitrogen consists of five outer electrons with an additional three electrons from each hydrogen atom, giving eight electrons in total, or pairs of four tetrahedral arranged electrons. This shape gives a dipole moment to the molecule, which makes it polar. The polarity of the molecule and, in particular, its capability to create hydrogen bonds makes ammonia very soluble in water. The chemical is naturally a base and a proton acceptor. Ammonia aqueous solution with a concentration 1.0 M has a pH of 11.6 at 298 K [51]. Even though accidental explosion or combustion is relatively low [16], the US-National Fire Protection Association (NFPA) has identified this as a hazardous material, putting it at high safety risk [52]. Safety concerns regarding ammonia storage for end-users arise due to the high vapour pressure of liquid ammonia. At high temperatures, ammonia can decompose, forming highly flammable hydrogen and toxic nitrogen oxide. Ammonia is also known to be corrosive with certain alloys such as copper, brass and bronze, as presented in Table 5.

Table 5. Material compatibility of ammonia and its derivative [16,53–55].

Materials	Ammonium Hydroxide (Ammonia in Water)	Ammonia Anhydrous
ABS plastic	-	D
Acetal (Delrin ®)	D	D
Aluminium	C	A
Brass	-	D
Bronze	-	D
Buna N (Nitrile)	D	B
Carbon graphite	-	A
Carbon steel	D	B
Carpenter 20	-	A
Cast iron	D	A
ChemRaz (FFKM)	-	B
Cooper	-	D
CPVC	-	A
EPDM	A	A
Epoxy	-	A
Fluorocarbon (FKM)	B	D
Hastelloy-C ®	B	B
Hypalon ®	A	D
Hytrel ®	-	D
Kalrez	-	A
Kel-F ®	-	A
LDPE	-	B
Leather		D
Natural rubber	-	D
Neoprene	B	A
NORYL ®	-	B
Nylon	C	A
Polycarbonate	-	D
PEEK	-	A
Polypropylene	A	A
Polyurethane	-	D
PPS (Ryton ®)	-	A
PTFE	A	A
PVC	-	A
PVDF (Kynar ®)	A	A
Santropene	A	A
Silicone	-	C
Stainless Steel 304	B	A
Stainless Steel 316	A	A
Titanium	-	C
Tygon ®	-	A
Viton ®	-	D

Note: A (excellent); B (good); C (fair); D (poor).

Ammonia is also fatal if inhaled, causing a lung injury. Some people may be somewhat irritated by 30 ppm for 10 min, while the remainder is sensitive to 50 ppm. At 500 ppm, the nose and throat get immediate and severe irritation. Short exposure to levels over

1500 ppm can lead to fluid accumulation in the lungs [56]. Immediately Dangerous to Life and Health (IDLH) is the degree whereby a healthy person can get 30 min of exposure without causing permanent health effects. Ammonia is also believed to be responsible for the depletion of ozone by an accumulation of nitrous compounds in the atmosphere [57]. Table 6 summarises the Health and Safety data, including their vapour pressure, IDLH and toxicity.

Table 6. Health and Safety data of ammonia and its derivatives [58–60].

Carriers	VP at 20 °C (Bar)	IDLH (ppm)	AT 20 °C (IDLH^{-1})
35 wt.% NH_3 solution	1.24	-	~4133
NH_3 (liquid)	8	300	~27,000
$Mg(NH_3)_6Cl_2$	1.4×10^{-3}	-	4.65

Note: VP (vapour pressure); AT (apparent toxicity).

3.2. Ammonia production Using Haber–Bosch Method

Fritz Haber and Carl Bosch invented the Haber–Bosch (H–B) process in the middle of the 20th century, replacing the Birkeland–Eyde process and Frank–Caro process to synthesise ammonia [37]. Since its discovery, the Haber–Bosch process dominating the industrial process and has undergone many substantial improvements and optimisations. The minimum energy consumption in the mid-1950s is reduced from more than 60 GJ t_{NH3}^{-1} to 27.4–31.8 GJ t_{NH3}^{-1} today [61]. Such improvements represent an improvement in overall energy efficiency from 36 to 65%. The most significant increase in productivity was made by replacing coal with CH_4 to produce H_2.

In short, the production cycle of ammonia can be broken down into two major phases; the first is synthetic mixture production, and the subsequent is the synthesis of ammonia synthesis by the H–B process. In the first phase, hydrogen production takes place via two-stage steam methane reforming (SMR) reactors before being transferred for water-gas-shifting (WGS) reaction, CO_2 subtraction and then methanation [61]. The primary SMR reactor works at 850–900 °C and 25–35 bar that travels through the catalyst. In this process, the energy needed for the endothermal reactions generated by the external combustion of methane fuel through the furnace [61]. Water-saturated methane is then fed into a catalytic converter that converts methane to hydrogen and carbon monoxide before it is transferred to a second SMR. In the secondary SMR, ambient air is compressed and transferred at 900–1000 °C to partially oxidise reagents in the next reactor. In this process, air–oxygen and steam convert unreacted methane into hydrogen and carbon monoxide, besides providing the stoichiometric nitrogen required for Haber–Bosch downstream reactions [58].

The next process involves converting carbon monoxide to carbon dioxide by steam injection in the WGS reactor to prevent coke from forming on the catalyst and side reactions. The feed is then sent to a CO_2 remover column, giving a nitrogen and hydrogen-rich feed with high purity for the next process [61]. The ammonia production then takes place on Haber–Bosch reactor when the process is usually performed in a reactor with two to four catalytic converter beds at 200 to 350 bar and 300 to 500 °C [58]. Since the process has low single-pass conversion efficiency (~15%), it is necessary to recycle the unreacted gas. The nitrogen fixation process for ammonia and the reaction process that occurs during the Haber–Bosch process is shown below:

$$N_2 + 3H_2 \rightarrow 2NH_3 \quad \Delta H_{298K} = -92 \text{ kJ mol}^{-1} \tag{1}$$

$$CH_4 + H_2O \rightarrow CO + 3H_2 \text{ (Primary Steam Reforming)} \tag{2}$$

$$4CH_4 + O_2 + 2H_2O \rightarrow 10H_2 + 4CO \text{ (Secondary Steam Reforming)} \tag{3}$$

$$CO + H_2O \rightarrow CO_2 + H_2 \text{ (Water} - \text{gas shift)} \tag{4}$$

3.3. Electrically-Driven Haber–Bosch Process

Notwithstanding the technological reliability of the existing Haber–Bosch process, the question remains as to whether or not such a process can be CO_2-free. To answer these questions, research and development are underway worldwide to substitute the centuries-old Haber–Bosch method for the production of ammonia, driven by renewable electricity. It includes switching H_2 obtained from the steam-reformed CH_4 to H_2 obtained from the electrolysed H_2O. Given the trend in renewable energy prices to compete with fossil fuels, the green Haber–Bosch process is no longer a vision. Figure 5 shows the electrically driven Haber–Bosch plant powered by renewable energy.

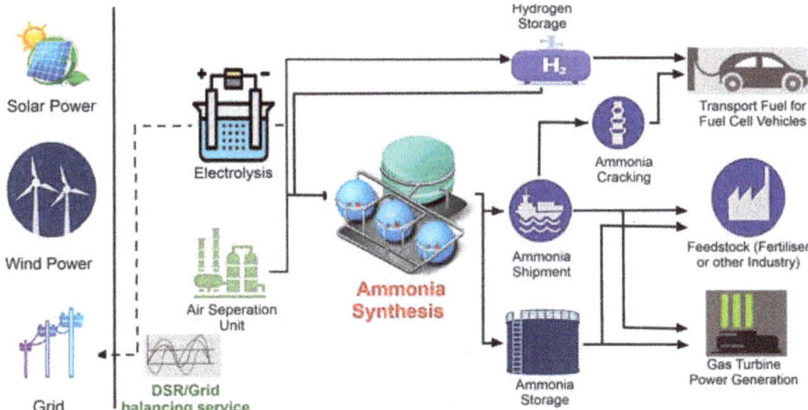

Figure 5. Schematic of renewable energy-driven Haber–Bosch plant [62].

Ammonia synthesis processes powered by renewable energy have been demonstrated in many countries. For example, in the USA, Schmuecker Pinehurst Farm LLC has built a solar ammonia plant and has been in operation for many years where H_2 and N_2 are generated from water and air by electrolysis and power swing system before pressurised and fed into the ammonia production facility [63]. In Zimbabwe, Africa, an ammonium nitrate plant was developed where ammonia is supplied from renewable energy-driven Haber–Bosch process. This production facility has been productive for years, with 240,000 tonnes of ammonium nitrate produced annually [64]. In Australia, several projects have been developed near Yara Pilbara, Western Australia [65] and Port Lincoln, South Australia [66] to evaluate the feasibility of renewable energy-driven ammonia plants. At the same time as optimisation of the renewable energy-driven Haber–Bosch process, the development of alternative methods to allow the N_2 reduction reaction at atmospheric pressure and moderate ambient temperature, such as photocatalysis, electrocatalysis and plasmacatalysis has attracted widespread interest in ammonia synthesis today.

4. Innovative Approaches for Ammonia Synthesis

In the absence of high temperatures and pressures, nature converts molecular N_2 to NH_3. This natural process uses enzyme nitrogenises containing metal ions (iron and molybdenum) to induce ammonia reactions from atmospheric nitrogen, electrons and protons. This phenomenon has aroused the researcher's interest in imitating nature.

In recent times, significant progress in understanding the process for nitrogenises reactions and in creating a modern synthetic method has been achieved. Photocatalysis, plasmacatalysis and electrocatalysis have been studied as alternative green routes for ammonia production. These new techniques offer distinct advantages compared to the old Haber–Bosch method.

Among those, the plasma-enabled ammonia synthesis is both energy and cost-effective in theory. The potential energy consumption of non-thermal plasma (NTP) ammonia production has been reported to be around 0.2 MJ/mol, which is lower than the Haber–Bosch technique (0.48 MJ/mol) [67]. This section details the method and provides a discussion on emerging technology for facilitating the N_2 reduction reaction to ammonia.

4.1. Photocatalysis

Photocatalytic ammonia production from water and air at low temperature and pressure show enormous potential, attracting increased research interest from scientists. The process is relatively safe, inexpensive and accessible to a free energy source (light). In general, photons are used in the photocatalytic mechanism to drive N_2 activation. The fixation process of N_2 into NH_3 by photocatalysts can be represented by the following equation:

$$N_2 + 3H_2O \overset{light}{\rightarrow} 2NH_3 + \frac{3}{2}O_2 \tag{5}$$

The N_2 fixation photocatalytic process involves several steps. In short, the electron generated by the photocatalyst effect is driven into the conduction band, creating an empty hole in the valance band. Some holes and electrons are then recombined together, while others transfer to the catalytic surface and take part in the redox reaction. H_2O is then oxidised to O_2 by holes, while N_2 is reduced to NH_3 by the reaction of water-derived protons and photo-generated electrons. Figure 6 illustrated the catalytic process of the photocatalyst.

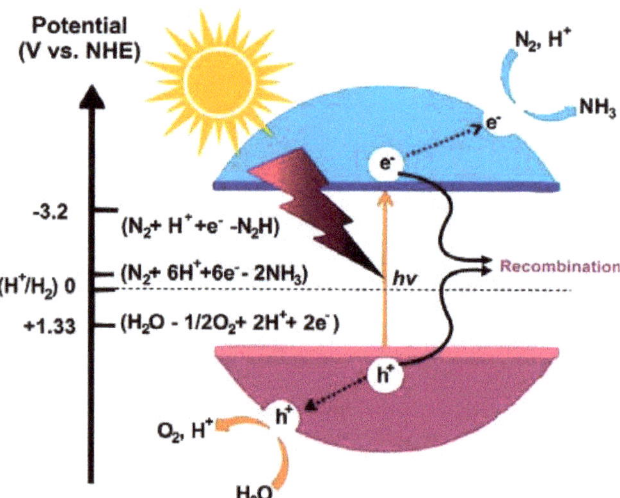

Figure 6. Schematic of photocatalyst reaction for ammonia synthesis [68].

TiO_2-based metal oxide photocatalysts were studied early in nitrogen fixation because of less expense and higher stability. Following the pioneering research in 1977 by Schrauzer and Guth [69], many semiconductors have been proposed for the process of photocatalysis viz. metal oxide, metal sulphide, oxyhalides and other graphitic nitride carbon materials.

In 1988, Bourgeois et al. [70] studied photocatalytic action of unmodified TiO_2 after annealing under atmospheric air pressure. Thermal pre-treatment is believed to trigger surface defects that caused defects or impurities in the semiconductor bandgap. In the most recent study, the photocatalytic N_2 fixation into NH_3 on TiO_2 surface oxygen vacancies was systematically investigated by Hirakawa et al. [71]. They found that the active sites for N_2 reduction are the oxygen vacancies of the Ti^{3+} species. The superficial Ti^{3+} provided an abundance of active N_2 fixing sites by acting as an electron donor, resulting in relatively easy dissociation of the $N \equiv N$ bond. In other studies, many other metal species are used as doping in the TiO_2 catalyst; however, the photocatalytic was unsatisfactory. The transition metal atoms loaded into TiO_2 can function as co-catalysts while acting as a dopant in the TiO_2 matrix. In addition, the Schottky junction formation on the semiconductor interface and the transition metal induces electrical fields and facilitates the separation of photo-induced electrons and holes.

In addition to metal doping, noble metals may also be embedded onto the surface of TiO_2. Ranjit et al. [72], in 1996, compared four noble metals (Ru, Rh, Pd, Pt) as TiO_2 co-catalysts. They found Ru > Rh > Pd > Pt to be the order of photoactivity between metals, which is closely related to the strength of the noble metals and the hydrogen bond. The experiment revealed that noble metals with higher barriers to H_2 evolution exhibit surprisingly higher NH_3 yields.

The development of metal-doped TiO_2 in N_2 fixation sparked interest in the development of binary metal oxide and ternary metal oxides as doping. The study by Lashgari and Zeinalkhani [73], on the synthesis of Fe_2O_3 by precipitation method, found that the ammonia is efficiently formed from N_2 at Fe_2O_3 nanoparticles. It was also reported that partially reduced Fe_2O_3 may be used for the reduction of photocatalytic N_2. More recently, bismuth monoxide (BiO) was employed for N_2 photoreduction. It has been stated that BiO quantum dots will significantly decrease N_2 at 1226 µmol $gcat^{-1}h^{-1}$ NH_3 with the simulated rate of sunlight [74].

In addition to metal oxides, the study of metal sulphides in the field of photocatalysis has also seen a significant increase in the scientific community. The metal sulphide band gap is conducive to intense visible light absorption, resulting in highly effective solar use. As a result of the growing multidisciplinary research in biology and material science, organic-sulphide has also been developed as a catalyst for NH_3 synthesis [68]. Researchers questioned whether sulphur vacancies could effectively promote the photocatalytic event. Motivated by this idea, Hu et al. [75] studied the sulphur vacancies effect on the efficiency of ternary metal sulphide for N_2 fixation. Sulphur vacancies have been found to introduce surface chemical adsorption sites, which have helped to activate N_2 molecules by widening the bonding space from 1.164 Å to 1.213 Å. In the recent development, $Cd_{0.5}Zn_{0.5}S$ solid solution loaded with Ni_2P was effectively reduced N_2 with the NH^3 yield of 253.8 µmol $g_{cat}^{-1} h^{-1}$ [76]. Other materials, such as Bismuth oxyhalides, are also found to be promising photocatalysts due to their layered crystal structure and suitable band gaps [68,77].

Other than the material mentioned above, graphitic nitride carbon photocatalyst for N_2 fixation has been designed and developed in recent years. Dong et al. [78] have shown that introducing vacancies in nitrogen will dramatically boost the photocatalytic behaviour of g-C_3N_4. It was because the nitrogen vacancies, which had the size and forms as the nitrogen atoms, are advantageous for adsorbing and activating the chemically inert N_2. In another study, honeycombed g-C_3N_4 doped with Fe^{3+} has shown enhanced photocatalytic performance [79]. After that, various forms of photocatalysts based on g-C_3N_4 were employed in photochemical ammonia synthesis. A summary of various photochemical catalysts by previous studies is presented in Table 7.

Table 7. Recent studies on photochemical catalysis systems for ammonia production.

Groups	Catalyst	T (K)	Scavenger	Catalyst Loading (g L^{-1})	NH$_3$ Yield (µmolg$_{cat}^{-1}$h^{-1})	Ref
Metal oxides	TiO_2	500	-	-	0.83	[70]
	0.4 wt.% Co-doped TiO_2	313	-	-	6.3	[69]
	0.4 wt% Cr-doped TiO_2	313	-	-	0.37	[69]
	0.4 wt% Mo-doped TiO_2	313	-	-	6.7	[69]
	2 wt% Mg-doped TiO_2	-	-	0.67	6.9 µM h^{-1}	[80]
	0.5 wt% Fe-doped TiO_2	353	-	-	6	[81]
	10 wt% Ce-doped TiO_2	-	-	0.8	3.4 µM h^{-1}	[82]
	10 wt% V-doped TiO_2	-	-	0.8	4.9 µM h^{-1}	[82]

Table 7. Cont.

Groups	Catalyst	T (K)	Scavenger	Catalyst Loading (g L^{-1})	NH$_3$ Yield (μmolg$_{cat}^{-1}$h^{-1})	Ref
	0.24 wt% Ru-loaded TiO$_2$	-	-	-	17.3	[72]
	0.8 wt% Pt-loaded TiO$_2$	-	-	-	4.8	[72]
	0.69 wt% Pd-loaded TiO$_2$	-	-	-	11.8	[72]
	0.2 wt% Rh-loaded TiO$_2$	-	-	-	12.6	[72]
	0.2 wt% Fe-doped TiO$_2$	313	-	-	10	[69]
	Fe-doped TiO$_2$	298	Ethanol	1	400 μM h^{-1}	[83]
	JRC-TIO-6 (rutile)	313	2-PrOH	1	2.5 μM h^{-1}	[71]
	Pt-loaded ZnO	-	Na-EDTA	-	860	[84]
	Fe$_2$O$_3$	298	Ethanol	0.5	1362.5 μM h^{-1}	[73]
	Partially reduced Fe$_2$O$_3$	303	-	-	10	[85]
	BiO quantum dots	298	-	-	1226	[74]
	H-Bi$_2$MoO$_6$	-	-	-	10	[86]
Metal sulphides	Zn$_{0.1}$Sn$_{0.1}$Cd$_{0.8}$S	303	Ethanol	0.4	105.2 μM h^{-1}	[75]
	CdS/Pt	311	-	-	3.26	[87]
	CdS/Pt/RuO$_2$		-	4	620 μM h^{-1}	[88]
	MoS$_2$	298	-	-	325	[89]
	Mo$_{0.1}$Ni$_{0.1}$Cd$_{0.8}$S	303	Ethanol	0.4	71.2 μM h^{-1}	[90]
	g-C$_3$N$_4$/ZnMoCdS	298	Ethanol	0.4	77.6 μM h^{-1}	[91]
	g-C$_3$N$_4$/ZnSnCdS	303	Ethanol	0.4	167.6 μM h^{-1}	[92]
	Ni$_2$P/Cd$_{0.5}$Zn$_{0.5}$S	293	-	-	253.8	[76]
Oxyhalides	Bi$_5$O$_7$I	293	Methanol	0.5	111.5 μM h^{-1}	[93]
	BiOCl	298	Methanol	0.67	46.2 μM h^{-1}	[94]
	BiOBr	298	-	0.5	1042 μM h^{-1}	[95]
	Bi$_5$O$_7$Br nanotubes	-	-	-	1380	[96]
Graphitic nitride carbon	Fe-doped g-C$_3$N$_4$	303	Ethanol	0.4	120 μM h^{-1}	[79]
	g-C$_3$N$_4$	-	Methanol	1	160 μM h^{-1}	[78]
	g-C$_3$N$_4$/MgAlFeO	303	Ethanol	0.4	166.8 μM h^{-1}	[97]
	g-C$_3$N$_4$/rGO	303	Na-EDTA	0.4	206 μM h^{-1}	[98]
	Ga$_2$O$_3$-DBD/g-C$_3$N$_4$	-	Ethanol	0.4	112.5 μM h^{-1}	[99]
	W$_{18}$O$_{49}$/g-C$_3$N$_4$	-	Ethanol	0.4	57.8 μM h^{-1}	[100]

4.2. Electrocatalysis

In addition to photocatalysis, the synthesis of ammonia by the electrocatalysis process is also currently being explored. In a traditional proton-conductive electrolyte cell, gaseous H$_2$ passes through the anode where it is changed to protons (H$^+$) while the nitrogen reduction reaction occurs at the cathode [101]. The H$^+$ then diffuses into the cathode, where it forms NH$_3$ in combination with the dissociated N. The following equations can describe the reaction:

$$3H_2 \rightarrow 6H^+ + 6e^- \tag{6}$$

$$N_2 + 6H^+ + 6e^- \rightarrow 2NH_3 \tag{7}$$

$$N_2 + 3H_2 \rightarrow 2NH_3 \tag{8}$$

The problem, though, is such cell configuration had to work at low temperatures where the kinetics of reaction were sluggish [102]. Moreover, an electrochemical cell is more advantageous when operating at higher temperatures since higher rates of reaction can be achieved in the same electrode area, and hydrazine development can be prevented [103]. As a result, proton (H$^+$) conductivity solid-state materials that operate at a temperature above 500 °C were developed [102].

Based on the working temperature, electrocatalytic ammonia synthesis can be broken down into high (higher than 500 °C), intermediate (between 100 °C and 500 °C) and low (lower than 100 °C) [36]. In the high-temperature electrocatalysis process, most of the studies occupy solid electrolytes with perovskite as the material in reactor configuration.

Among the studies, quite high ammonia rates were reported by Wang et al. [104,105] using doped Ceria-$Ca_3(PO_4)_2$-K_3PO_4 composite electrolyte in combination with a Ag-Pd electrode. Ammonia yields up to 9.5×10^{-9} mol s^{-1}cm^{-2} can be achieved using N_2/H_2 as feedstock and up to 6.95×10^{-9} mol s^{-1}cm^{-2} using N_2/natural gas. In 2009, the novel Solid State Ammonia Synthesis (SSAS) configuration using the Ag-Ru/MgO cathode developed by Skodra and Staukides [106] was able to directly use water as a source of hydrogen. Other than the configurations mentioned above, the oxygen-ion (O^{2-}) conductor has also been shown to be used for ammonia synthesis in SSAS where both processes of electrolysis and ammonia synthesis occur at the cathode but have suffered very low ammonia production rates [106,107].

In intermediate electrocatalysis processes with operating temperatures ranging from 500 °C to 100 °C, molten salts are typically used as electrolytes. Murakami et al. [108] made the earliest study in this temperature range in 2003, using a molten salt mixture electrolyte and porous nickel as electrodes. Other sources of hydrogen, such as water steam [109,110], hydrogen chloride [111], methane [112] and hydrogen sulphide [113], have also been tested. More recently, Licht et al. [114] used a similar configuration for the experiment with NaOH/KOH as electrolyte and nickel as the electrode but added a Fe_2O_3 nanoparticle catalyst to the molten salt. The maximum forming rate of ammonia 1×10^{-8} mols^{-1}cm^{-2} could be achieved by this setup, although it is much lower compared to the works of Murakami's group.

In low-temperature electrochemical ammonia synthesis below 100 °C, Nafion and Sulfonate Polysulfone (SPSF) are commonly used as proton electrolyte conductors [103]. Kordali et al. [115] in 1999 reported a novel configuration that could synthesise ammonia below 100 °C using a combination of the Nafion membrane and KOH solution. The anode was Pt, while the anode was carbon cloth on which Ru had been deposited. The hydrogen source was either hydrogen gas or water. A summary of the selected electrochemical system by previous studies is presented in Table 8.

Table 8. A summary of the selected electrocatalytic ammonia synthesis by previous studies.

Temperature	Electrolyte	Cathode/Anode	NH$_3$ Yield ($\times 10^{-9}$ mol/s.cm^2)	FE (%)	Ref
High (T > 500 °C)	SCY	Pd	4.50	78	[116]
	BCN	Ag-Pd	1.42	-	[117]
	BCZN	Ag-Pd	1.82	-	[117]
	BCNN	Ag-Pd	2.16	-	[117]
	BCS	Ag-Pd	5.23	-	[118]
	BCGS	Ag-Pd	5.82	-	[118]
	BZCY	Ni- BZCY/Rh	2.86	6.2	[119]
	LSGM	Ag-Pd	2.37	70	[120]
	LCZ	Ag-Pd	2.00	-	[121]
	LCC	Ag-Pd	1.30	-	[121]
	LCZ	Ag-Pd	1.76	80	[122]
	BCG	Ag-Pd	3.09	-	[123]
	BCY1	Ag-Pd	2.10	60	[124]
	BZCY	Ni- BZCY/Cu	1.70	2.7	[125]
	BZCY	Ni- BZCY/Cu	4.10	10	[125]
	BCY2 − ZnO	Ag-Pd	2.60	45	[126]
	CYO − Ca$_3$(PO$_4$)$_2$/K$_3$PO$_4$	Ag-Pd	9.50	-	[105]
	CYO − Ca$_3$(PO$_4$)$_2$/K$_3$PO$_4$	Ag-Pd	6.95	-	[104]
	CSO	Ag-Pd	7.20	-	[127]
	CGO	Ag-Pd	7.50	-	[127]
	CYO	Ag-Pd	7.70	-	[127]
	CLO	Ag-Pd	8.20	-	[127]
	LCGM	Ag-Pd	1.63	47	[128]
	LSGM	Ag-Pd	2.53	73	[128]
	LBGM	Ag-Pd	2.04	60	[128]
	LBGM	Ag-Pd	1.89	60	[129]

Table 8. *Cont.*

Temperature	Electrolyte	Cathode/Anode	NH$_3$ Yield ($\times 10^{-9}$ mol/s.cm^2)	FE (%)	Ref
	BCG2	Ag-Pd/Ni- BCG	5.00	70	[130]
	BCZS	Ag-Pd	2.67	50	[131]
	BCC	Ag-Pd	2.69	50	[132]
Intermediate (100 °C < T < 500 °C)	(Li, Na, K)$_2$CO$_3$-CSO	LSFC/Ni- CSO	5.39	7.5	[133]
	LiCl-KCl-CsCl	Porous Ni Plate	3.33	72	[108]
	LiCl-KCl-CsCl	Al/Porous Ni Plate	33.3	72	[134]
	LiCl-KCl-CsCl	Porous Ni Plate/Baron-doped diamond	5.80	80	[111]
	LiCl-KCl-CsCl	Porous Ni Plate/Glassy carbon rod	20	23	[109]
	NaOH/KOH/Nano-Fe$_2$O$_3$	Ni	10	35	[114]
	Na$_{0.5}$K$_{0.5}$OH/Nano-Fe$_2$O$_3$	Monel/Ni	16.2	76	[135]
	NaOH-KOH molten salt	(Fe$_2$O$_3$/AC)/Ni	8.27	13.7	[136]
Low (T < 100 °C)	Nafion	SBCF/Ni-CSO	6.90	-	[137]
	Nafion	SBCF/Ni-CSO	7.20	-	[137]
	Nafion	SBCC/Ni-CSO	8.70	-	[137]
	Nafion	Pt	3.13	2.2	[138]
	Nafion	Pt	3.50	0.7	[138]
	Nafion	Pt	1.14	0.55	[138]
	SPSF	SSC/NiO−CSO	6.50	-	[139]
	SPSF	NiO−CSO	2.40	-	[139]
	Nafion	SSN/NiO−CSO	10.5	-	[140]
	SPSF	SSN/NiO−CSO	10.3	-	[141]
	Nafion	SFCN/NiO−CSO	11.3	90.4	[141]
	Nafion	MOF (Fe)/Pt	2.12	1.43	[142]
	Nafion	MOF (Co)/Pt	1.64	1.06	[142]
	Nafion	MOF (Cu)/Pt	1.24	0.96	[142]
	Nafion	MOF (Fe)/Pt	1.52	0.88	[142]
	Nafion	Rh NNs/Carbon rod	6.24	0.7	[143]
	Nafion	Carbon nanospikes/Pt	1.59	11.56	[144]

Note: SCY denotes ScCe$_{0.95}$Yb$_{0.05}$O$_{3-\alpha}$, BCN denotes Ba$_3$(Ca$_{1.18}$Nb$_{1.82}$)O$_{9-\delta}$, BCZN denotes Ba$_3$CaZr$_{0.5}$Nb$_{1.5}$O$_{9-\delta}$, BCNN denotes Ba$_3$Ca$_{0.9}$Nd$_{0.28}$Nb$_{1.82}$O$_{9-\delta}$, BCS denotes BaCe$_{0.9}$Sm$_{0.1}$O$_{3-\delta}$, BCGS denotes BaCe$_{0.8}$Gd$_{0.1}$Sm$_{0.1}$O$_{3-\delta}$, BZCY denotes BaZr$_{0.7}$Ce$_{0.2}$Y$_{0.1}$O$_{3-\delta}$, LSGM denotes La$_{0.9}$Sr$_{0.1}$Ga$_{0.8}$Mg$_{0.2}$O$_{3-\alpha}$, LCZ denotes La$_{1.95}$Ca$_{0.05}$Zr$_2$O$_{7-\delta}$, LCC denotes La$_{1.95}$Ca$_{0.05}$Ce$_2$O$_{7-\delta}$, BCG1 denotes BaCe$_{0.8}$Gd$_{0.2}$O$_{3-\delta}$, BCY1 denotes BaCe$_{0.85}$Y$_{0.15}$O$_{3-\alpha}$, BCY2 denotes Ba$_{0.98}$Ce$_{0.8}$Y$_{0.2}$O$_{3-\alpha}$, CYO denotes Ce$_{0.8}$Y$_{0.2}$O$_{1.9}$, CSO denotes Ce$_{0.8}$Sm$_{0.2}$O$_{1.9}$, CGO denotes Ce$_{0.8}$Gd$_{0.2}$O$_{1.9}$, CLO denotes Ce$_{0.8}$La$_{0.2}$O$_{1.9}$, LCGM denotes La$_{0.9}$Ca$_{0.1}$Ga$_{0.8}$Mg$_{0.2}$O$_{3-\alpha}$, LSGM denotes La$_{0.9}$Sr$_{0.1}$Ga$_{0.8}$Mg$_{0.2}$O$_{3-\alpha}$, LBGM denotes La$_{0.9}$Ba$_{0.1}$Ga$_{0.8}$Mg$_{0.2}$O$_{3-\alpha}$, BCG2 denotes BaCe$_{0.85}$Gd$_{0.15}$O$_{3-\alpha}$, BCZS denotes BaCe$_{0.7}$Zr$_{0.2}$Sm$_{0.1}$O$_{3-\alpha}$, BCC denotes BaCe$_{0.9}$Ca$_{0.1}$O$_{3-\alpha}$, LSFC denotes La$_{0.6}$Sr$_{0.4}$Fe$_{0.8}$Cu$_{0.2}$O$_{3-\delta}$, SBCF denotes SmBaCuFeO$_{5+\delta}$, SBCC denotes SmBaCuCoO$_{5+\delta}$, SSC denotes Sm$_{0.5}$Sr$_{0.5}$CoO$_{3-\delta}$, SSN denotes Sm$_{1.5}$Sr$_{0.5}$NiO$_4$, SFCN denotes SmFe$_{0.7}$Cu$_{0.1}$Ni$_{0.2}$O$_3$.

4.3. Plasmacatalysis

The plasmacatalysis process was described as a possible alternative to many chemicals' high-temperature and pressure synthesis systems. In addition to positive and negative ions, plasma often contains a large number of neutral particles, such as atoms, molecules, radicals and excited particles, resulting in highly reactive physical and chemical reactions when used in chemical synthesis. The plasma's ionised and excited species concentration is considerably higher than the traditional thermally heated gas phases. These properties can, therefore, benefit from attaining further effective interaction even without a catalyst [145]. In the interaction of plasma and catalyst, plasma creates a more active spot yielding to higher catalytic activity. When the beneficial effect of plasma and catalyst is combined effectively, it is likely to produce a much higher yield [146].

Depending on the thermal equilibrium or not, plasma could be classified into thermal and non-thermal plasma (NTP). The temperature of plasma, like that of any other gas, is determined by the average energies of the plasma particles (neutral and charged) and their degrees of freedom (translational, rotational, vibrational and those related to electronic excitation) [147,148]. Plasmas can thus exhibit multiple temperatures as a multi-component system. In common electrical discharge for plasma generation, energy is transferred to heavy articles by collision with electron. In thermal plasma, electron and heavy particles achieved thermal equilibrium due to joule heating. Joule heating or ohmic heating define the process in which the energy of an electric current is converted into heat as it flows through a resistance. The temperature of the gas in thermal plasma is extremely high, typically ranging from 4000 K to 20,000 K. On the other hand, non-thermal plasma is characterised by multiple different temperatures related to different plasma particles and different degrees of freedom. In non-thermal plasma, thermal equilibrium between electron and heavy particles is not achieved, and the temperature of the NTP may be as low as room temperature, although the electron, the excited and the ionised species have a high temperature (Te >> T_0) [147].

The temperature of the gas in thermal plasma is extremely high, typically ranging from 4000 K to 20,000 K and is equivalent to that of the electron, which has achieved a thermodynamic equilibrium between the electron and other species. On the other hand, the temperature of the NTP may be as low as room temperature, although the electron, the excited and the ionised species have a high temperature. Since NTP offers less power input, this plasma is a more attractive option for chemical synthesis.

Many researchers have explored the mechanism of plasmacatalytic synthesis of ammonia since the 1900s. The first attempt to utilise plasma to synthesis ammonia can be traced back to 1929 when Brewer et al. [149] successfully synthesised ammonia using glow discharge plasma and achieved an energy yield of 3.03 g-NH$_3$/kWh. The system used was complex in that high voltage, vacuum and magnetic fields were applied. The magnetic field was reported to have no significant impact on the yield of ammonia. Since these experiments demonstrated the principle of using plasma to produce ammonia, researchers have begun to conduct detailed investigations into the process of plasma-assisted ammonia synthesis using other types of plasma. Table 9 summarises previous research on plasmacatalytic ammonia synthesis.

Table 9. Summary of previous research on plasma-chemical for ammonia synthesis.

Groups	Catalyst	Pressure (Torr)	H_2/N_2	Flow Rate (mL/min)	NH$_3$ Yield (%)	Energy Yield (g-NH$_3$/kWh)	Ref
GD	-	3	3	-	-	3.5	[149]
	Pt	50	3	833.3	2	-	[150]
	Pt	50	3	833.3	7.9	-	[151]
	Pt	7	2.5	-	8.1	0.12	[152]
	MgCl$_2$	10	3	-	-	-	[153]
	Ag	5	0.6	78	80.8	-	[154]
RF	-	4.97	4	20	0.27	-	[155]
	Fe	4.97	4	20	0.35	0.01	[156]
	Au	0.26	4	20	0.2	0.1	[157]
	Ga-In	0.26	4	20	0.3	0.3	[158]
	Ni-MOF-74	0.26	4	20	0.23	0.23	[159]
MW	Rh	2.25	0.56	7.5	11.25	0.01	[160]
	-	760	0.11	15,000	3.1	0.04	[161]
	Co/Al$_2$O$_3$	760	1.167	120	0.112	0.01	[162]
DBD	Pd	760	3	-	3.13	-	[163]
	MgO	760	0.8	2266.7	0.33	-	[164]
	-	760	3.56	730	1.36	1.83	[165]
	Ru/alumina	760	3	40	4.36	0.37	[166]
	Ru/alumina	760	3	30	4.62	0.40	[167]
	-	760	6	500	0.74	0.69	[168]
	PZT	760	1	11.5	5.9	0.7	[169]
	Alumina	760	3	60	0.67	0.18	[170]
	Cs-Ru/MgO	760	3	4000	2.41	2.3	[171]

Table 9. Cont.

Groups	Catalyst	Pressure (Torr)	H_2/N_2	Flow Rate (mL/min)	NH_3 Yield (%)	Energy Yield (g-NH_3/kWh)	Ref
	PZT	760	3	11.5	0.5	0.75	[172]
	Ru-Mg/alumina	760	4	2000	2.55	35.7	[173]
	Cu	760	1	100	3.5	3.3	[174]
	Ru/alumina	760	1.5	-	-	0.64	[175]
	Ni/silica with BaTiO$_3$	760	3	25	12	0.75	[176]
	Ru/alumina	760	3	1000	0.05	1.9	[176]
	Ru/Si-MCM-41	760	1	-	-	1.7	[177]
	Ni/alumina	760	2	100	2	0.89	[178]
	MgCl$_2$	760	1	4000	-	20.5	[179]

Note: GD: glow discharge, RF: radiofrequency, MW: microwave; DBD: dielectric barrier discharge.

Following the initial study, Eremin et al. [151] revealed that ammonia is formed by surface reactions. Afterwards, Venugopalan et al. [152] achieved high productivity of ammonia on Ag coated quartz. Uyama et al. [156,180] found the formation of nitride in addition to hydrazine and ammonia in their study. Nakajima and Sekiguchi [161] found that when plasma is generated by H_2/N_2 gas mixture, the nitrogen gas activation in the plasma has been depressed by hydrogen while hydrogen injection into the afterglow area increased the production of ammonia. In addition to the plasma mentioned above, dielectric barrier discharge was also widely explored. In 2000, researchers [164] tested MgO as a catalyst in combination with DBD. The result shows that the catalyst could increase the ammonia yield by up to 75% more than a plasma-assisted reaction. The group also studied the synthesis of ammonia from methane and nitrogen without any catalyst and achieved an energy yield of as much as 0.69 g-NH_3/kWh [168]. The use of porous materials for ammonia absorption was also studied by Peng et al. [171]. They found that by using a porous material for ammonia absorption, the rate of ammonia synthesis increases due to the lower gas phase of ammonia. The group also examined some metal catalysts as a promoter for improving Ru loaded on magnesium-oxide particles. Caesium (Cs) was reported as the best promoter and capable of achieving energy yield as high as 2.41 g-NH_3/kWh [171]. Akay and Zhang [181] performed research where barium titanate enhanced by Ni/SiO$_2$ was used as the catalyst. The overall energy yield of 1.9 g-NH_3/kWh was achieved by this configuration. More recently, in a pulsed DBD plasma reactor, Peng et al. [179] used MgCl$_2$ as a catalyst and ammonia absorber. The study discovered that MgCl$_2$ was effective to store generated ammonia for later extraction. The configuration also achieved a very high energy yield with 20.5 g-NH_3/kWh.

More recently, an attempt was made to synthesis NH_3 in the catalyst-free plasma–water interfaces system based on the batch reactor process. Breakthroughs have recently occurred in which plasmas in contact with water surfaces have achieved significant results, putting the interfaces between plasma and water as an NH_3 reaction locus by using a combination of electrochemical and plasma. In these studies, the metal anode was replaced with N_2-plasma gas and successfully produced up to 0.44 mg/hour of NH_3 on 1 mm^2 plasma–liquid interfaces [182].

5. Ammonia as a Renewable Fuel

5.1. Direct Fuel Cell

Fuel cells are currently intensively examined as a breakthrough candidate for carbon-free power generation. The device provides a highly efficient conversion directly from chemical to electricity and has a low environmental footprint. In an early study, ammonia, which has 17% hydrogen by weight, was proposed for use in PEM fuel cells. Though, due to the low operating temperature of the PEM fuel cell, the thermodynamics decomposition of ammonia cannot occur [183]. On top of that, ammonia is lethal to the Nafion membrane utilised in PEM fuel cells. Thus, external cracking reactors are required to completely convert ammonia into hydrogen, giving an extra energy input and additional costs [184].

Based on electrolyte and reaction, a direct ammonia-fed fuel cell can be divided into three major systems, alkaline fuel cell (AFC), alkaline membrane fuel cell (AMFC) and solid oxide fuel cells (SOFC). The discussion of each configuration is provided below. Ammonia has been reported as a feed for a fuel cell as early as the 1960s based on the alkaline fuel cell developed by Francis Thomas Bacon [185]. The cells use alkaline electrolytes such as potassium hydroxide (KOH) and platinum cathodes. Most recently, Hejze et al. [186] reported the potential of molten hydroxide (NaOH/KOH) as an electrolyte. Unfortunately, the use of KOH and NaOH is not favourable for air-intake fuel cells since it reacts with CO_2 to form K_2CO_3 and Na_2CO_3 and degrades the performance of the alkaline electrolyte.

Recently, alkaline membrane fuel cells (AMFCs) gained attention from the fuel cell society due to the compatibility with CO_2. As reported by Unlu et al. [187], CO_2 introduced in the cathode has a positive effect on improving fuel cell performance. In a recent development, room temperature AMFC has been developed by Lan and Tao [185]. Compared to fuel cells based on acidic polymer electrolytes, low-cost non-precious catalysts, including MnO_2, silver or nickel, may be used for AMFCs [37]. Moreover, Pt/C, PtRu/C and Ru/C were recently investigated AMFCs and can also be used as anodes [188].

Other types of ammonia fuel cells, namely SOFC, are initially developed to prevent NO_X formation [189]. However, the number of scientists who studied SOFC becomes more intense caused by the potencies of the cell to operate at high temperature, thus overcoming the disadvantage suffered by PEMs. At high temperatures, ammonia can be directly decomposed into hydrogen, normally ranging between 500 and 1000 °C, and hence the need for an external cracking reactor is negated. In addition, there was no evidence of ammonia having a bad effect on the ceramic electrolytes used in SOFCs [190]. Nonetheless, because of the fragility of porcelain materials, SOFCs are usually not appropriate for transport use [191].

Research on SOFC fuel cells can be separated into Oxygen Ion-Conducting Electrolytes (SOFC-O) and Hydrogen Ion-Conducting Electrolyte (SOFC-H), which is also known as proton-conducting electrolytes. The schematic of SOFC-O and SOFC-H fuel cells is illustrated in Figure 7.

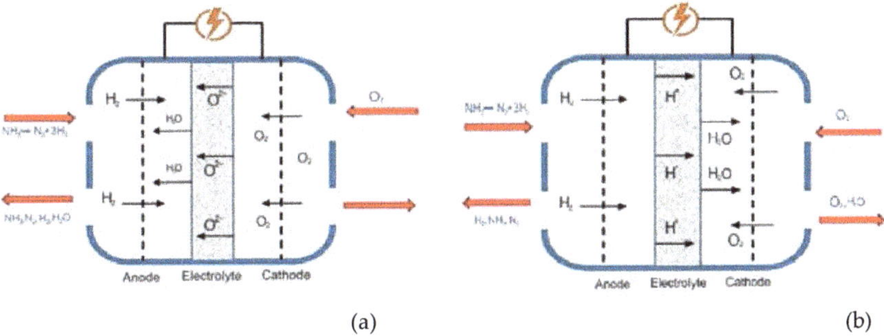

Figure 7. Schematic representation of (**a**) SOFC-O and (**b**) SOFC-H [27].

The SOFC-O operating principle lies in the transportation of oxygen anions across the electrolyte while the charge carrier in SOFC-H is a proton [192]. For both types, ammonia is fed into the anodic site, where it thermally decomposes into nitrogen and hydrogen [193]. In SOFC-O, the oxygen in the cathode compartment is reduced into oxygen ions at the cathode–electrolyte interface and transported across the solid electrolyte, which then reacts with hydrogen electrochemically to produce water [27,194]. The reactions that occur at the anode and cathode are stated below:

$$\text{Anode}: H_2 + O^{2-} \rightarrow H_2O + 2e^- \tag{9}$$

$$\text{Cathode}: \frac{1}{2}O_2 + 2e^- \rightarrow O_2^- \tag{10}$$

The SOFC-O electrolytes tend to be built based on solid ceramics with metal oxide. YSZ is most widely used because of its high ionic conversion, which facilitates the efficient movement of oxygen anions through the electrolyte [195]. These solid electrolytes also show strong chemical and thermal stability, which is crucial for the treatment of high temperatures. Samarium doped ceria (SDC)-based electrolytes also sparked interest due to its capability to have high ionic conductivities at lower temperatures [196].

In SOFC-H, the hydrogen in the anode compartment is oxidised into a proton which is then transported across solid electrolyte into the cathode [197]. This later reacts with oxygen to produce water. The reactions that occur at the anode and cathode are given below:

$$\text{Anode}: H_2 \rightarrow 2H^+ + 2e^- \tag{11}$$

$$\text{Cathode}: O_2 + 2H^+ + 2e^- \rightarrow H_2O \tag{12}$$

An SOFC-H system electrolyte is selected based on the conductivity of a proton as well as chemical and mechanical stability. The extraordinarily high proton conductivities of doped $BaCeO_3$(BCO) and $BaZrO_3$(BZO) have been shown over a wide 300 to 1000 °C temperature range [198].

In addition to the qualities of the employed material, electrolyte thickness has a direct effect on fuel cell performance. When a thinner electrolyte has been used, the internal resistance of SOFC decreases. However, reducing electrolyte thickness can affect the mechanical strength and consequently stability over the long term [198]. Table 10 summarises the preceding SOFC fuel cell work that turns ammonia into electricity.

Table 10. Summarization of previous research on SOFC for ammonia synthesis.

Groups	Electrolyte	Thickness (µm)	Cathode	Anode	T(°C)	Power Density (mW/cm^2)	Ref
SOFC-O	YSZ	200	Ag	Pt-YSZ	800–1000	50–125	[199]
	YSZ	400	Ag planar	Ni-YSZ	800	75	[200]
	YSZ	1000	Ag tubular	Ni-YSZ	800	10	[200]
	YSZ	400	Ag	NiO-YSZ	800	60	[196]
	YSZ	150	LSM	NiO-YSZ	700	55	[201]
	YSZ	30	YSZ -LSM	Ni-YSZ	750–850	299	[201]
	YSZ	15	YSZ -LSM	Ni-YSZ	800	526	[202]
	SDC	24	SSC-SDC	NiO-SDC	650	467	[203]
	SDC	10	BSCF	Ni-SDC	700	1190	[204]
SOFC-H	BCGP	1300	Pt	Pt	700	35	[205]
	BCG	1300	Pt	Pt	700	25	[206]
	BCE	1000	Pt	Pt	700	32	[207]
	BCGP	1000	Pt	Ni-BCE	600	23	[208]
	BCG	50	LSC	Ni-BCG	700	355	[209]
	BZCY	35	BSCF	Ni- BZCY	450–700	135–420	[210]
	BCG	30	BSCF	Ni-CGO	600	147	[211]
	BCN	20	LSC	NiO-BCN	700	315	[212]

Note: YSZ denotes yttria-stabilized zirconia, SDC denotes samarium doped ceria, LSM denotes $La_{0.5}Sr_{0.5}MnO_3$, SSC denotes $Sm_{0.5}Sr_{0.5}Co_{3-\delta}$, BSCF denotes $Ba_{0.5}Sr_{0.5}Co_{0.8}Fe_{0.2}O_{3-\delta}$, BCGP denotes $BaCe_{0.8}Gd_{0.19}Pr_{0.01}O_{3-\delta}$, BCG denotes $BaCe_{0.8}Gd_{0.2}O_{3-\delta}$, BCE denotes $BaCe_{0.85}Eu_{0.15}O_3$, LSC denotes $La_{0.5}Sr_{0.5}CoO_{3-\delta}$, BZCY denotes $BaZr_{0.1}Ce_{0.7}Y_{0.2}O_{3-\delta}$, BSCF denotes $Ba_{0.5}Sr_{0.5}Co_{0.8}Fe_{0.2}O_{3-\delta}$, CGO denotes $Ce_{0.8}Gd_{0.2}O_{1.9}$, BCN denotes $BaCe_{0.9}Nd_{0.1}O_{3-\delta}$.

In addition to all of the above forms, microbial fuel cells (MFC) are also seen as an alternate technique for generating electricity directly from ammonia. MFC uses microorganisms in the oxidation process for the conversion of chemical energy from bio-degradable material, for example, ammonia contaminated wastewater. The electrons flow from the anodic side of the external circuit to the cathode, where they combine with the proton and oxygen to form water [198]. The schematic diagram of MFC is shown in Figure 8.

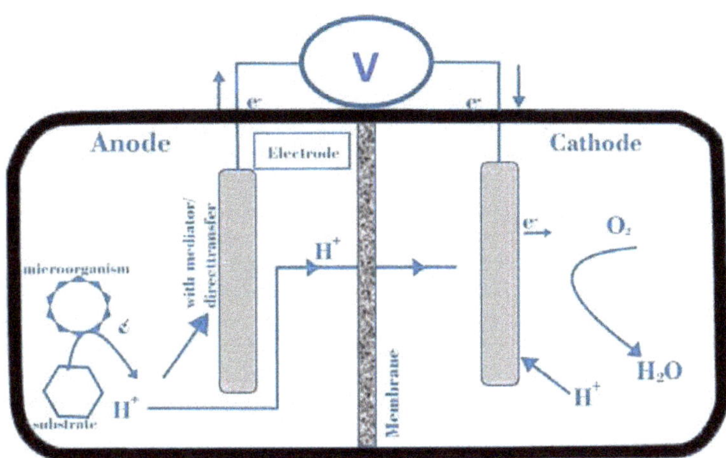

Figure 8. Schematic diagram of MFC [213].

According to Li et al. [214], MFC has been deemed a potential technique for treating wastewater while producing energy, but low power, high cost and reactor scalability issues severely limit its advancement. In addition to wastewater treatment, MFC technology has been proposed as a feasible alternative for air cleaning by removing ammonia from the environment. Yan and Liu [215], in 2020, found Sn-doped V_2O_5 nanoparticles to be a good catalyst for the rapid removal of ammonia in the air using photo-electrocatalysis (PEC) MFCs.

5.2. IC Engine

One of the main industries contributing to GHG emissions worldwide is the transport industry. The search for the right alternative energy source to reduce fossil fuel addiction has been a long, challenging journey. Ammonia-fuelled vehicles, which have received a lot of publicity recently, are one of the solutions to reduce GHG emissions and fossil fuel dependency. Ammonia can be used as fuel in both spark ignition (SI) and compression ignition (CI). Numerous companies and research groups have tested these engines in the last decades. However, ammonia poses undesirable combustion properties, which require further study of its combustion properties. Table 11 compares the characteristics of ammonia with other IC engine fuels.

Table 11. Comparison of NH_3 combustion properties with other fuels at 27 °C [216–220].

	NH₃	H₂	MeOH	DME	Gasoline	Diesel
Storage	L	C	L	L	L	L
Storage pressure (MPa)	1.1	70	0.1	0.5	0.1	0.1
Density (kg m⁻³)	600	39.1	784.6	668	740	820
Laminar burning velocity (m s⁻¹)	0.07	3.51	0.36	0.54	0.58	1.28
Low heating value (MJ/kg)	18.8	120	19.92	28.43	42.9	44.41
Latent heat of vaporization (kJ/kg)	1369	0	1100	467	71.78	47.86
Minimum ignition energy (MJ)	8	0.011	0.14	0.29	0.24	0.24
Auto-ignition temperature (K)	930	773–850	712	598–623	530.37	588.7
Octane/cetane number	130	>100	119	55–65	90–98	40–55
AFT (K)	1850	2483	1910	-	2138	-
HCR	1.32	1.41	1.20	-	1.28	-
Explosion limit in air (% vol)	15–28	4.7–75	6.7–36	3.2–18.6	0.6–8	0.6–5.5
Gravimetric hydrogen density (%)	17.8	100	12.5	13	13	12.75

Note: L (liquid); C (compressed); AFT (adiabatic flame temperature); HCR (heat capacity ratio).

Ammonia has low flame velocity, very high auto-ignition temperatures, narrow flammability limits and high vaporisation heat compared to other fuels [217]. Narrow flammability limits and high auto-ignition temperatures create problems for NH_3 to be used in the engine [221]. Although NH_3 can be used as single fuel in the CI engine, extremely higher CR is required to auto-ignite the fuel [222,223]. In addition, high latent heat of vaporisation at the time of injection decreases the gas temperature in the engine, which further complicates it [221]. In the SI engine, the use of ammonia is restricted by low flame velocity and narrow explosion limits resulting in incomplete combustion [224]. Ammonia combustion in the SI-engine can be improved by providing stronger igniters such as plasma jet igniters, smaller combustion chambers to overcome the resistance of NH_3 combustion [223]. Supercharging can also achieve improved combustion [225].

In addition to the problems mentioned above, ammonia also shows low flame speed and specific energy in combination with high ignition energies and high auto-ignition temperatures, resulting in a relatively low propagation rate from the combustion [16]. Although ammonia has been successfully used as mono-fuel both in SI and CI engines, such a low ammonia combustion rate induces inconsistency in combustion under conditions of low engine load or high engine speed [225]. Thereby, it is essential to mix with secondary fuel to overcome its disadvantages as a fuel. In addition to ammonia, potential fuels to be used in SI engines are hydrogen, methanol, ethanol, ethane and gasoline. For CI engines, fuels with higher cetane numbers are preferred as a combustion promoter due to the better ignition characteristics [223]. However, these approaches require some special features [16].

On a dual fuel CI engine, ammonia could be used up to 95% of the fuel energy basis with diesel as a combustion promoter. An optimal mixture, however, is 60% of ammonia on an energy basis because a smaller amount of ammonia would limit the flammability of ammonia [226]. Other studies suggested that an optimal content of ammonia is between 60–80% based on the mass basis [225]. A demonstration of biodiesel as a combustion promoter by Kong et al. [227] revealed that the fuel performed similar engine performance characteristics with ammonia/diesel blends. The operating characteristics are, however, different when Dimethyl Ether (DME) is used as the ammonia fuel combustion promoter. The study shows that ammonia could be used up to 80% only when DME is used as a combustion promoter [217]. Moreover, the study also revealed that the fuel mix of ammonia and DME has a competitive energy cost with current diesel fuel. Even the addition of NH_3 in the blend has been shown to significantly raise emissions of CO, HC and NOx [228].

On the SI engine, gasoline is used as a combustion promoter in most of the studies. A compression ratio of 10:1 is required to get optimal operation of the engine with ammonia content 70% [229]. Gaseous fuels are chosen for SI-engines due to the same phase with ammonia gas, while anhydrous ammonia would lower the temperature of the in-cylinder, thus adversely affecting subsequent turbulence triggering deteriorated combustion and misfire [228]. In the SI engine powered by ammonia/gasoline, gasoline is port-injected while ammonia gas is direct-injected. Direct injection of ammonia gas substantially lowers the cylinder temperature due to the high latent heat of ammonia. Thus, creating turbulence in the combustion chamber will enhance the combustion of the fuel. However, too small swirls do not affect the combustion, while too large swirls have a negative effect on the combustion by blowing out the flames due to the slow propagation of the ammonia flames [223].

Another alternative is by using hydrogen as a secondary fuel by installing an on-board reformer to split ammonia into hydrogen and nitrogen [230]. Morch et al. [219] give a complete database of SI engine performance with ammonia/hydrogen mixture as a fuel. Ammonia and hydrogen have, in this investigation, been incorporated into the CFR engine intake manifold. A series of studies were undertaken with different excess air ratios and hydrogen ammonia ratios. The results revealed that a fuel mixture with 10 vol.% hydrogens has the best performance in terms of efficiency and power. In a comparison study with gasoline, there is a high possibility that an increase in efficiency and power is caused by a greater compression ratio. The analysis of the system has also shown that most of the heat required can be covered by the exhaust heat. The diagram showing the fuel system setup for the experiments is given in Figure 9.

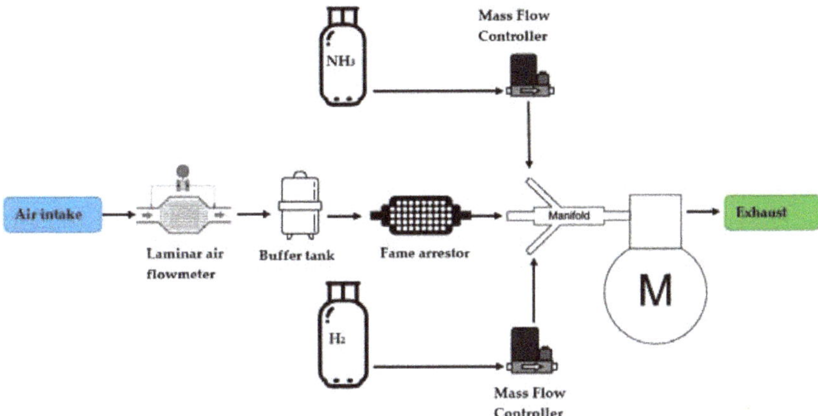

Figure 9. Diagram showing the fuel system setup for the experiments [219].

Much farther trials of ammonia/hydrogen-fuelled engines were also applied to more commercial applications in Italy, where a prototype for electric vehicles with a 15 kW engine was built. The study found that the optimal performance in full load is achieved with 7% of hydrogen content while only 5% in half-load [219]. More recently, Ezzat and Dincer [231] proposed, thermodynamically analysed and compared two different integrated systems of an ammonia/hydrogen-fuelled engine. The first system is made up of hydrogen production. In the second system, an ammonia fuel cell is added to complement the IC engine. The study shows that the first system has higher energy and exergy efficiency with 61.89% and 63.34%, compared to the second system with 34.73% and 38.44%, respectively. The study also shows that when compared with pure ammonia injection, the use of hydrogen from cracked ammonia is extremely beneficial. In addition to road transportations, ammonia is also a favourable fuel for marine industries [232,233]. Unlike the automotive industry, marine systems are not space-constrained, so that catalytic equipment can be deployed for NOx reduction solutions. In the recent development, MAN Energy Solutions replaces the 3000 B&W double-fuel engines operating in the field of LPG and diesel engines [234].

5.3. Gas Turbine

The ammonia-fuelled gas turbine seems destined to become one of the key technologies in the sustainable energy economy of the future. In the 1960s, an early attempt was made to use ammonia as fuel in the gas turbine. However, due to nitrogen molecule in its structure, ammonia combustion is always associated with the formation of nitrogen oxides which exceed current standards [235,236]. Moreover, a longer residence time in the combustion chamber is required for ammonia to be completely combustible due to low laminar burning velocity, which makes it difficult to achieve stable combustion [16]. It also might cause an ammonia slip [237]. Therefore, ammonia-fuelled gas turbines were poorly studied in early development. However, the shift towards carbon-free alternative energy carriers has returned interest in ammonia, including the utilisation in power industries. Burning NH_3 in turbines is the most promising direction of using ammonia as a carrier of surplus electricity generated from renewable energy to balance seasonal energy demand. Thus, many efforts have been devoted to overcoming these shortcomings.

Kobayashi et al. [238] have produced a review article covering the current ammonia combustion research and future directions. The study emphasised that the final product of ammonia combustion is not nitric oxides because the overall reaction is $4NH_3 + 3O_2 \rightarrow 2N_2 + 6H_2O$. This implies that the configuration of the parameters of the combustion system is crucial to perform the reaction according to this direction. Karabeyoglu et al. [239] developed a test rig and conducted a series of trials with a pre-burner system to partially crack ammonia into H_2. The study revealed that ammonia combustion could be self-sustained when 10% cracking is applied. In 2014, Iki et al. [240] developed a 50 kWe micro gas-turbine system that enables a bi-fuel supply of kerosene and ammonia. The system was able to achieve over 25-kWe power generation by supplying about 10% heat from ammonia gas. In 2015, Iki et al. [241] were able to achieve 21 kWe by replacing the standard combustor with a prototype combustor. The gas turbine performance showed an efficiency of combustion up to 96% with NOx emission above 1000 ppm at 16% O_2. Hayakawa et al. [242] investigated stretching limits for high-pressure flames and observed lower NO formation with higher mixture pressure. Okafor et al. [243], in their experiment and numerical calculations, concluded that predominant rate-limiting reactions in methane–ammonia flames are belonging to the ammonia oxidation path, which controls H and OH radicals. These radicals influence the burning velocity.

$$NO + NH_2 \rightarrow NNH + OH \tag{13}$$

$$NH_2 + O \rightarrow HNO + H \tag{14}$$

$$HNO + H \rightarrow NO + H_2 \tag{15}$$

The study also revealed that the NO concentration decreases when ammonia increases under rich conditions.

In 2016, Ito et al. [244] developed a gas-turbine combustion system with controlled emissions, which uses a mixture of NH_3 and natural gas as the fuel. Combustion properties have been explored via the use of a swirl burner, generally employed in gas turbines, experimentally and numerically. Detailed compositions of the burner exhaust gas were measured under atmospheric pressure and lean fuel circumstances. The results show that with this system, the combustion efficiency above 97% can be achieved for an ammonia mixing ratio below 50%. The study also shows that with an increase of equivalency ratio, unburned species such as NH_3, CO and THC decrease while NO and N_2O emissions increase. The study concludes that low emissions and good combustion efficiency are difficult to obtain in a single-stage reactor.

In 2017, Onishi et al. [245] attempted to create novel ways for reducing NOx emissions while burning an ammonia-natural gas combination in a gas turbine combustor. The concept of low-emission combustion in two-stage combustion was examined numerically and experimentally. In the main zone, methane and ammonia are used as fuel. The secondary zone is then only supplied ammonia. The results indicated two methods for attaining low NOx combustion: rich-lean combustion and a combination of lean combustion and extra ammonia delivery. In the first technique, NOx is created only in the main zone when the fuel is abundant, and the burnt gas is diluted in the secondary zone by secondary air. As a result, primary zone NOx production dominates emission. A lean combustion state in the primary zone results in a low temperature and oxygen concentration in the secondary zone in the second approach. The NOx concentration at the combustor outlet is low as a result of these circumstances. These expected combustion qualities have been experimentally validated. The experimental findings showed that the NOx emission behaviour corresponded to the numerical results.

In 2018, Ito et al. [246] studied a mixture of ammonia and natural gas fuel in a 2 MWe gas turbine. Their results indicated that ammonia is suitable for use in large turbines. Before ammonia is fed to the combustor, the gas turbine power is raised up to 2 MWe using natural gas. The ammonia's heat input ratio to total fuel is used to calculate the ammonia feed to the engine. The result shows that the gas turbine engine's operation was shown to be steady across a wide variety of ammonia mixing ratios ranging from 0% to 20%. As the ratio of ammonia in the mixture increases, the concentration of NOx at the turbine outlet rises significantly up to 5% mixing ratio, then remains steady until it reaches 20%.

6. Conclusions

Ammonia is among the most commonly shipped bulk-produced chemicals, marketed for more than a decade in mass all over the globe. Originally used in the chemical industries and as an intermediate for the production of fertilisers, ammonia has also been explored recently as a hydrogen storage media and a substitute fuel for hydrocarbon. Unlike the conventional ammonia production process that used natural gas as a feedstock and is responsible for carbon emission, ammonia is a means of renewable energy storage formulated from H_2 generated by an electrically driven electrolyser and N_2 separate from the atmospheric air. In addition to that, innovative approaches, such as photocatalysis, electrocatalysis and plasmacatalysis, have attracted widespread interest in ammonia synthesis today. Thus, the application of ammonia as a renewable energy carrier not only plays a key role to lower GHG emissions but also allows transporting H_2 efficiently and economically, permits the direct conversion to electricity by fuel cell and provides versatility in its use, as fuel for the IC engine and power generation. Thus, the ease of processing, transporting and using NH_3 makes it an appealing choice to serve as the link between renewable energy production and demands. However, work is still needed to improve the efficiency of the conversion process for the chemical to compete with hydrocarbon fuel.

Author Contributions: Original draft preparation, M.H.H.; Supervision, T.M.I.M. and M.M.; Review and Editing, H.C.O. and I.M.R.F. Revision, A.S.S.; and F.H. All authors have read and agreed to the published version of the manuscript.

Funding: This research received no external funding.

Acknowledgments: This study was carried out under the International Research Scholarship (IRS) and UTS Presidential Scholarship (UTSP) program, funded by the University of Technology Sydney, Australia.

Conflicts of Interest: The authors declare no conflict of interest.

References

1. Cavicchioli, R.; Ripple, W.J.; Timmis, K.N.; Azam, F.; Bakken, L.R.; Baylis, M.; Behrenfeld, M.J.; Boetius, A.; Boyd, P.W.; Classen, A.T. Scientists' warning to humanity: Microorganisms and climate change. *Nat. Rev. Microbiol.* **2019**, *17*, 569–586. [CrossRef] [PubMed]
2. Edenhofer, O. *Climate Change 2014: Mitigation of Climate Change*; Cambridge University Press: Cambridge, UK, 2015; Volume 3.
3. Lüthi, D.; Le Floch, M.; Bereiter, B.; Blunier, T.; Barnola, J.-M.; Siegenthaler, U.; Raynaud, D.; Jouzel, J.; Fischer, H.; Kawamura, K. High-resolution carbon dioxide concentration record 650,000–800,000 years before present. *Nature* **2008**, *453*, 379–382. [CrossRef] [PubMed]
4. Pachauri, R.K.; Allen, M.R.; Barros, V.R.; Broome, J.; Cramer, W.; Christ, R.; Church, J.A.; Clarke, L.; Dahe, Q.; Dasgupta, P. *Climate Change 2014: Synthesis Report. Contribution of Working Groups I, II and III to the Fifth Assessment Report of the Intergovernmental Panel on Climate Change*; IPCC: Geneva, Switzerland, 2014.
5. United Nations Framework Convention on Climate Change. 1992. Available online: https://unfccc.int/files/essential_background/background_publications_htmlpdf/application/pdf/conveng.pdf (accessed on 13 January 2021).
6. Olivier, G.J.; Peters, J.A. *Trends In Global CO$_2$ and Total Greenhouse Gas Emissions: 2017 Report*; PBL Netherlands Environmental Assessment Agency: The Hague, The Netherland, 2017.
7. Jiang, X.; Guan, D. The global CO$_2$ emissions growth after international crisis and the role of international trade. *Energy Policy* **2017**, *109*, 734–746. [CrossRef]
8. Protocol, K. United Nations framework convention on climate change. *Kyoto Protoc.* **1997**, *19*. Available online: https://unfccc.int/kyoto_protocol (accessed on 18 January 2021).
9. International Cooperation on Climate Change; United Nations Framework Convention on Climate Change (UNFCCC). Available online: https://www.dfat.gov.au/international-relations/themes/climate-change/Pages/international-cooperation-on-climate-change (accessed on 19 May 2020).
10. Appl, M. The Haber-Bosch heritage: The ammonia production technology. In Proceedings of the 50th Anniversary of the IFA Technical Conference, Seville, Spain, 25–26 September 1997.
11. Klein, D.; Carazo, M.P.; Doelle, M.; Bulmer, J.; Higham, A. *The Paris Agreement on Climate Change: Analysis and Commentary*; Oxford University Press: Oxford, UK, 2017.
12. Myllyvirta, L. Analysis: Coronavirus Temporarily Reduced China's CO2 Emissions by a Quarter; Carbon Brief. 2020. Available online: https://www.carbonbrief.org/analysis-coronavirus-has-temporarily-reduced-chinas-co2-emissions-by-a-quarter (accessed on 8 January 2021).

13. Zhang, R.; Zhang, Y.; Lin, H.; Feng, X.; Fu, T.-M.; Wang, Y. NOx Emission Reduction and Recovery during COVID-19 in East China. *Atmosphere* **2020**, *11*, 433. [CrossRef]

14. Viswanathan, B. Chapter 9—Hydrogen as an Energy Carrier. In *Energy Sources*; Viswanathan, B., Ed.; Elsevier: Amsterdam, The Netherlands, 2017; pp. 161–183.

15. Edwards, P.; Kuznetsov, V.; David, W. Hydrogen energy. *Philos. Trans. R. Soc. A Math. Phys. Eng. Sci.* **2007**, *365*, 1043–1056. [CrossRef]

16. Valera-Medina, A.; Xiao, H.; Owen-Jones, M.; David, W.I.; Bowen, P.J. Ammonia for power. *Prog. Energy Combust. Sci.* **2018**, *69*, 63–102. [CrossRef]

17. Paris 2015: Tracking Country Climate Pledges. 2017. Available online: https://www.carbonbrief.org/paris-2015-tracking-country-climate-pledges (accessed on 24 June 2020).

18. Innovation Promotion Program Energy Carriers. SIP 2016. Available online: https://www8.cao.go.jp/cstp/panhu/sip_english/20-23.pdf (accessed on 8 January 2021).

19. Crolius, S.H. Ammonia Included in Japan's International Resource Strategy. 2020. Available online: https://www.ammoniaenergy.org/articles/ammonia-included-in-japans-international-resource-strategy/ (accessed on 18 May 2020).

20. Brown, T. Green Ammonia Plants Win Financing in Australia and New Zealand. 2020. Available online: https://www.ammoniaenergy.org/articles/green-ammonia-plants-win-financing-in-australia-and-new-zealand/ (accessed on 19 May 2020).

21. Service, R.F. Ammonia—A Renewable Fuel Made From Sun, Air, and Water—Could Power the Globe Without Carbon. 2018. Available online: https://www.sciencemag.org/news/2018/07/ammonia-renewable-fuel-made-sun-air-and-water-could-power-globe-without-carbon# (accessed on 17 May 2020).

22. Pattabathula, V.; Richardson, J. Introduction to Ammonia Production. 2016. Available online: https://www.aiche.org/resources/publications/cep/2016/september/introduction-ammonia-production (accessed on 15 May 2020).

23. Nitrogen Statistics and Information. 2020. Available online: https://www.usgs.gov/centers/nmic/nitrogen-statistics-and-information (accessed on 20 May 2021).

24. Mineral. Commodity Summaries. 2020. Available online: https://pubs.usgs.gov/periodicals/mcs2020/mcs2020,pdf (accessed on 18 June 2020).

25. Egenhofer, C.; Schrefler, L.; Rizos, V.; Marcu, A.; Genoese, F.; Renda, A.; Wieczorkiewicz, J.; Roth, S.; Infelise, F.; Luchetta, G. The Composition and Drivers of Energy Prices and Costs in Energy-Intensive Industries: The Case of Ceramics, Glass and Chemicals. 2014. Available online: http://aei.pitt.edu/50255/1/CEPS_Energy_Prices_Study_Consolidated_version.pdf (accessed on 14 January 2021).

26. Patil, B.; Wang, Q.; Hessel, V.; Lang, J. Plasma N2-fixation: 1900–2014. *Catal. Today* **2015**, *256*, 49–66. [CrossRef]

27. Afif, A.; Radenahmad, N.; Cheok, Q.; Shams, S.; Kim, J.H.; Azad, A.K. Ammonia-fed fuel cells: A comprehensive review. *Renew. Sustain. Energy Rev.* **2016**, *60*, 822–835. [CrossRef]

28. Davis, B.L.; Dixon, D.A.; Garner, E.B.; Gordon, J.C.; Matus, M.H.; Scott, B.; Stephens, F.H. Efficient regeneration of partially spent ammonia borane fuel. *Angew. Chem. Int. Ed.* **2009**, *48*, 6812–6816. [CrossRef]

29. Leigh, G. Haber-Bosch and Other Industrial Processes. In *Catalysts for Nitrogen Fixation*; Springer: Dordrecht, The Netherlands, 2004; pp. 33–54. [CrossRef]

30. Boerner, L.K. Industrial ammonia production emits more CO_2 than any other chemical-making reaction. *Chem. Eng. News* **2019**, *97*. Available online: https://cen.acs.org/environment/green-chemistry/Industrial-ammonia-production-emits-CO2/97/i24#:~{}:text=It%20belched%20up%20to%20about,reaction%20(see%20page%2023) (accessed on 12 February 2021).

31. Snoeckx, R.; Bogaerts, A. Plasma technology–a novel solution for CO 2 conversion? *Chem. Soc. Rev.* **2017**, *46*, 5805–5863. [CrossRef]

32. MacFarlane, D.R.; Choi, J.; Suryanto, B.H.; Jalili, R.; Chatti, M.; Azofra, L.M.; Simonov, A.N. Liquefied sunshine: Transforming renewables into fertilizers and energy carriers with electromaterials. *Adv. Mater.* **2020**, *32*, 1904804. [CrossRef]

33. Yan, Z.; Ji, M.; Xia, J.; Zhu, H. Recent Advanced Materials for Electrochemical and Photoelectrochemical Synthesis of Ammonia from Dinitrogen: One Step Closer to a Sustainable Energy Future. *Adv. Energy Mater.* **2020**, *10*, 1902020. [CrossRef]

34. Xu, H.; Ithisuphalap, K.; Li, Y.; Mukherjee, S.; Lattimer, J.; Soloveichik, G.; Wu, G. Electrochemical Ammonia Synthesis through N_2 and H_2O under Ambient Conditions: Theory, Practices, and Challenges for Catalysts and Electrolytes. *Nano Energy* **2020**, *69*, 104469. [CrossRef]

35. MacFarlane, D.R.; Cherepanov, P.V.; Choi, J.; Suryanto, B.H.R.; Hodgetts, R.Y.; Bakker, J.M.; Vallana, F.M.F.; Simonov, A.N. A Roadmap to the Ammonia Economy. *Joule* **2020**, *4*, 1186–1205. [CrossRef]

36. Wang, L.; Xia, M.; Wang, H.; Huang, K.; Qian, C.; Maravelias, C.T.; Ozin, G.A. Greening ammonia toward the solar ammonia refinery. *Joule* **2018**, *2*, 1055–1074. [CrossRef]

37. Lan, R.; Irvine, J.T.; Tao, S. Ammonia and related chemicals as potential indirect hydrogen storage materials. *Int. J. Hydrog. Energy* **2012**, *37*, 1482–1494. [CrossRef]

38. Kojima, Y. High Purity Hydrogen Generation from Ammonia. 2017. Available online: https://ep70.eventpilotadmin.com/web/page.php?page=IntHtml&project=ACS17FALL&id=2747276 (accessed on 22 January 2021).

39. Makepeace, J.W.; He, T.; Weidenthaler, C.; Jensen, T.R.; Chang, F.; Vegge, T.; Ngene, P.; Kojima, Y.; de Jongh, P.E.; Chen, P. Reversible ammonia-based and liquid organic hydrogen carriers for high-density hydrogen storage: Recent progress. *Int. J. Hydrog. Energy* **2019**, *44*, 7746–7767. [CrossRef]

40. Bartels, R.J.; Pate, M.B.; Olson, N.K. An economic survey of hydrogen production from conventional and alternative energy sources. *Int. J. Hydrog. Energy* **2010**, *35*, 8371–8384. [CrossRef]
41. Schultz, K.R. *Use of the Modular Helium Reactor for Hydrogen Production*; General Atomics (US): San Diego, CA, USA, 2003. Available online: https://www.osti.gov/biblio/821808-use-modular-helium-reactor-hydrogen-production (accessed on 3 February 2021).
42. Orhan, M.F.; Dincer, I.; Naterer, G.F. Cost analysis of a thermochemical Cu–Cl pilot plant for nuclear-based hydrogen production. *Int. J. Hydrog. Energy* **2008**, *33*, 6006–6020. [CrossRef]
43. Charvin, P.; Stéphane, A.; Florent, L.; Gilles, F. Analysis of solar chemical processes for hydrogen production from water splitting thermochemical cycles. *Energy Convers. Manag.* **2008**, *49*, 1547–1556. [CrossRef]
44. Bartels, J.R. A Feasibility Study of Implementing an Ammonia Economy. Master's Theses, Iowa State University, Iowa, IA, USA, 2008.
45. Andersson, J.; Grönkvist, S. Large-scale storage of hydrogen. *Int. J. Hydrog. Energy* **2019**, *44*, 11901–11919. [CrossRef]
46. Anhydrous Ammonia Transportation Information. 2007. Available online: http://www.mda.state.mn.us/chemicals/spills/ammoniaspills/transportation.htm (accessed on 15 June 2020).
47. Paster, M. Hydrogen Deliver Options and Issues. Available online: http://www1.eere.energy.gov/hydrogenandfuelcells/analysis/pdfs/paster_h2_delivery.pdf (accessed on 15 June 2020).
48. Appl, M. *Ammonia: Principles and Industrial Practice*; Vch Verlagsgesellschaft Mbh: Weinheim, Germany, 1999. Available online: https://trove.nla.gov.au/work/32581098 (accessed on 15 June 2020).
49. Christensen, C.H.; Johannessen, T.; Sørensen, R.Z.; Nørskov, J.K. Towards an ammonia-mediated hydrogen economy? *Catal. Today* **2006**, *111*, 140–144. [CrossRef]
50. Christensen, C.H.; Sørensen, R.Z.; Johannessen, T.; Quaade, U.J.; Honkala, K.; Elmøe, T.D.; Køhler, R.; Nørskov, J.K. Metal ammine complexes for hydrogen storage. *J. Mater. Chem.* **2005**, *15*, 4106–4108. [CrossRef]
51. Haynes, W.M. *CRC Handbook of Chemistry and Physics*; CRC Press: Boca Raton, FL, USA, 2014.
52. Karabeyoglu, A.; Evans, B. Fuel conditioning system for ammonia fired power plants. In Proceedings of the NH3 Congress, Ames, IA, USA, 1 October 2012; Available online: https://www.ammoniaenergy.org/wp-content/uploads/2021/01/evans-brian.pdf (accessed on 23 January 2021).
53. Cole-Parmer Instrument Company. Chemical Compatibility Database, Ammonia, Anhydrous. 2017. Available online: https://www.coleparmer.co.uk/chemical-resistance (accessed on 10 June 2021).
54. Pincha, M.E.; Heizer, B.L.; McHale, M.P. *Material Compatibility Problems for Ammonia Systems*; SAE Technical Paper: Washington, DC, USA, 1988.
55. Graco, Chemical Compatibility Guide. 2013. Available online: https://www.graco.com/content/dam/graco/ipd/literature/misc/chemical-compatibility-guide/Graco_ChemCompGuideEN-B.pdf (accessed on 9 June 2021).
56. CDC. Ammonia. Available online: https://www.cdc.gov/niosh/idlh/7664417 (accessed on 26 May 2020).
57. Ravishankara, A.; Daniel, J.S.; Portmann, R.W. Nitrous oxide (N_2O): The dominant ozone-depleting substance emitted in the 21st century. *Science* **2009**, *326*, 123–125. [CrossRef]
58. Klerke, A.; Christensen, C.H.; Nørskov, J.K.; Vegge, T. Ammonia for hydrogen storage: Challenges and opportunities. *J. Mater. Chem.* **2008**, *18*, 2304–2310. [CrossRef]
59. Organization, W.H. *Ammonia Health and Safety Guide-Health and Safety Guide 37*; International Programme on Chemical Safety: Geneva, Switzerland, 1990.
60. Ryer-Powder, J.E. Health effects of ammonia. *Plant/Oper. Prog.* **1991**, *10*, 228–232. [CrossRef]
61. Smith, C.; Hill, A.K.; Torrente-Murciano, L. Current and future role of Haber-Bosch ammonia in a carbon-free energy landscape. *Energy Environ. Sci.* **2020**, *13*, 331–344. [CrossRef]
62. Wilkinson, I. Green Ammonia. 2017. Available online: Siemens.com (accessed on 23 January 2021).
63. Schmuecker, J. Schmuecker Pinehurst Farm, l.L.C. Available online: http://solarhydrogensystem.com/ (accessed on 16 June 2020).
64. Cholteeva, Y. Sable Chemicals and Tatanga Energy to Construct Solar Energy Plant in Zimbabwe. Available online: https://www.power-technology.com/news/sable-chemicals-and-tatanga-energy-to-construct-solar-energy-plant-in-zimbabwe/ (accessed on 18 June 2020).
65. Solar Ammonia. 2017. Available online: https://www.yara.com/ (accessed on 13 June 2020).
66. Brown, T. Renewable Ammonia Demonstration Plant Announced in South Australia. 2018. Available online: https://ammoniaindustry.com/renewable-ammonia-demonstration-plant-announced-in-south-australia/ (accessed on 14 June 2020).
67. Rusanov, V.D.; Fridman, A.A.; Sholin, G.V. The physics of a chemically active plasma with nonequilibrium vibrational excitation of molecules. *Sov. Phys. Uspekhi* **1981**, *24*, 447. [CrossRef]
68. Chen, X.; Li, N.; Kong, Z.; Ong, W.-J.; Zhao, X. Photocatalytic fixation of nitrogen to ammonia: State-of-the-art advancements and future prospects. *Mater. Horiz.* **2018**, *5*, 9–27. [CrossRef]
69. Schrauzer, G.; Guth, T. Photolysis of water and photoreduction of nitrogen on titanium dioxide. *J. Am. Chem. Soc.* **2002**, *99*, 7189–7193. [CrossRef]
70. Bourgeois, S.; Diakite, D.; Perdereau, M. A study of TiO2 powders as a support for the photochemical synthesis of ammonia. *React. Solids* **1988**, *6*, 95–104. [CrossRef]

71. Hirakawa, H.; Hashimoto, M.; Shiraishi, Y.; Hirai, T. Photocatalytic conversion of nitrogen to ammonia with water on surface oxygen vacancies of titanium dioxide. *J. Am. Chem. Soc.* **2017**, *139*, 10929–10936. [CrossRef] [PubMed]
72. Ranjit, K.T.; Varadarajan, T.K.; Viswanathan, B. Photocatalytic reduction of dinitrogen to ammonia over noble-metal-loaded TiO_2. *J. Photochem. Photobiol. A Chem.* **1996**, *96*, 181–185. [CrossRef]
73. Lashgari, M.; Zeinalkhani, P. Photocatalytic N2 conversion to ammonia using efficient nanostructured solar-energy-materials in aqueous media: A novel hydrogenation strategy and basic understanding of the phenomenon. *Appl. Catal. A Gen.* **2017**, *529*, 91–97. [CrossRef]
74. Sun, S.; An, Q.; Wang, W.; Zhang, L.; Liu, J.; Goddard, W.A., III. Efficient photocatalytic reduction of dinitrogen to ammonia on bismuth monoxide quantum dots. *J. Mater. Chem. A* **2017**, *5*, 201–209. [CrossRef]
75. Hu, S.; Chen, X.; Li, Q.; Zhao, Y.; Mao, W. Effect of sulfur vacancies on the nitrogen photofixation performance of ternary metal sulfide photocatalysts. *Catal. Sci. Technol.* **2016**, *6*, 5884–5890. [CrossRef]
76. Ye, L.; Han, C.; Ma, Z.; Leng, Y.; Li, J.; Ji, X.; Bi, D.; Xie, H.; Huang, Z. Ni_2P loading on $Cd_{0.5}Zn_{0.5}S$ solid solution for exceptional photocatalytic nitrogen fixation under visible light. *Chem. Eng. J.* **2017**, *307*, 311–318. [CrossRef]
77. Wei, P.; Yang, Q.; Guo, L. Bismuth oxyhalide compounds as photocatalysts. *Prog. Chem.* **2009**, *21*, 1734–1741.
78. Dong, G.; Ho, W.; Wang, C. Selective photocatalytic N_2 fixation dependent on gC_3N_4 induced by nitrogen vacancies. *J. Mater. Chem. A* **2015**, *3*, 23435–23441. [CrossRef]
79. Hu, S.; Chen, X.; Li, Q.; Li, F.; Fan, Z.; Wang, H.; Wang, Y.; Zheng, B.; Wu, G. Fe^{3+} doping promoted N_2 photofixation ability of honeycombed graphitic carbon nitride: The experimental and density functional theory simulation analysis. *Appl. Catal. B Environ.* **2017**, *201*, 58–69. [CrossRef]
80. Ileperuma, O.A.; Tennakone, K.; Dissanayake, W.D.D.P. Photocatalytic behaviour of metal doped titanium dioxide: Studies on the photochemical synthesis of ammonia on Mg/TiO_2 catalyst systems. *Appl. Catal.* **1990**, *62*, L1–L5. [CrossRef]
81. Soria, J.; Conesa, J.; Augugliaro, V.; Palmisano, L.; Schiavello, M.; Sclafani, A. Dinitrogen photoreduction to ammonia over titanium dioxide powders doped with ferric ions. *J. Phys. Chem.* **1991**, *95*, 274–282. [CrossRef]
82. Ileperuma, O.A.; Thaminimulla, C.T.K.; Kiridena, W.C.B. Photoreduction of N_2 to NH_3 and H_2O to H_2 on metal doped TiO_2 catalysts (M = Ce, V). *Sol. Eergy Mater. Sol. Cells* **1993**, *28*, 335–343. [CrossRef]
83. Zhao, W.; Zhang, J.; Zhu, X.; Zhang, M.; Tang, J.; Tan, M.; Wang, Y. Enhanced nitrogen photofixation on Fe-doped TiO_2 with highly exposed (1, 0 1) facets in the presence of ethanol as scavenger. *Appl. Catal. B Environ.* **2014**, *144*, 468–477. [CrossRef]
84. Janet, C.; Navaladian, S.; Viswanathan, B.; Varadarajan, T.; Viswanath, R. Heterogeneous wet chemical synthesis of superlattice-type hierarchical ZnO architectures for concurrent H_2 production and N_2 reduction. *J. Phys. Chem. C* **2010**, *114*, 2622–2632. [CrossRef]
85. Khader, M.M.; Lichtin, N.N.; Vurens, G.H.; Salmeron, M.; Somorjai, G.A. Photoassisted catalytic dissociation of water and reduction of nitrogen to ammonia on partially reduced ferric oxide. *Langmuir* **1987**, *3*, 303–304. [CrossRef]
86. Hao, Y.; Dong, X.; Zhai, S.; Ma, H.; Wang, X.; Zhang, X. Hydrogenated bismuth molybdate nanoframe for efficient sunlight-driven nitrogen fixation from air. *Chem.–A Eur. J.* **2016**, *22*, 18722–18728. [CrossRef]
87. Miyama, H.; Fujii, N.; Nagae, Y. Heterogeneous photocatalytic synthesis of ammonia from water and nitrogen. *Chem. Phys. Lett.* **1980**, *74*, 523–524. [CrossRef]
88. Khan, T.M.M.; Bhardwaj, R.C.; Bhardwaj, C. Catalytic Fixation of Nitrogen by the Photocatalytic $CdS/Pt/RuO_2$ Particulate System in the Presence of Aqueous [Ru (Hedta) N_2]⊖ Complex. *Angew. Chem. Int. Ed. Engl.* **1988**, *27*, 923–925. [CrossRef]
89. Sun, S.; Li, X.; Wang, W.; Zhang, L.; Sun, X. Photocatalytic robust solar energy reduction of dinitrogen to ammonia on ultrathin MoS_2. *Appl. Catal. B Environ.* **2017**, *200*, 323–329. [CrossRef]
90. Cao, Y.; Hu, S.; Li, F.; Fan, Z.; Bai, J.; Lu, G.; Wang, Q. Photofixation of atmospheric nitrogen to ammonia with a novel ternary metal sulfide catalyst under visible light. *RSC Adv.* **2016**, *6*, 49862–49867. [CrossRef]
91. Zhang, Q.; Hu, S.; Fan, Z.; Liu, D.; Zhao, Y.; Ma, H.; Li, F. Preparation of $gC_3N_4/ZnMoCdS$ hybrid heterojunction catalyst with outstanding nitrogen photofixation performance under visible light via hydrothermal post-treatment. *Dalton Trans.* **2016**, *45*, 3497–3505. [CrossRef] [PubMed]
92. Hu, S.; Li, Y.; Li, F.; Fan, Z.; Ma, H.; Li, W.; Kang, X. Construction of $g\text{-}C_3N_4/Zn_{0.11}Sn_{0.12}Cd_{0.88}S_{1.12}$ hybrid heterojunction catalyst with outstanding nitrogen photofixation performance induced by sulfur vacancies. *ACS Sustain. Chem. Eng.* **2016**, *4*, 2269–2278. [CrossRef]
93. Bai, Y.; Ye, L.; Chen, T.; Wang, L.; Shi, X.; Zhang, X.; Chen, D. Facet-dependent photocatalytic N2 fixation of bismuth-rich Bi5O7I nanosheets. *ACS Appl. Mater. Interfaces* **2016**, *8*, 27661–27668. [CrossRef]
94. Li, H.; Shang, J.; Shi, J.; Zhao, K.; Zhang, L. Facet-dependent solar ammonia synthesis of BiOCl nanosheets via a proton-assisted electron transfer pathway. *Nanoscale* **2016**, *8*, 1986–1993. [CrossRef]
95. Li, H.; Shang, J.; Ai, Z.; Zhang, L. Efficient visible light nitrogen fixation with BiOBr nanosheets of oxygen vacancies on the exposed facets. *J. Am. Chem. Soc.* **2015**, *137*, 6393–6399. [CrossRef]
96. Wang, S.; Hai, X.; Ding, X.; Chang, K.; Xiang, Y.; Meng, X.; Yang, Z.; Chen, H.; Ye, J. Light-Switchable Oxygen Vacancies in Ultrafine Bi5O7Br Nanotubes for Boosting Solar-Driven Nitrogen Fixation in Pure Water. *Adv. Mater.* **2017**, *29*, 1701774. [CrossRef]
97. Wang, Y.; Wei, W.; Li, M.; Hu, S.; Zhang, J.; Feng, R. In situ construction of Z-scheme $gC_3N_4/Mg_{1.1}Al_{0.3}Fe_{0.2}O_{1.7}$ nanorod heterostructures with high N_2 photofixation ability under visible light. *RSC Adv.* **2017**, *7*, 18099–18107. [CrossRef]

98. Hu, S.; Zhang, W.; Bai, J.; Lu, G.; Zhang, L.; Wu, G. Construction of a 2D/2D gC$_3$N$_4$/rGO hybrid heterojunction catalyst with outstanding charge separation ability and nitrogen photofixation performance via a surface protonation process. *RSC Adv.* **2016**, *6*, 25695–25702. [CrossRef]

99. Cao, S.; Zhou, N.; Gao, F.; Chen, H.; Jiang, F. All-solid-state Z-scheme 3, 4-dihydroxybenzaldehyde-functionalized Ga$_2$O$_3$/graphitic carbon nitride photocatalyst with aromatic rings as electron mediators for visible-light photocatalytic nitrogen fixation. *Appl. Catal. B Environ.* **2017**, *218*, 600–610. [CrossRef]

100. Liang, H.; Zou, H.; Hu, S. Preparation of the W$_{18}$O$_{49}$/gC$_3$N$_4$ heterojunction catalyst with full-spectrum-driven photocatalytic N2 photofixation ability from the UV to near infrared region. *New J. Chem.* **2017**, *41*, 8920–8926. [CrossRef]

101. Cong, L.; Yu, Z.; Liu, F.; Huang, W. Electrochemical synthesis of ammonia from N$_2$ and H$_2$O using a typical non-noble metal carbon-based catalyst under ambient conditions. *Catal. Sci. Technol.* **2019**, *9*, 1208–1214. [CrossRef]

102. Kyriakou, V.; Garagounis, I.; Vasileiou, E.; Vourros, A.; Stoukides, M. Progress in the electrochemical synthesis of ammonia. *Catal. Today* **2017**, *286*, 2–13. [CrossRef]

103. Garagounis, I.; Vourros, A.; Stoukides, D.; Dasopoulos, D.; Stoukides, M. Electrochemical Synthesis of Ammonia: Recent Efforts and Future Outlook. *Membranes* **2019**, *9*, 112. [CrossRef]

104. Wang, B.H.; de Wang, J.; Liu, R.; Xie, Y.H.; Li, Z.J. Synthesis of ammonia from natural gas at atmospheric pressure with doped ceria–Ca$_3$(PO$_4$)$_2$–K$_3$PO$_4$ composite electrolyte and its proton conductivity at intermediate temperature. *J. Solid State Electrochem.* **2007**, *11*, 27–31. [CrossRef]

105. Wang, B.; Liu, R.; Wang, J.; Li, Z.; Xie, Y. Doped ceria-Ca-3 (PO4) 2-K3PO4 composite electrolyte: Proton conductivity at intermediate temperature and application in atmospheric pressure ammonia synthesis. *Chin. J. Inorg. Chem.* **2005**, *21*, 1551–1555.

106. Skodra, A.; Stoukides, M. Electrocatalytic synthesis of ammonia from steam and nitrogen at atmospheric pressure. *Solid State Ion.* **2009**, *180*, 1332–1336. [CrossRef]

107. Jeoung, H.; Kim, J.N.; Yoo, C.-Y.; Joo, J.H.; Yu, J.H.; Song, K.C.; Sharma, M.; Yoon, H.C. Electrochemical synthesis of ammonia from water and nitrogen using a Pt/GDC/Pt cell. *Korean Chem. Eng. Res.* **2014**, *52*, 58–62. [CrossRef]

108. Murakami, T.; Nishikiori, T.; Nohira, T.; Ito, Y. Electrolytic synthesis of ammonia in molten salts under atmospheric pressure. *J. Am. Chem. Soc.* **2003**, *125*, 334–335. [CrossRef]

109. Murakami, T.; Nohira, T.; Goto, T.; Ogata, Y.H.; Ito, Y. Electrolytic ammonia synthesis from water and nitrogen gas in molten salt under atmospheric pressure. *Electrochim. Acta* **2005**, *50*, 5423–5426. [CrossRef]

110. Murakami, T.; Nohira, T.; Araki, Y.; Goto, T.; Hagiwara, R.; Ogata, Y.H. Electrolytic Synthesis of Ammonia from Water and Nitrogen under Atmospheric Pressure Using a Boron-Doped Diamond Electrode as a Nonconsumable Anode. *Electrochem. Solid State Lett.* **2007**, *10*, E4. [CrossRef]

111. Murakami, T.; Nishikiori, T.; Nohira, T.; Ito, Y. Electrolytic ammonia synthesis from hydrogen chloride and nitrogen gases with simultaneous recovery of chlorine under atmospheric pressure. *Electrochem. Solid-State Lett.* **2005**, *8*, D19–D21. [CrossRef]

112. Murakami, T.; Nohira, T.; Ogata, Y.H.; Ito, Y. Electrolytic ammonia synthesis in molten salts under atmospheric pressure using methane as a hydrogen source. *Electrochem. Solid-State Lett.* **2005**, *8*, D12–D14. [CrossRef]

113. Murakami, T.; Nohira, T.; Ogata, Y.H.; Ito, Y. Electrochemical synthesis of ammonia and coproduction of metal sulfides from hydrogen sulfide and nitrogen under atmospheric pressure. *J. Electrochem. Soc.* **2005**, *152*, D109–D112. [CrossRef]

114. Licht, S.; Cui, B.; Wang, B.; Li, F.-F.; Lau, J.; Liu, S. Ammonia synthesis by N$_2$ and steam electrolysis in molten hydroxide suspensions of nanoscale Fe$_2$O$_3$. *Science* **2014**, *345*, 637–640. [CrossRef]

115. Kordali, V.; Kyriacou, G.; Lambrou, C. Electrochemical synthesis of ammonia at atmospheric pressure and low temperature in a solid polymer electrolyte cell. *Chem. Commun.* **2000**, *17*, 1673–1674. [CrossRef]

116. Marnellos, G.; Stoukides, M. Ammonia synthesis at atmospheric pressure. *Science* **1998**, *282*, 98–100. [CrossRef]

117. Li, Z.; Liu, R.; Xie, Y.; Feng, S.; Wang, J. A novel method for preparation of doped Ba$_3$(Ca$_{1.18}$Nb$_{1.82}$)O$_{9-\delta}$: Application to ammonia synthesis at atmospheric pressure. *Solid State Ion.* **2005**, *176*, 1063–1066. [CrossRef]

118. Li, Z.; Liu, R.; Wang, J.; Xu, Z.; Xie, Y.; Wang, B. Preparation of double-doped BaCeO$_3$ and its application in the synthesis of ammonia at atmospheric pressure. *Sci. Technol. Adv. Mater.* **2007**, *8*, 566. [CrossRef]

119. Vasileiou, E.; Kyriakou, V.; Garagounis, I.; Vourros, A.; Stoukides, M. Ammonia synthesis at atmospheric pressure in a BaCe$_{0.2}$Zr$_{0.7}$Y$_{0.1}$O$_{2.9}$ solid electrolyte cell. *Solid State Ion.* **2015**, *275*, 110–116. [CrossRef]

120. Zhang, F.; Yang, Q.; Pan, B.; Xu, R.; Wang, H.; Ma, G. Proton conduction in La$_{0.9}$Sr$_{0.1}$Ga$_{0.8}$Mg$_{0.2}$O$_{3-\alpha}$ ceramic prepared via microemulsion method and its application in ammonia synthesis at atmospheric pressure. *Mater. Lett.* **2007**, *61*, 4144–4148. [CrossRef]

121. Wang, J.-D.; Xie, Y.-H.; Zhang, Z.-F.; Liu, R.-Q.; Li, Z.-J. Protonic conduction in Ca$_{2+}$-doped La$_2$M$_2$O$_7$ (M= Ce, Zr) with its application to ammonia synthesis electrochemically. *Mater. Res. Bull.* **2005**, *40*, 1294–1302. [CrossRef]

122. Xie, Y.-H.; Wang, J.-D.; Liu, R.-Q.; Su, X.-T.; Sun, Z.-P.; Li, Z.-J. Preparation of La$_{1.9}$Ca$_{0.1}$Zr$_2$O$_{6.95}$ with pyrochlore structure and its application in synthesis of ammonia at atmospheric pressure. *Solid State Ion.* **2004**, *168*, 117–121. [CrossRef]

123. Li, Z.-J.; Liu, R.-Q.; Wang, J.-D.; Xie, Y.-H.; Yue, F. Preparation of BaCe$_{0.8}$Gd$_{0.2}$O$_{3-\delta}$ by the citrate method and its application in the synthesis of ammonia at atmospheric pressure. *J. Solid State Electrochem.* **2005**, *9*, 201–204. [CrossRef]

124. Guo, Y.; Liu, B.; Yang, Q.; Chen, C.; Wang, W.; Ma, G. Preparation via microemulsion method and proton conduction at intermediate-temperature of BaCe$_{1-x}$YxO$_{3-\alpha}$. *Electrochem. Commun.* **2009**, *11*, 153–156. [CrossRef]

125. Vasileiou, E.; Kyriakou, V.; Garagounis, I.; Vourros, A.; Manerbino, A.; Coors, W.; Stoukides, M. Electrochemical enhancement of ammonia synthesis in a $BaZr_{0.7}Ce_{0.2}Y_{0.1}O_{2.9}$ solid electrolyte cell. *Solid State Ion.* **2016**, *288*, 357–362. [CrossRef]

126. Zhang, M.; Xu, J.; Ma, G. Proton conduction in $Ba_xCe_{0.8}Y_{0.2}O_{3-\alpha}$ + 0.04ZnO at intermediate temperatures and its application in ammonia synthesis at atmospheric pressure. *J. Mater. Sci.* **2011**, *46*, 4690–4694. [CrossRef]

127. Liu, R.-Q.; Xie, Y.-H.; Wang, J.-D.; Li, Z.-J.; Wang, B.-H. Synthesis of ammonia at atmospheric pressure with $Ce_{0.8}M_{0.2}O_{2-\delta}$ (M= La, Y, Gd, Sm) and their proton conduction at intermediate temperature. *Solid State Ion.* **2006**, *177*, 73–76. [CrossRef]

128. Chen, C.; Wang, W.; Ma, G. Proton conduction in $La_{0.9}M_{0.1c}Ga_{0.8}Mg_{0.2}O_{3-\alpha}$ at intermediate temperature and its application to synthesis of ammonia at atmospheric pressure. *Acta Chim. Sin.* **2009**, *67*, 623–628.

129. Chen, C.; Ma, G. Preparation, proton conduction, and application in ammonia synthesis at atmospheric pressure of $La_{0.9}Ba_{0.1}Ga_{1-x}Mg_xO_{3-\alpha}$. *J. Mater. Sci.* **2008**, *43*, 5109–5114. [CrossRef]

130. Chen, C.; Ma, G. Proton conduction in $BaCe_{1-x}Gd_xO_{3-\alpha}$ at intermediate temperature and its application to synthesis of ammonia at atmospheric pressure. *J. Alloy. Compd.* **2009**, *485*, 69–72. [CrossRef]

131. Wang, X.; Yin, J.; Xu, J.; Wang, H.; Ma, G. Chemical stability, ionic conductivity of $BaCe_{0.9-x}Zr_xSm_{0.10}O_{3-\alpha}$ and its application to ammonia synthesis at atmospheric pressure. *Chin. J. Chem.* **2011**, *29*, 1114–1118. [CrossRef]

132. Liu, J.; Li, Y.; Wang, W.; Wang, H.; Zhang, F.; Ma, G. Proton conduction at intermediate temperature and its application in ammonia synthesis at atmospheric pressure of $BaCe_{1-x}Ca_xO_{3-\alpha}$. *J. Mater. Sci.* **2010**, *45*, 5860–5864. [CrossRef]

133. Amar, I.A.; Petit, C.T.; Zhang, L.; Lan, R.; Skabara, P.J.; Tao, S. Electrochemical synthesis of ammonia based on doped-ceria-carbonate composite electrolyte and perovskite cathode. *Solid State Ion.* **2011**, *201*, 94–100. [CrossRef]

134. Murakami, T.; Nishikiori, T.; Nohira, T.; Ito, Y. Investigation of anodic reaction of electrolytic ammonia synthesis in molten salts under atmospheric pressure. *J. Electrochem. Soc.* **2005**, *152*, D75–D78. [CrossRef]

135. Li, F.-F.; Licht, S. Advances in understanding the mechanism and improved stability of the synthesis of ammonia from air and water in hydroxide suspensions of nanoscale Fe_2O_3. *Inorg. Chem.* **2014**, *53*, 10042–10044. [CrossRef]

136. Cui, B.; Zhang, J.; Liu, S.; Liu, X.; Xiang, W.; Liu, L.; Xin, H.; Lefler, M.J.; Licht, S. Electrochemical synthesis of ammonia directly from N_2 and water over iron-based catalysts supported on activated carbon. *Green Chem.* **2017**, *19*, 298–304. [CrossRef]

137. Zhang, Z.; Zhong, Z.; Ruiquan, L. Cathode catalysis performance of $SmBaCuMO_{5+\delta}$ (M= Fe, Co, Ni) in ammonia synthesis. *J. Rare Earths* **2010**, *28*, 556–559. [CrossRef]

138. Lan, R.; Irvine, J.T.; Tao, S. Synthesis of ammonia directly from air and water at ambient temperature and pressure. *Sci. Rep.* **2013**, *3*, 1–7. [CrossRef]

139. Wang, J.; Liu, R.-Q. Property research of SDC and SSC in ammonia synthesis at atmospheric pressure and low temperature. *Acta Chim. Sin* **2008**, *66*, 717–721.

140. Liu, R.; Xu, G. Comparison of electrochemical synthesis of ammonia by using sulfonated polysulfone and nafion membrane with $Sm_{1.5}Sr_{0.5}NiO_4$. *Chin. J. Chem.* **2010**, *28*, 139–142. [CrossRef]

141. Xu, G.; Liu, R.; Wang, J. Electrochemical synthesis of ammonia using a cell with a Nafion membrane and $SmFe_{0.7}Cu_{0.3-x}Ni_xO_3$ (x= 0−0.3) cathode at atmospheric pressure and lower temperature. *Sci. China Ser. B Chem.* **2009**, *52*, 1171–1175. [CrossRef]

142. Zhao, X.; Yin, F.; Liu, N.; Li, G.; Fan, T.; Chen, B. Highly efficient metal–organic-framework catalysts for electrochemical synthesis of ammonia from N 2 (air) and water at low temperature and ambient pressure. *J. Mater. Sci.* **2017**, *52*, 10175–10185. [CrossRef]

143. Liu, H.-M.; Han, S.-H.; Zhao, Y.; Zhu, Y.-Y.; Tian, X.-L.; Zeng, J.-H.; Jiang, J.-X.; Xia, B.Y.; Chen, Y. Surfactant-free atomically ultrathin rhodium nanosheet nanoassemblies for efficient nitrogen electroreduction. *J. Mater. Chem. A* **2018**, *6*, 3211–3217. [CrossRef]

144. Song, Y.; Johnson, D.; Peng, R.; Hensley, D.K.; Bonnesen, P.V.; Liang, L.; Huang, J.; Yang, F.; Zhang, F.; Qiao, R. A physical catalyst for the electrolysis of nitrogen to ammonia. *Sci. Adv.* **2018**, *4*, e1700336. [CrossRef] [PubMed]

145. Carreon, M.L. Plasma catalytic ammonia synthesis: State of the art and future directions. *J. Phys. D Appl. Phys.* **2019**, *52*, 483001. [CrossRef]

146. Chen, G.; Wang, L.; Godfroid, T.; Snyders, R. *Progress in Plasma-Assisted Catalysis for Carbon Dioxide Reduction, In Plasma Chemistry and Gas Conversion*; IntechOpen: London, UK, 2018.

147. Fridman, A. *Plasma Chemistry*; Cambridge University Press: Cambridge, UK, 2008.

148. Bruggeman, P.; Sadeghi, N.; Schram, D.; Linss, V. Gas temperature determination from rotational lines in non-equilibrium plasmas: A review. *Plasma Sources Sci. Technol.* **2014**, *23*, 023001. [CrossRef]

149. Brewer, K.A.; Westhaver, J.W. The Synthesis of Ammonia in the Glow Discharge. *J. Phys. Chem.* **1928**, *33*, 883–895. [CrossRef]

150. Maltsev, A.; Churina, L.; Eremin, E. Activity of heterogeneous catalysts in synthesis of ammonia in glow discharge. *Russ. J. Phys. Chem. USSR* **1968**, *42*, 1235.

151. Eremin, E.N.; Maltsev, A.N.; Belova, V.M. Behaviour of a Catalyst in a Glow-discharge Plasma. *Russ. J. Phys. Chem.* **1969**, *43*, 443.

152. Yin, S.K.; Venugopalan, M. Plasma chemical synthesis. I. *Effect of electrode material on the synthesis of ammonia. Plasma Chem. Plasma Process.* **1983**, *3*, 343–350. [CrossRef]

153. Sugiyama, K.; Akazawa, K.; Oshima, M.; Miura, H.; Matsuda, T.; Nomura, O. Ammonia synthesis by means of plasma over MgO catalyst. *Plasma Chem. Plasma Process.* **1986**, *6*, 179–193. [CrossRef]

154. Touvelle, M.; Licea, J.M.; Venugopalan, M. Plasma chemical synthesis. II. *Effect of wall surface on the synthesis of ammonia. Plasma Chem. Plasma Process.* **1987**, *7*, 101–108. [CrossRef]

155. Uyama, H.; Uchikura, T.; Niijima, H.; Matsumoto, O. Synthesis of ammonia with RF discharge. Adsorption of products on zeolite. *Chem. Lett.* **1987**, *16*, 555–558. [CrossRef]
156. Uyama, H.; Matsumoto, O. Synthesis of ammonia in high-frequency discharges. *Plasma Chem. Plasma Process.* **1989**, *9*, 13–24. [CrossRef]
157. Shah, J.; Wang, W.; Bogaerts, A.; Carreon, M.L. Ammonia synthesis by radio frequency plasma catalysis: Revealing the underlying mechanisms. *ACS Appl. Energy Mater.* **2018**, *1*, 4824–4839. [CrossRef]
158. Shah, R.J.; Harrison, J.M.; Carreon, M.L. Ammonia plasma-catalytic synthesis using low melting point alloys. *Catalysts* **2018**, *8*, 437. [CrossRef]
159. Shah, J.; Wu, T.; Lucero, J.; Carreon, M.A.; Carreon, M.L. Nonthermal Plasma Synthesis of Ammonia over Ni-MOF-74. *ACS Sustain. Chem. Eng.* **2018**, *7*, 377–383. [CrossRef]
160. Siemsen, L.G. The Synthesis of Ammonia from Hydrogen and Atomic Nitrogen on the Rh Surfac. 1990. Available online: https://lib.dr.iastate.edu/cgi/viewcontent.cgi?article=12220&context=rtd (accessed on 9 January 2021).
161. Nakajima, J.; Sekiguchi, H. Synthesis of ammonia using microwave discharge at atmospheric pressure. *Thin Solid Film.* **2008**, *516*, 4446–4451. [CrossRef]
162. Bai, X.; Tiwari, S.; Robinson, B.; Killmer, C.; Li, L.; Hu, J. Microwave catalytic synthesis of ammonia from methane and nitrogen. *Catal. Sci. Technol.* **2018**, *8*, 6302–6305. [CrossRef]
163. Eremin, E.; Maltsev, A.; Syaduk, V. Catalytic synthesis of ammonia in a barrier discharge. *Russ. J. Phys. Chem. USSR* **1971**, *45*, 635.
164. Mingdong, B.; Xiyao, B.; Zhitao, Z.; Mindi, B. Synthesis of ammonia in a strong electric field discharge at ambient pressure. *Plasma Chem. Plasma Process.* **2000**, *20*, 511–520. [CrossRef]
165. Bai, M.; Zhang, Z.; Bai, X.; Bai, M.; Ning, W. Plasma synthesis of ammonia with a microgap dielectric barrier discharge at ambient pressure. *IEEE Trans. Plasma Sci.* **2003**, *31*, 1285–1291.
166. Mizushima, T.; Matsumoto, K.; Sugoh, J.-I.; Ohkita, H.; Kakuta, N. Tubular membrane-like catalyst for reactor with dielectric-barrier-discharge plasma and its performance in ammonia synthesis. *Appl. Catal. A Gen.* **2004**, *265*, 53–59. [CrossRef]
167. Mizushima, T.; Matsumoto, K.; Ohkita, H.; Kakuta, N. Catalytic effects of metal-loaded membrane-like alumina tubes on ammonia synthesis in atmospheric pressure plasma by dielectric barrier discharge. *Plasma Chem. Plasma Process.* **2007**, *27*, 1–11. [CrossRef]
168. Bai, M.; Zhang, Z.; Bai, M.; Bai, X.; Gao, H. Synthesis of ammonia using CH_4/N_2 plasmas based on micro-gap discharge under environmentally friendly condition. *Plasma Chem. Plasma Process.* **2008**, *28*, 405–414. [CrossRef]
169. Bai, M.; Zhang, Z.; Bai, M.; Bai, X.; Gao, H. Conversion of methane to liquid products, hydrogen, and ammonia with environmentally friendly condition-based microgap discharge. *J. Air Waste Manag. Assoc.* **2008**, *58*, 1616–1621. [CrossRef]
170. Hong, J.; Aramesh, M.; Shimoni, O.; Seo, D.H.; Yick, S.; Greig, A.; Charles, C.; Prawer, S.; Murphy, A.B. Plasma catalytic synthesis of ammonia using functionalized-carbon coatings in an atmospheric-pressure non-equilibrium discharge. *Plasma Chem. Plasma Process.* **2016**, *36*, 917–940. [CrossRef]
171. Peng, P.; Li, Y.; Cheng, Y.; Deng, S.; Chen, P.; Ruan, R. Atmospheric pressure ammonia synthesis using non-thermal plasma assisted catalysis. *Plasma Chem. Plasma Process.* **2016**, *36*, 1201–1210. [CrossRef]
172. Gómez-Ramírez, A.; Montoro-Damas, A.M.; Cotrino, J.; Lambert, R.M.; González-Elipe, A.R. About the enhancement of chemical yield during the atmospheric plasma synthesis of ammonia in a ferroelectric packed bed reactor. *Plasma Process. Polym.* **2017**, *14*, 1600081. [CrossRef]
173. Kim, H.H.; Teramoto, Y.; Ogata, A.; Takagi, H.; Nanba, T. Atmospheric-pressure nonthermal plasma synthesis of ammonia over ruthenium catalysts. *Plasma Process. Polym.* **2017**, *14*, 1600157. [CrossRef]
174. Aihara, K.; Akiyama, M.; Deguchi, T.; Tanaka, M.; Hagiwara, R.; Iwamoto, M. Remarkable catalysis of a wool-like copper electrode for NH 3 synthesis from N 2 and H 2 in non-thermal atmospheric plasma. *Chem. Commun.* **2016**, *52*, 13560–13563. [CrossRef]
175. Xie, D.; Sun, Y.; Zhu, T.; Fan, X.; Hong, X.; Yang, W. Ammonia synthesis and by-product formation from H_2O, H_2 and N_2 by dielectric barrier discharge combined with an Ru/Al_2O_3 catalyst. *RSC Adv.* **2016**, *6*, 105338–105346. [CrossRef]
176. Patil, B. *Plasma (Catalyst)-Assisted Nitrogen Fixation: Reactor Development for Nitric Oxide and Ammonia Production*; Micro Flow Chemistry and Synthetic Methodology: Eindhoven, The Netherlands, 2017.
177. Peng, P.; Cheng, Y.; Hatzenbeller, R.; Addy, M.; Zhou, N.; Schiappacasse, C.; Chen, D.; Zhang, Y.; Anderson, E.; Liu, Y. Ru-based multifunctional mesoporous catalyst for low-pressure and non-thermal plasma synthesis of ammonia. *Int. J. Hydrog. Energy* **2017**, *42*, 19056–19066. [CrossRef]
178. Mehta, P.; Barboun, P.; Herrera, F.A.; Kim, J.; Rumbach, P.; Go, D.B.; Hicks, J.C.; Schneider, W.F. Overcoming ammonia synthesis scaling relations with plasma-enabled catalysis. *Nat. Catal.* **2018**, *1*, 269–275. [CrossRef]
179. Peng, P.; Chen, P.; Addy, M.; Cheng, Y.; Anderson, E.; Zhou, N.; Schiappacasse, C.; Zhang, Y.; Chen, D.; Hatzenbeller, R. Atmospheric plasma-assisted ammonia synthesis enhanced via synergistic catalytic absorption. *ACS Sustain. Chem. Eng.* **2018**, *7*, 100–104. [CrossRef]
180. Uyama, H.; Nakamura, T.; Tanaka, S.; Matsumoto, O. Catalytic effect of iron wires on the syntheses of ammonia and hydrazine in a radio-frequency discharge. *Plasma Chem. Plasma Process.* **1993**, *13*, 117–131. [CrossRef]
181. Akay, G.; Zhang, K. Process intensification in ammonia synthesis using novel coassembled supported microporous catalysts promoted by nonthermal plasma. *Ind. Eng. Chem. Res.* **2017**, *56*, 457–468. [CrossRef]

182. Hawtof, R.; Ghosh, S.; Guarr, E.; Xu, C.; Sankaran, R.M.; Renner, J.N. Catalyst-free, highly selective synthesis of ammonia from nitrogen and water by a plasma electrolytic system. *Sci. Adv.* **2019**, *5*, eaat5778. [CrossRef]
183. Cheddie, D. *Ammonia as a Hydrogen Source for Fuel Cells: A Review*; InTechOpen: London, UK, 2012.
184. Yu, X.; Ye, S. Recent advances in activity and durability enhancement of Pt/C catalytic cathode in PEMFC: Part, I. Physico-chemical and electronic interaction between Pt and carbon support, and activity enhancement of Pt/C catalyst. *J. Power Sources* **2007**, *172*, 133–144. [CrossRef]
185. Lan, R.; Tao, S. Direct ammonia alkaline anion-exchange membrane fuel cells. *Electrochem. Solid-State Lett.* **2010**, *13*, B83–B86. [CrossRef]
186. Hejze, T.; Besenhard, J.; Kordesch, K.; Cifrain, M.; Aronsson, R. Current status of combined systems using alkaline fuel cells and ammonia as a hydrogen carrier. *J. Power Sources* **2008**, *176*, 490–493. [CrossRef]
187. Unlu, M.; Zhou, J.; Kohl, P.A. Anion exchange membrane fuel cells: Experimental comparison of hydroxide and carbonate conductive ions. *Electrochem. Solid-State Lett.* **2009**, *12*, B27–B30. [CrossRef]
188. Suzuki, S.; Muroyama, H.; Matsui, T.; Eguchi, K. Fundamental studies on direct ammonia fuel cell employing anion exchange membrane. *J. Power Sources* **2012**, *208*, 257–262. [CrossRef]
189. Carrette, L.; Friedrich, K.A.; Stimming, U. Fuel cells: Principles, types, fuels, and applications. *ChemPhysChem* **2000**, *1*, 162–193. [CrossRef]
190. Meng, G.; Ma, G.; Ma, Q.; Peng, R.; Liu, X. Ceramic membrane fuel cells based on solid proton electrolytes. *Solid State Ion.* **2007**, *178*, 697–703. [CrossRef]
191. Li, T.; Rabuni, M.; Kleiminger, L.; Wang, B.; Kelsall, G.; Hartley, U.; Li, K. A highly-robust solid oxide fuel cell (SOFC): Simultaneous greenhouse gas treatment and clean energy generation. *Energy Environ. Sci.* **2016**, *9*, 3682–3686. [CrossRef]
192. Siddiqui, O.; Dincer, I. A review and comparative assessment of direct ammonia fuel cells. *Therm. Sci. Eng. Prog.* **2018**, *5*, 568–578. [CrossRef]
193. Lan, R.; Tao, S. Ammonia as a suitable fuel for fuel cells. *Front. Energy Res.* **2014**, *2*, 35. [CrossRef]
194. Ishak, F.; Dincer, I.; Zamfirescu, C. Thermodynamic analysis of ammonia-fed solid oxide fuel cells. *J. Power Sources* **2012**, *202*, 157–165. [CrossRef]
195. Vayenas, C.; Farr, R. Cogeneration of electric energy and nitric oxide. *Science* **1980**, *208*, 593–594. [CrossRef]
196. Fournier, G.; Cumming, I.; Hellgardt, K. High performance direct ammonia solid oxide fuel cell. *J. Power Sources* **2006**, *162*, 198–206. [CrossRef]
197. Hossain, S.; Abdalla, A.M.; Jamain, S.N.B.; Zaini, J.H.; Azad, A.K. A review on proton conducting electrolytes for clean energy and intermediate temperature-solid oxide fuel cells. *Renew. Sustain. Energy Rev.* **2017**, *79*, 750–764. [CrossRef]
198. Jeerh, G.; Zhang, G.M.; Tao, S. Recent progress in ammonia fuel cells and their potential applications. *J. Mater. Chem. A* **2021**.
199. Wojcik, A.; Middleton, H.; Damopoulos, I. Ammonia as a fuel in solid oxide fuel cells. *J. Power Sources* **2003**, *118*, 342–348. [CrossRef]
200. Choudhary, V.T.; Goodman, D. CO-free fuel processing for fuel cell applications. *Catal. Today* **2002**, *77*, 65–78. [CrossRef]
201. Ma, Q.; Ma, J.; Zhou, S.; Yan, R.; Gao, J.; Meng, G. A high-performance ammonia-fueled SOFC based on a YSZ thin-film electrolyte. *J. Power Sources* **2007**, *164*, 86–89. [CrossRef]
202. Zhang, L.; You, C.; Weishen, Y.; Liwu, L. A direct ammonia tubular solid oxide fuel cell. *Chin. J. Catal.* **2007**, *28*, 749–751. [CrossRef]
203. Liu, M.; Peng, R.; Dong, D.; Gao, J.; Liu, X.; Meng, G. Direct liquid methanol-fueled solid oxide fuel cell. *J. Power Sources* **2008**, *185*, 188–192. [CrossRef]
204. Meng, G.; Jiang, C.; Ma, J.; Ma, Q.; Liu, X. Comparative study on the performance of a SDC-based SOFC fueled by ammonia and hydrogen. *J. Power Sources* **2007**, *173*, 189–193. [CrossRef]
205. Pelletier, L.; McFarlan, A.; Maffei, N. Ammonia fuel cell using doped barium cerate proton conducting solid electrolytes. *J. Power Sources* **2005**, *145*, 262–265. [CrossRef]
206. Maffei, N.; Pelletier, L.; Charland, J.; McFarlan, A. An intermediate temperature direct ammonia fuel cell using a proton conducting electrolyte. *J. Power Sources* **2005**, *140*, 264–267. [CrossRef]
207. Maffei, N.; Pelletier, L.; Charland, J.; McFarlan, A. An ammonia fuel cell using a mixed ionic and electronic conducting electrolyte. *J. Power Sources* **2006**, *162*, 165–167. [CrossRef]
208. Maffei, N.; Pelletier, L.; McFarlan, A. A high performance direct ammonia fuel cell using a mixed ionic and electronic conducting anode. *J. Power Sources* **2008**, *175*, 221–225. [CrossRef]
209. Ma, Q.; Peng, R.; Lin, Y.; Gao, J.; Meng, G. A high-performance ammonia-fueled solid oxide fuel cell. *J. Power Sources* **2006**, *161*, 95–98. [CrossRef]
210. Lin, Y.; Ran, R.; Guo, Y.; Zhou, W.; Cai, R.; Wang, J.; Shao, Z. Proton-conducting fuel cells operating on hydrogen, ammonia and hydrazine at intermediate temperatures. *Int. J. Hydrog. Energy* **2010**, *35*, 2637–2642. [CrossRef]
211. Zhang, L.; Yang, W. Direct ammonia solid oxide fuel cell based on thin proton-conducting electrolyte. *J. Power Sources* **2008**, *179*, 92–95. [CrossRef]
212. Xie, K.; Ma, Q.; Lin, B.; Jiang, Y.; Gao, J.; Liu, X.; Meng, G. An ammonia fuelled SOFC with a $BaCe_{0.9}Nd_{0.1}O_{3-\delta}$ thin electrolyte prepared with a suspension spray. *J. Power Sources* **2007**, *170*, 38–41. [CrossRef]

213. Khan, M.R.; Amin, M.; Rahman, M.; Akbar, F.; Ferdaus, K. Factors affecting the performance of double chamber microbial fuel cell for simultaneous wastewater treatment and power generation. *Pol. J. Chem. Technol.* **2013**, *15*, 7–11. [CrossRef]

214. Li, W.-W.; Yu, H.-Q.; He, Z. Towards sustainable wastewater treatment by using microbial fuel cells-centered technologies. *Energy Environ. Sci.* **2014**, *7*, 911–924. [CrossRef]

215. Yan, C.; Liu, L. Sn-doped V_2O_5 nanoparticles as catalyst for fast removal of ammonia in air via PEC and PEC-MFC. *Chem. Eng. J.* **2020**, *392*, 123738. [CrossRef]

216. Zacharakis-Jutz, G. Performance Characteristics of Ammonia Engines Using Direct Injection Strategies. 2013, p. 13032. Available online: https://lib.dr.iastate.edu/etd/13032 (accessed on 15 February 2021).

217. Gross, W.C.; Kong, S.-C. Performance characteristics of a compression-ignition engine using direct-injection ammonia–DME mixtures. *Fuel* **2013**, *103*, 1069–1079. [CrossRef]

218. Zamfirescu, C.; Dincer, I. Ammonia as a green fuel and hydrogen source for vehicular applications. *Fuel Process. Technol.* **2009**, *90*, 729–737. [CrossRef]

219. Mørch, C.S.; Bjerre, A.; Gøttrup, M.P.; Sorenson, S.C.; Schramm, J. Ammonia/hydrogen mixtures in an SI-engine: Engine performance and analysis of a proposed fuel system. *Fuel* **2011**, *90*, 854–864. [CrossRef]

220. Zamfirescu, C.; Dincer, I. Using ammonia as a sustainable fuel. *J. Power Sources* **2008**, *185*, 459–465. [CrossRef]

221. Brohi, E. Ammonia as Fuel for Internal Combustion Engines? *2014*. Available online: https://publications.lib.chalmers.se/records/fulltext/207145/207145.pdf (accessed on 23 December 2020).

222. Reiter, J.A.; Kong, S.-C. Diesel engine operation using ammonia as a carbon-free fuel. In *ASME 2010 Internal Combustion Engine Division Fall Technical Conference*; American Society of Mechanical Engineers Digital Collection: San Antonio, TX, USA, 2010.

223. Pearsall, J.T.; Garabedian, C.G. Combustion of anhydrous ammonia in diesel engines. *SAE Trans.* **1968**, 3213–3221.

224. Garabedian, G.C.; Johnson, J.H. *The Theory of Operation of an Ammonia Burning Internal Combustion Engine*; Army Tank-Automotive Center: Warren, MI, USA, 1966; pp. 333–348.

225. Cooper, J.; Crookes, R.; Mozafari, A.; Rose, J. Ammonia as a Fuel for the IC Engine. In Proceedings of the International Conference on Environmental Pollution, Zurich, Switzerland, 31 December 1991.

226. Reiter, J.A.; Kong, S.-C. Demonstration of compression-ignition engine combustion using ammonia in reducing greenhouse gas emissions. *Energy Fuels* **2008**, *22*, 2963–2971. [CrossRef]

227. Kong, S.-C. A study of natural gas/DME combustion in HCCI engines using CFD with detailed chemical kinetics. *Fuel* **2007**, *86*, 1483–1489. [CrossRef]

228. Ryu, K.; Zacharakis-Jutz, G.E.; Kong, S.-C. Effects of gaseous ammonia direct injection on performance characteristics of a spark-ignition engine. *Appl. Energy* **2014**, *116*, 206–215. [CrossRef]

229. Grannell, S.M.; Assanis, D.N.; Bohac, S.V.; Gillespie, D.E. The operating features of a stoichiometric, ammonia and gasoline dual fueled spark ignition engine. In *ASME 2006 International Mechanical Engineering Congress and Exposition*; American Society of Mechanical Engineers Digital Collection: New York, NY, USA, 2006; Available online: https://nh3fuelassociation.org/wp-content/uploads/2012/05/grannell_nh3.pdf (accessed on 18 January 2021).

230. Duynslaegher, C. *Experimental and Numerical Study of Ammonia Combustion*; University of Leuven: Leuven, Belgium, 2011; pp. 1–314. Available online: https://dial.uclouvain.be/pr/boreal/object/boreal:89103 (accessed on 12 January 2021).

231. Ezzat, M.; Dincer, I. Comparative assessments of two integrated systems with/without fuel cells utilizing liquefied ammonia as a fuel for vehicular applications. *Int. J. Hydrog. Energy* **2018**, *43*, 4597–4608. [CrossRef]

232. Hansson, J.; Fridell, E.; Brynolf, S. *On the Potential of Ammonia as Fuel for Shipping: A Synthesis of Knowledge*; Lighthouse Swedish Maritime Competence Centre: Göteborg, Sweden, 2020.

233. De Vries, N. *Safe and Effective Application of Ammonia as a Marine Fuel*; Delft University of Technology: Delft, The Netherland, 2019.

234. Laursen, R.S. Ship operation using LPG and ammonia as fuel on MAN B&W dual fuel ME-LGIP engines. In Proceedings of the 15th Annual NH3 Fuel Conference, Pittsburgh, PA, USA, 31 October–1 November 2018.

235. Shrestha, K.P.; Seidel, L.; Zeuch, T.; Mauss, F. Detailed kinetic mechanism for the oxidation of ammonia including the formation and reduction of nitrogen oxides. *Energy Fuels* **2018**, *32*, 10202–10217. [CrossRef]

236. Jojka, J.; Ślefarski, R. Dimensionally reduced modeling of nitric oxide formation for premixed methane-air flames with ammonia content. *Fuel* **2018**, *217*, 98–105. [CrossRef]

237. Nakamura, H.; Shindo, M. Effects of radiation heat loss on laminar premixed ammonia/air flames. *Proc. Combust. Inst.* **2019**, *37*, 1741–1748. [CrossRef]

238. Kobayashi, H.; Hayakawa, A.; Somarathne, K.K.A.; Okafor, E.C. Science and technology of ammonia combustion. *Proc. Combust. Inst.* **2019**, *37*, 109–133. [CrossRef]

239. Karabeyoglu, A.; Evans, B.; Stevens, J.; Cantwell, B.; Micheletti, D. Development of ammonia based fuels for environmentally friendly power generation. In Proceedings of the 10th International Energy Conversion Engineering Conference, Atlanta, GA, USA, 30 July–1 August 2012.

240. Iki, N.; Kurata, O.; Matsunuma, T.; Inoue, T.; Suzuki, M.; Tsujimura, T.; Furutani, H. Micro gas turbine operation with kerosene and ammonia. In Proceedings of the 11th annual NH3 Fuel Conference, Iowa, IA, USA, 21–24 September 2014.

241. Iki, N.; Kurata, O.; Matsunuma, T.; Inoue, T.; Suzuki, M.; Tsujimura, T.; Furutani, H. Micro gas turbine firing kerosene and ammonia. In *ASME Turbo Expo 2015: Turbine Technical Conference and Exposition*; American Society of Mechanical Engineers Digital Collection: Montreal, QC, Canada, 2015.

242. Hayakawa, A.; Goto, T.; Mimoto, R.; Arakawa, Y.; Kudo, T.; Kobayashi, H. Laminar burning velocity and Markstein length of ammonia/air premixed flames at various pressures. *Fuel* **2015**, *159*, 98–106. [CrossRef]
243. Okafor, E.C.; Naito, Y.; Colson, S.; Ichikawa, A.; Kudo, T.; Hayakawa, A.; Kobayashi, H. Experimental and numerical study of the laminar burning velocity of CH$_4$–NH$_3$–air premixed flames. *Combust. Flame* **2018**, *187*, 185–198. [CrossRef]
244. Ito, S.; Kato, S.; Saito, T.; Fujimori, T.; Kobayashi, H. Development of ammonia/natural gas dual fuel gas turbine combustor. In Proceedings of the NH3 Fuel Conference, Los Angeles, CA, USA, 18–21 September 2016.
245. Onishi, S.; Ito, S.; Uchida, M.; Kato, S.; Saito, T.; Fujimori, T.; Kobayashi, H. Methods for Low NOx Combustion in Ammonia/Natural Gas Dual Fuel Gas Turbine Combustor. In Proceedings of the 2017 AIChE Annual Meeting, AIChE, Minneapolis Convention Center, Minneapolis, MN, USA, 29 October–3 November 2017.
246. Ito, S.; Uchida, M.; Onishi, S.; Fujimor, T.; Kobayashi, T. Performance of Ammonia–Natural Gas Co-fired Gas Turbine for Power Generation. In Proceedings of the 15th Annu. NH3 Fuel Conference, Pittsburgh, PA, USA, 31 October–1 November 2018.

Review

Barriers and Solutions for Increasing the Integration of Solar Photovoltaic in Kenya's Electricity Mix

Dominic Samoita [1], Charles Nzila [2], Poul Alberg Østergaard [3,*] and Arne Remmen [4]

[1] Department of Electrical and Communications Engineering, Moi University, P.O. Box 3900 Eldoret, Kenya; dsamoita@gmail.com

[2] Department of Manufacturing, Industrial and Textiles Engineering, Moi University, P.O. Box 3900 Eldoret, Kenya; cnzila@gmail.com

[3] Department of Planning, Aalborg University, Rendsburggade 14, 9000 Aalborg, Denmark

[4] Department of Planning, Aalborg University, A.C. Meyers Vænge 15, 2450 Copenhagen, Denmark; ar@plan.aau.dk

* Correspondence: poul@plan.aau.dk; Tel.: +45-9940-8424

Received: 4 September 2020; Accepted: 13 October 2020; Published: 20 October 2020

Abstract: Currently, Kenya depends mainly on oil, geothermal energy and hydro resources for electricity production, however all three have associated issues. Oil-based electricity generation is environmentally harmful, expensive and a burden to the national trade balance. The rivers for hydropower and their tributaries are found in arid and semi-arid areas with erratic rainfall leading to problems of supply security, and geothermal exploitation has cost and risk issues amongst others. Given these problems and the fact that Kenya has a significant yet underexploited potential for photo voltaic (PV)-based power generation, the limited—although growing—exploitation of solar PV in Kenya is explored in this paper as a means of diversifying and stabilising electricity supply. The potential for integration of PV into the Kenyan electricity generation mix is analysed together with the sociotechnical, economic, political, and institutional and policy barriers, which limit PV integration. We argue that these barriers can be overcome with improved and more robust policy regulations, additional investments in research and development, and improved coordination of the use of different renewable energy sources. Most noticeably, storage solutions and other elements of flexibility need to be incorporated to balance the intermittent character of electricity generation based on solar PV.

Keywords: technical; economic; institutional; policy; pumped hydro storage

1. Introduction

Nations in the developed world are transitioning towards the use of renewable energy sources (RES) as the main resource for meeting energy needs because of its potential to address issues of climate change [1]. As applications in the economically developed countries have helped to drive down cost, low and middle-income countries are also increasingly looking towards RES. Up until 2008, however, most countries had not included PV technology into their electricity generation mix [2]. One of the reasons is that PV technology lacked cost competitiveness when compared to other RES like wind power as well as when compared to power generation based on fossil fuels.

In general, in terms of installations, fossil fuel-based technologies have had an upper hand over PV on a global scale. Concerns, however, have mounted over the steady increase of greenhouse gas emissions, which has been prompting governments to adopt planning practices and policies [3] as well as subsidies favouring RES [4]. Such concerns as well as reductions in cost of PV technology have resulted in increased uptake and technology development in parallel.

The cost of PV systems in Germany for example, has been declining steadily and significantly over the past decade; this is attributed to the rapid technological development spurred by government subsidies [5]. Installed costs between 2006 and 2013 declined by an average of 16% per year. By the first quarter of 2017, the typical cost of solar PV in typical roof top applications had fallen to 1640 EUR per kWp from over 5000 EUR per kWp in 2006 [6].

This decrease in installation cost took place as installed capacity increased in the same time period. By the end of 2018, 1.5 million rooftop installations had thus been installed in Germany [7]—a country of approximately 42 million households [8].

Concurrently, the development of PV technology that resulted in increased efficiency from 15% to over 30% [9]. The decreasing module costs combined with increasing efficiencies have resulted in a compound decrease in the cost of electricity from PV modules. Consequently, in the global context, PV has become much more competitive and the cumulative capacity of PV technology has increased significantly [10].

The development has been uneven across the globe though. In a study by [11], the current status and outlook of RES in Morocco was assessed. Morocco has exceptional good potential for the exploitation of particularly PV, but also good prospects for wind power—two technologies already adopted in the country. In the study, challenges and barriers to the development of RES and the national strategy for energy security and how the challenges will be met was evaluated using time series method. Results of the study showed that in the long term, towards 2030, wind and solar power can be injected without creating constraints of transit on the solar and wind power.

Kenya, on the other hand is neither as ambition nor as successful in terms of PV development in spite of a good solar radiation. Data show that the total installed PV capacity in Kenya was only about 50.25 MWp as of 2019 [12]. This capacity is marginal compared to the total installed power production capacity of approximately 5000 MW in Kenya. PV is projected to grow at 15% annually [12] in Kenya mainly attributed to the decreasing prices and PV thus becoming more competitive, but this is still marginal compared to potentials, and in many case developments is on off-grid systems. For these applications, in addition to wishing to save money, consumers like the idea of being autonomous [13]—a motivational factor also seen in e.g., Denmark, where PV owners may even opt for costly storage systems to increase their level of electricity autonomy [14].

While a 15% annual increase in other areas would seem high, it does, however, not suffice for a transition as this expansion rate would require many decades of installations. Also, concurrently, an increase in income and a process of urbanization is generating increases in fossil fuel usage in Kenya [15] that also needs to be balanced through increased RES exploitation. Besides, the peak demand for electricity increased by 3.6% from 2018 to 2019, and the peak is projected to increase from 1802 MW in 2018 to 15,000 MW by 2030 [12]. More RES development is thus needed and any barriers have to be overcome.

In fact, the Kenyan electricity generation mix (See Figure 1) shows that fossil fuels only provided approximately one quarter of the electricity while hydropower and geothermal had even more significant shares of approximately 30% each. Thus Kenya is heavily supplied by RES as it is.

Kenya has a high potential for the use of geothermal energy, with potentials to increase from currently about 200 MW up towards 10 GW [16]. The exploitation faces several challenges however including rising investment charges, increasing resource exploration and expansion risks, land-use conflicts, inadequate expertise, and high investment in grid infrastructure due to long distances from geothermal sites to existing load centres [17].

Hydropower in Kenya is mainly in the form of dammed hydro power plants with production susceptible to drought. This results in a less-than-optimal robustness and outright load shedding [18]. Pumped hydro storage could provide more flexibility by also enabling also excess generation from wind and PV to be stored for later use. Wind and PV could thus assist Kenyan hydropower and making it more robust against drought.

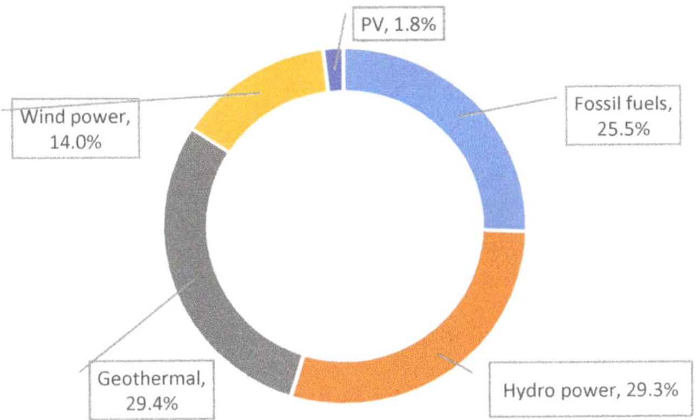

Figure 1. Kenyan electricity production mix shares in 2019. Based on data from [12].

Also, both hydropower and geothermal projects are characterised by long lead-times, thus in spite of potentials, these may not be adequate to fill the projected production gap.

Thus, as it is—and also combined with other constraints—the electricity system in Kenya suffers from frequent power outages. In a typical month, firms and homesteads connected to the grid experience on an average about six power outages, each lasting approximately five hours [19]. The economic cost of power interruptions is assessed to be about 7.1% of the power distribution companies' sales. Power outages therefore have a significant economic cost on businesses [19] and in turn on the Kenyan society.

Kenya has a large potential for PV since it is located near the equator, which provides it with a high insolation [18]. The insolation levels in Kenya and the large rural population is a stimulant for the penetration of solar power. According to [20] about 70% of the land area in Kenya has the potential of receiving approximately 5 kWh/m^2/day throughout the year with an annual mean radiation of 6.98 kWh/m^2.

A literature survey on the integration of PV in the electricity generation mix reveals, however, that focus is predominantly on Europe and United States of America. Little attention has been paid to emerging economies such as Kenya where electricity production and demand is expected to grow considerably in the coming years. Besides, there is room for adopting an infrastructure capable of meeting the future power demands using RES from the outset—and given the resource availability—notably PV.

Also, no study has highlighted the potential of and possible barriers to solar PV generation where hydropower already exists. Studies on PV in Kenya have a leaning towards ensuring access to electricity in areas located far away from the national grid as opposed to grid-connected projects [21,22]. This of course limits the analyses of interplay with hydropower. The only exceptions to off-grid analyses to date are viability studies in South Africa [23] and a review focussing on the development of mini-grids [24]. This paper therefore focuses on the prospects of balancing grid-connected PV systems with hydropower generation.

In spite of the demonstrated large potential of PV utilisation in Kenya, current exploitation is still limited, and projections show a modest growth that may not even match the increase in electricity and general energy consumption. Also, a predominant focus in the existing studies on Kenya is on the potential of PV as a source of renewable energy from a technical perspective and with a particular focus on stand-alone applications. While other countries—especially more economically developed—already target an increased PV exploitation even with poorer solar insulation conditions than Kenya, the country is still lagging behind in this respect. To exploit the potential more thoroughly, there is need for proper analysis of the opportunities and barriers for integration of PV in Kenya's electricity generation mix. The scope of this paper is therefore to analyse the potentials and barriers for deployment of PV technology in Kenya's electricity generation mix.

The paper is based on a review of the existing knowledge within the field, which is subsequently synthesized to provide a multifaceted perspective on opportunities and barriers to grid-connected PV in Kenya. The paper does not investigate off-grid PV systems.

2. Materials and Methods

The paper is based on existing literature as well as primary data collection from selected cases of grid-connected solar PV that have either recently been added to the national grid or is in the process of being connected in Kenya.

2.1. Literature Survey

The literature survey included a systematic four-step literature review where Scopus was used to locate the published literature.

In the first step, a combination of keywords was identified as follows: ("barriers" OR "opportunities") AND "PV". This combination was searched for in the abstracts, titles and keywords of the publications from 2012 to 2020 yielding 264 journal papers. Note that the search was not restricted to Kenya but yielded results from analyses worldwide to capture barriers and opportunities more generally.

In the second step, each publication was evaluated for its relevance to the integration of PV systems in the electricity generation mix, based on the abstract, title and keywords. If a publication had relevance, it was eligible for inclusion in the third step. This resulted in 102 journal papers.

In the third step, the full texts were evaluated in detail for their relevance to the barriers and opportunities for integration of PV technology in Kenya's electricity generation mix. This procedure ensured that no relevant information was missed from the publications. If a study addressed any kind of opportunity or barrier to the diffusion, it was included for analysis in the fourth step. Consequently, this resulted in 46 journal papers.

2.2. Analytical Framework

Barriers to renewable energy technology penetration in general and to PV technology specifically are global phenomena, and appear quite similarly in the literature, with only some country or technology-specific differences. Child et al. [25] has categorized major barriers to renewable energy technology penetration in six categories: "market failure/imperfection, market distortions, economic and financial, institutional, technical, and social, cultural and behavioural".

Some of the barriers noted by [25] include limited access to information, a bias for established energy resources, financial barriers, the circumstance that externalities are not factored into the decision-main process, lack of training, and a lacking acceptance [26].

Hvelplund has paid extensive attention to ownership [27,28] as a barrier and as an incentive for the advancement of energy technologies, however in our assessment we found little indication of the impacts of ownership in terms of acting as a barrier or opportunity. Some anecdotal evidence exists in terms of what type of actors have adapted PV technology in Kenya—but mainly on off-grid applications.

Also, not all the barriers identified by [25] are relevant or applicable in the Kenyan context. These categories of barriers and opportunities for success were therefore revisited and updated to make them relevant in the Kenyan context. Consequently, the range of barriers constraining the deployment of PV was categorized into technological, economic, institutional and policy as shown in Figure 2. This categorization is applied as the analytical framework.

As the fourth step in this analysis, the four categories of barriers and opportunities identified in the analytical framework presented in Figure 2 were evaluated. In addition to the literature search and analysis of other secondary materials such as government reports, conference papers and peer reviewed articles, this paper draws on in-sights from case studies on the Garissa solar PV project and the UNDP solar PV project—both in Kenya. The first author also draws on first-hand experience with installation of PV systems in the western part of Kenya. The first author reflects on his practical experiences on barriers that limit the uptake of PV and the existing opportunities and how they apply on a national scale to enrich the content of this paper.

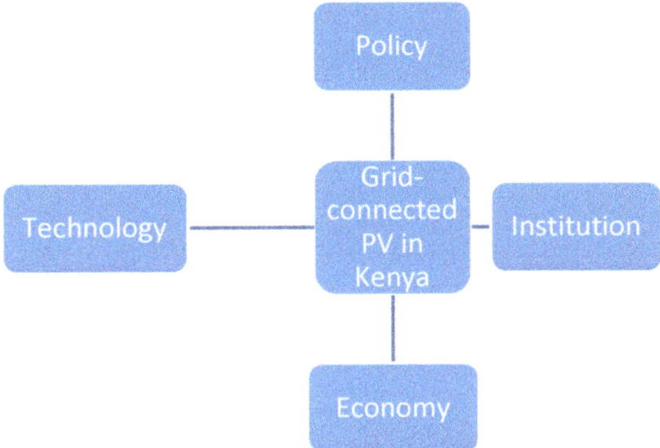

Figure 2. Analytical framework for assessing barriers and opportunities for increased integration of solar PV in Kenya.

As a last step in the analyses, international experience with pumped hydro storage is explored in order to show that it is technically and economically feasible to roll out this technology in Kenya.

3. Barriers to Increasing Integration of PV in Kenya

The barriers identified in the literature review are presented in this section—organised according to the analytical framework present in Section 2. These barriers are categorized into three: technological, economic and institutional barriers.

The advent of the PV technology in Kenya in the 1970s was mainly facilitated by donors and particularly in the 1980s, PV systems were donated to facilities such as health clinics that were off-grid [17]. During the implementation of PV in rural off-grid areas it was established that there was a market for PV technology beyond the scope of health clinics and off-grid missions [21]. Thus, within two years from 2000 to 2002, households increased the purchases of PV systems from 20% to 40% of total annual PV purchases. However, the benefits from the solar PV systems were mainly accrued by the rural middle class [28]. Consequentially, poorer rural households were sidestepped in terms of receiving subsidized PV systems [29].

With time, a focus on the rural affluent allowed the PV sector to be commercially viable without reliance on donors or subsidies. Specifically, in the 2000s, focus was on the middle-income segment, which was propelled by reduced costs of PV systems and a desire for the middle class to watch television [28]. These advances have led to the development of a considerable amount of PV systems being installed [30], however most of the development in Kenya has been in the form of stand-alone (i.e., off-grid) systems that are not connected to national grid. Thus, the PV market in Kenya has evidenced use of Solar Home Systems, which has been considered the most successful off-grid solar market in the developing economies [31].

3.1. Technological Barriers

Although PV technology has advanced significantly in the last decades, there are still several technical barriers to its adoption as highlighted in the literature. The quality of PV systems is of vital importance for its integration. Lack of adequate knowledge is a crucial barrier that may result in improper usage and inability to maintain the systems [13]. This may create a negative perception and prevent potential customers making a decision to adopt the systems.

A large technological challenge for PV in Kenya has been the lack of energy storage systems [32]. Meanwhile since Kenya relies largely on reservoir/dammed hydroelectric power supply, a PV-based pumped storage hydropower could offer an even more flexible solution to the variability of the residual production (demand minus non-dispatchable power production). Battery storage solutions also exist but the accompanying initial cost is high (though falling) for grid-connected systems hence research has pointed the need to look at development of hybrid systems combining solar PV with hydropower as a viable alternative option [33]. The storage requirement is of course dependent on both the composition of the rest of the electricity systems and on the share of PV. While a small share of PV can be integrated into the grid with no technical issues, large-scale integration of PV sets higher demands for the flexibility of energy systems.

Mathiesen at al. [34] identifies three phases of implementing RES. The first stage is the introduction phase where RES only marginally replaces production based on fossil fuels. The second stage is the large-scale integration phase where the integration of fluctuating renewables in the system becomes complex and where grid stability becomes an issue in the power system. The third stage is the full renewable energy phase with a complex system with suitable storage and conversion technologies to maintain the temporal balance of the energy system.

As the national Kenyan system is in the second phase—similar to most other countries in the world—the integration issues of large-scale integration are not faced yet, but a strong focus on PV will eventually bring the country there.

In this connection, the key technical issues that demand to be addressed when integrating solar PV on the grid include voltage level and point of common coupling, network voltage variations, power quality, voltage ride-through capability, reactive power compensation capability, frequency regulation capability and protection issues.

3.2. Economic Barriers

The main economic barrier for PV integration has been the high upfront cost and an unwillingness of banks to fund such investments [35]. The higher construction costs have previously made financial institutions more likely to perceive renewables as risky, lending money at higher rates and making it harder for utilities or developers to justify the investment. Perceptions of the (high) costs associated with solar PV can still be a barrier even if prices have come down considerably in recent years.

There have been efforts towards establishing the economic viability of solar PV in different regions of the world. A study focusing on the Middle East and North Africa regions concluded that rooftop PV systems were competitive with other energy producing plants in the region [33], however development has not followed suit. Nevertheless, a study in the Kenyan context confirms this. In [36] the authors used a levelised cost of energy evaluation and established that solar power in combination with other renewable energy generation technologies is competitive in relation to non-renewable energy. In spite of the high insolation levels, solar PV has mainly been used for off-grid application such as solar lanterns because of the high upfront investment costs of grid-connected solar PV systems [37]. However, currently, solar PV has been highly incentivized and prices have dropped significantly thus making it a viable option compared to diesel-fuel generators that are expensive to run [38].

Cost comparisons and competitiveness of solar PV with the conventional fossil fuels to electricity production are among the most significant barriers for adoption of solar PV in Kenya both for the utility company and for individual investors [17]. From the perspective of both the national utility company Kenya Power and Lighting Company (Nairobi, Kenya) and independent power producers, large-scale PV electricity generation without a well-functioning energy storage system maybe too expensive considering the fact these companies are also required to maintain the grid infrastructure and manage other running costs [31].

To mitigate these challenges, the incumbent could consider selling or leasing out PV energy storage systems, provide financing and grid connections or build a service relationship with individual investors. Once the domestic market grows, the installation costs will become cheaper; this could further be facilitated by training and certification of solar PV installers at the national level [39].

3.3. Institutional Barriers

The institutional barriers vary largely and are here grouped into four categories; (a) grid access, (b) research and development programmes, (c) university linkages and (d) policy experience from other African nations.

3.3.1. Grid Access

Kenya has an interconnected national electricity grid operated and run by the monopoly Kenya Power and Lighting Company Limited (Nairobi, Kenya). Unfortunately, the process of grid connection for new PV systems is long and complicated (that requires up to fourteen licensing steps) which discourages potential investment in PV power generation [18].

3.3.2. Role of Education, Training, Research and Development Programmes

Globally, China is a powerhouse for solar technology [40]. In 2001, China had thirty research institutes and universities that were working collaboratively to develop the materials used in PV cells [41]. Notably, key Chinese producers of PV manufacturing equipment have emerged with a number of firms being tasked with PV system design, technology research and development, manufacturing of the components as well as sales and after-sales service.

In Kenya, most solar PV firms have been involved in the government led rural electrification programme [42]. However, there is minimal research and development within the firms since most of them rarely devote a portion of their annual turnover to facilitate their technological capabilities and the competitiveness regarding PV [43]. Hence, in view of the extant literature, instances of research and development in the PV in Kenya are close to none.

In 2015, Maclean and Brass [44] noted that the Kenya Industrial Research and Development Institute is engaged in researching the development of low-carbon and climate-resilient technologies such as PV. The authors noted that Migori, Bungoma, Kirinyaga, Embu and Samburu were among the counties that were surveyed for low-carbon technologies. This is referred as a niche development for Kenya relevant to make affordable PV systems for off-grid rural areas. This study concluded that a policy brief should be developed to spur a pico solar market in Kenya. Pico refers to the smallest portable photovoltaic systems mainly typified by rechargeable battery.

3.3.3. University-Industry Linkages

Institutions of higher learning are instrumental players in both national and regional innovation systems [45]. In the recent past, universities have received substantial attention with regard to their role in innovation and spurring economic development. Universities are credited with major advances in scientific research and the creation of innovations with impact on the society.

Universities in Kenya have historically played a role in its National Innovation System also in development of RES. Strathmore Energy Research Centre (SERC)—a brainchild of Strathmore University in Nairobi—offers professional training, project development, and technical research in the renewable energy sector [36]. Since its inception in 2012, SERC has been at the forefront in implementing innovative pilot projects with the intention to promote RES into Africa. Trained technicians from SERC are able to do PV installation, repair and maintenance thereby raising awareness and contributing towards uptake of PV technology.

Furthermore, the centre for research on New and Renewable Energies at Maseno University has also been instrumental in promoting RES exploitation locally and regionally with special emphasis on rural application [44]. The centre focuses on harnessing bio-energy, geothermal, solar and wind energy.

The energy technology programmes offered in Kenyan universities are created to develop individuals with the capacity to address the national and global challenges in the energy sector. Therefore, this indicates that university-industry linkages exist and maybe utilized to foster increased integration of PV in the electricity generation mix.

3.3.4. Policy Experience from Africa

The African continent has a rich source of solar energy and in the recent years, PV has been becoming a viable alternative source of electricity for both small and large-scale application in Africa [42]. Like Kenya, solar PV deployment in most other countries in Africa has mainly been driven by rural (off-grid) electrification. As opposed to most of the African nations, Morocco saw the need for spurring PV on a larger scale for electricity generation at an early stage. In fact, a Moroccan integrated solar project was launched that comprised of solar and wind technologies that complement each other [46]. In addition, the Moroccan government commissioned a 500 MWp PV plant in 2018 [46].

In 2015, Rwanda was at an early stage of solar power integration according to [1]. At the time, there were at most eight companies which were mainly donor-driven and whose scope was to install solar systems in government hospitals and schools. According to the authors, there was a growing market for solar PV mainly among the private households.

Despite this, Rwanda possesses the highest grid connection in East Africa ahead of Kenya and Tanzania [1]. Nonetheless, there has been a tremendous growth in PV for off-grid Rwandan rural electrification since 2009 [28]. Further development of the technology was due to a strong focus on the Rwandan vision 2020 [47], which puts emphasis on renewable energy technology.

Besides, the Monetary Growth and Poverty reduction policy in Rwanda states that the government in collaboration with the private sector should facilitate the distribution and the sale of solar PV systems and further provide a regulatory environment that is conducive for the rapid integration of PV [47]. Experience from Rwanda and Morocco could be used to inspire policies in Kenya to help foster increased integration of solar PV in the country.

3.4. Political Barriers

Policy measures are of vital importance for large-scale introduction of PV. A lack of stability of incentives for the adoption of PV can be a significant barrier—for instance a sudden removal of existing subsidies or inconsistencies in policy measures.

Kenya has instituted policies meant to increase the integration of PV [1]. Particularly, among these policies is the Kenya Rural Electrification Master Plan [48], a Feed-in Tariff policy [49] and the Vision 2030 [50]. The government has also made a move towards ensuring that entrepreneurs willing to invest in solar PV are catered for [47].

As early as 2008, Kenya developed a feed-in tariff policy meant to ensure market stability for investors in PV. The feed-in tariff made it possible for independent power producers to deliver power from wind and hydro sources to the national grid. In 2012, the feed-in tariff policy was revised to also include solar power [37]. However, these policies have not translated to higher installed grid-connected PV capacity largely because the policies are not well coordinated during implementation or at worst, they are not implemented at all [1].

Under- or over-prioritization of investments in certain sectors in relation to others is usually not based on technical decisions but rather involves political choices and prioritizations. Large-scale PV projects are essentially large infrastructure projects that are typically highly political and that involve a multitude of actors with competing interests and negotiations across various levels. For example, [51] argue that the push for RE in Kenya is not necessarily being driven by environmental concerns, but rather by the need to provide access to electricity to the highest number of people within the shortest time possible. These authors highlight the tensions that come from pursuing the multiple objectives of 'growth', 'inclusiveness' and 'sustainability'.

4. Pumped Hydro Storage Solution

This section discusses pumped hydro storage solution by reflecting on experiences from various countries which have similar conditions as Kenya. The importance of pumped hydro as a possible solution that can be replicated locally to meet electricity demand can therefore be proven since it has been applied in various countries with similar conditions as Kenya.

The shift from fossil fuel-fired power production to RES such as PV requires some form of storage or flexibility to cope with intermittency of the sources. A limitation with batteries is that they cannot offer multiple-hour storage capacity and discharge over a long time period. Moreover, batteries have a specific lifespan beyond which their performance is not guaranteed. Batteries also require ample housing space for large power output.

Pumped hydro storage (PHS) is a form of energy storage whereby gravitational potential energy of water is pumped from a lower reservoir to a higher one serve as a dispatchable reservoir to feed turbines on request [52]. PHS is the largest form of grid energy storage available; currently, PHS accounts for about 95% of all active and tracked storage [53]. The first PHS systems were commissioned in Alpine Switzerland, Austria and Italy in the 1890's. World-wide installed capacity stands at over 181 GW, of which about 29 GW are in the USA [52].

Modern PHS plants have a cycle efficiency of about 80% [54,55], and allow the utilization of excess electricity from base-load power sources (such as coal or nuclear) or fluctuating RES to be saved for use during periods of higher demand. Reservoirs used in these systems are generally smaller compared to conventional hydroelectric dams of similar power capacity and their generating periods are often less than half a day.

Chile has vast amounts of solar and wind resources which are increasingly being harnessed to replace fossil fuel generation. The impact of large scale integration of these variable RES for grid-level electricity storage was evaluated by [55]. In this study, a cost-based linear optimization of the Chilean electricity system was developed to analyse and optimize various RES generation, transmission and energy storage scenarios until 2050. Results of this study showed that for the base scenario of decommissioning of aging coal plants and no new coal and large hydro generation, the generation gap can be filled by PV, concentrated solar power and flexible gas generation resulting in a drop of 78% in carbon dioxide emissions. Integration of PV for on-grid storage increases the solar PV fraction which consequently leads to a 6% reduction in operation and investment costs by 2050 [55].

Switzerland was a pioneer in electricity generation and a European leader in power storage [56]. The country's mountains, extensive snows and glaciers favour development of hydro technology which is well developed and mature. As a result, hydro power generation from Switzerland is critical in western European electric power system supply backup and means of storage of reserve power [5,55]. Power management and storage are also critical towards achieving 100% renewable energy-based system in Switzerland. The country utilizes two different hydro storage systems; PHS and hydro systems with natural inflows to reservoirs that are subsequently feed to turbines on demand

A steady increase of the RES share in Australia's electricity mix is causing a move away from dependency on fossil fuels. A study by [57] focusing on the South West Interconnected System in Western Australia modelled several high penetration scenarios for renewables comprising wind and PV and PHS. The scenarios were examined using a chronological dispatch model restricting to technologies that were already deployed on a large scale i.e., greater than 150GW were utilized. Results obtained demonstrated that 100% penetration of wind and PV electricity is compatible with a balanced grid—though requiring PHS. Furthermore, with the integration of PHS, a RES share over 90% will still be allowed at a competitive electricity supply cost.

A case study by [58] based on Ometepe island, Nicaragua simulated a PHS and geothermal plant using HOMER software. The island was chosen because it has wind, solar and geothermal resources as well as an extinct volcano with a crater on its top that can serve as the upper reservoir for the PHS system. Different system configurations were demonstrated and the results obtained revealed that PHS technology is able to serve the base load of the system, therefore reducing the required installed capacity of other power resources as well as decreasing the storage requirements and excess electricity production.

Zimbabwe is among the African countries which rely on power imports to meet its energy needs which endangers the energy security of the nation. Several studies have been conducted to assess the feasibility of hybrid power generating systems that incorporate intermittent power sources such as PV with or without storage in order to maximize technical and economic feasibility. One such study by [59] made a techno-economic comparison between standalone wind or PV and hybrid PV/wind. The hybrid system was based on maximizing intermittent energy sources. Results obtained showed that the levelized cost of electricity was less than or equivalent to the local grid tariff. The study further revealed that the utilization of RES would boost energy security and reduce dependency on imported energy.

A 2000 MW PHS plant supported by a 300 MW PV plant is being constructed on the Osborne dam, river Odzi in the Manicaland province of Zimbabwe [60]. This is the first project of its kind in the country which will provide backup for the national power grid during peak hours when the available network capacity is insufficient. Peak demand in Zimbabwe is observed for 8.5 h in a day [61,62].

Evidently from the foregoing, PHS is a mature technology that has been tried and tested. Through coordination, a higher penetration of solar PV may be achieved by using hydropower to compensate for the fluctuations in PV output in Kenya. Consequently, there will be less disruptions in electricity supply since there is reliance on both solar and hydropower.

5. Modelling of Kenya's Future Electricity Grid

This section reviews modelling studies that have been conducted relating to the future of electricity grid in Kenya. Similar studies in selected African countries and outside of Africa are also reviewed.

PV as a technology for electrification of rural Kenyan communities was modelled in [21]. The results showed that interconnecting a number of solar minigrids into one common grid leads to better technical performance. Moreover, the more minigrids connected, the better the performance and the more the power available for the national grid for supply to consumers.

A system-level model of Kenya was presented in [2]. This was used to assess the combination of PV and hydropower to displace diesel-based power generation. The research tested various generation mixes for the years 2012 to, and results obtained showed the value of high penetrations of PV exceeded expected feed-in-tariff payments.

In the work of [62], a spatially explicit supply model was developed to seek for a least-cost PV electrification model to connect consumers that are not served by the national electric grid. Information from individual consumer demand was used to develop this model. It was concluded from the results obtained that PV minigrids can serve up to 17% of the country's population. This includes due consideration for latent demands—i.e., demands currently unmet but which either grid expansion or access to PV electricity may stimulate.

In another study by [63] a rural electrification spatial model for Kenya was developed to identify various approaches to meet electricity needs for various places in the country. The analyses considered both diesel-based generators, various renewable energy sources and expansion of the national grid. Results obtained showed that renewable energy sources can play a key role in meeting energy demand for rural consumers.

Another study conducted in Morocco by [64], an energy management algorithm was modelled and simulated using MATLAB/SIMULINK to serve the load. The system is a grid-connected PV-battery which can manage its energy flows via an optimal management algorithm. Results of this modelling showed that the load was well served in all cases by instant solar production.

In Tunisia, research by [65], a grid connected PV system was modelled using a command approach to function under normal conditions and Symmetrical Grid Voltage Dips (SGVD). In normal operation mode, the command developed increased the low solar voltage to a suitable level corresponding to the Maximum Power Point Tracking. Under the SGVD, the control strategy should ensure stable connection as long as possible and inject more reactive power to support the grid faults. The modelled control scheme in the various operation modes presented high performances in transient and permanent phased. The system improved safety of the overall system and increased the connection time.

Experience from outside Africa has demonstrated how a decentralisation of power production on cogeneration of heat and power plants [66,67] and wind power [68,69] reduce grid loading thus enabling weaker grids, leave capacity for transit if so needed while at the same time reduces grid losses [70]. These analyses are based on a combination of energy systems analyses [71] using the EnergyPLAN model [72,73] and transmission grid analyses taking into consideration hour-step analyses of the 150 and 400 kV transmission grid.

All in all, experience from the literature demonstrate that a decentral deployment of grid connected PV systems could benefit the energy system—though due concern of ancillary services is still required.

6. Discussion

This section discusses the key challenges facing integration of PV into Kenya's electricity generation mix, and also puts some of the issues into a wider perspective.

A key technical challenge for integration of PV technology is the temporal match with the demand; solar power is available during the day and there must be a viable way of harnessing it for use when the sun is not up. In Kenya, electricity generation in the form of pumped hydroelectric power generation is seen as a viable option to replace run-of-river hydroelectric power plants whose production in susceptible to drought which result in reduced electricity generation capacity as well as outright power outages.

As it is, the electricity system in Kenya suffers from frequent power outages that are well documented. In a typical month, firms and homesteads connected to the grid experience on average 6.3 power outages, each lasting approximately five hours [19]. The economic cost of power interruptions is approximately 7.1% of the firms' sales; power outages therefore have a significant economic cost on businesses [19] and in turn on the Kenyan society.

There has been less emphasis on solar PV in Kenya. Specifically, no study has highlighted the potential of and possible barriers to solar PV generation where hydropower already exists. Studies on PV in Kenya have a leaning to integrate PV with other technologies. For policy-makers and international organizations keen to reduce carbon emissions and dependence on imported fuels, the deployment of hydro resources alongside solar PV is a viable option for many sub-Saharan African countries. It is here important to note the role played by hydropower sources, which supply approximately 30% of the end-use electricity demand in Kenya.

Kenya has a great potential for pumped-hydro storage for the integration of PV [2]. Dammed hydro or even pumped hydro storage are both dispatchable and thus offering regulating capability to the power system. For pumped hydro storage, low-cost surplus off-peak electric power can be used to fill reservoirs which is subsequently released through turbines during periods of high electrical demand. The cost of pumping water can easily be met from the revenue generated from selling power during peak demand periods—even when factoring in the cycle losses. At any rate, with cycle efficiencies in the 70% to 80% bracket [74] and modest cost per size [75], pumped hydro systems are a good storage candidate for large-scale PV integration.

In colder climates where heating systems offer flexibility through system-integration, options are more diverse with the use of cogeneration of heat and power [70,76–78], heat pumps [79,80], heat storage [75] and smart energy systems [81–84] for the integration of renewables. The potential for this sectorial integration between heat and electricity is not so pronounced in warmer climates. Also, in the future, electric vehicles provide an opportunity for integrating PV power [35,85–87], however this solution comes with a cost barrier.

There are conflicting standpoints and expectations of policy makers and utilities that constitute institutional and policy barriers. Incumbent electricity companies in Kenya are likely to favour maintaining the status quo since they have made investments in the existing electricity generation system. This creates path dependency and lock-in effect [87]. Kenya Power has the sole monopoly of managing costs of connecting PV systems to the grid and they manage the grid single-handedly. At the same time, new forms of electricity generation such as PV generation try to break the lock-in and this clashes with the current Kenyan electricity regime that is mainly based on hydro, geothermal and fossil fuels. Resistance from the utility or other industry players can be sensed in the context of path dependence and lock-in, and therefore undermines integration of renewable sources of energy [36].

On the other hand, the supportive policies advanced by the Kenyan government towards renewable energy have contributed towards a growing interest among citizens related to solar energy. There is a range of different kinds of support instruments in use, such as a conducive environment for investment, innovative financing schemes, exemptions from value added tax and import taxes, standardized power purchase agreements and feed-in-tariffs which have led to the growth of PV market in Kenya [49].

These policies have raised the point that dynamic support structures for renewable energy technologies can aid in increasing their market penetration. A period of high subsidy may be particularly important to establish early growth in market share, but should be followed by adjustments in subsidies to prevent markets from growing too quickly. At the same time, [87] reminds that support must go beyond financial measures to be sustainable by offering training programs on operation and maintenance. Also, [87] found that one-off investment support or tax rebates were preferable to feed-in tariffs, as they were deemed more cost efficient and were likely to instil greater confidence in investors.

In addition, research institutions play a critical role in integration of PV systems through building of local capacity to handle installation, operation and maintenance of PV systems. As highlighted earlier, research and development has played a key in the advancement of PV technology in countries such as China. Kenya needs to invest more in research and development in order to scale up electricity generation from PV technology and therefore increase economic competitiveness of solar PV.

These linkages can be extended to collaborations between Kenya and the international community. In most African countries integration of solar PV systems (mainly small-scale) has been driven by donor-supported projects aimed at serving specific needs for electricity [1]. Historically, development of PV systems has been aided by the donor community which has facilitated acquisition of solar systems to local communities and institutions by providing the requisite resources. Arguably, availability of resources from PV actors from outside Kenya may also help scaling up of PV technology in the country. PV development in Kenya is also shaped by the market forces of demand and supply.

In terms of the proposed storage solution, PHS is already a well-established technology in several countries—and as our analyses show, it is also a technology that is applied or considered in conjunction with PV. The non-technical prospects of PHS in a Kenyan context have not been considered, but it is a technology that is traditionally owned and operated by larger players in electricity markets—power producers or transmission system operators e.g., Kenya Generating Company Limited and Kenya Transmission Company Limited could thus fit institutionally well into a system as the Kenyan with a strong central player in Kenyan Power and Lighting Company.

7. Conclusions

This paper has discussed the barriers and possibilities for overcoming these for the integration of PV technology in Kenya based on a literature review to achieve high installed capacities. Significant changes must occur in the Kenyan electricity sector. Most noticeably, storage solutions and other elements of flexibility need to be incorporated to balance the intermittent nature of electricity generation based on solar PV. This is particularly eminent for large-scale deployment of PV technology in Kenya. A complement between hydro and solar PV to address the storage challenge was proposed in this investigation.

Such a hybrid system represents a complete transformation from the current scenario. A variety of technical, economic, institutional, political barriers have been pointed out which currently restrict further increase of PV technology.

Some of the technological barriers identified include lack of adequate knowledge of PV technology and lack of energy storage systems or energy system flexibility to integrate PV. An important economic barrier identified is the high upfront costs and unwillingness of banks to fund PV investments. Some of the institutional and policy barriers identified are lack of stability incentives for adoption of PV and the long and complicated grid connection process.

These barriers can be overcome with robust policy regulations, additional investments in education, training, research and development, better regulation of the electricity sector and improved coordination between key actors.

While this analysis focused on the barriers for solar PV in the Kenyan grid system, the results may be applicable to other sub-Saharan African countries, many of whom are faced with the same challenges: growing demand for electricity, insufficient generating capacity, and long lead times and extensive financial investments required for planned generation projects. As a result, many countries have turned to short-term expensive solutions such as diesel plants. Further, the other characteristic—that may make solar PV a favourable option in Kenya such as abundant solar resource—is also present across the continent.

Author Contributions: Conceptualization, D.S.; methodology, D.S.; investigation, D.S.; writing—original draft preparation, D.S.; writing—review and editing, C.N., P.A.Ø., A.R.; supervision, C.N., P.A.Ø., A.R.; funding acquisition, C.N. All authors have read and agreed to the published version of the manuscript.

Funding: This article was prepared with support from the Danish Ministry of Foreign Affairs through the Innovation and Renewable Electrification in Kenya (IREK) project—Grant DFC 14-09AAU.

Acknowledgments: The present paper is a substantially revised and improved version of a working paper published at the Aalborg University repository VBN [87].

Conflicts of Interest: The authors declare no conflict of interest.

References

1. Hansen, U.E.; Pedersen, M.B.; Nygaard, I. Review of solar PV policies, interventions and diffusion in East Africa. *Renew. Sustain. Energy Rev.* **2015**, *46*, 236–248. [CrossRef]
2. Rose, A.; Stoner, R.; Pérez-Arriaga, I. Prospects for grid-connected solar PV in Kenya: A systems approach. *Appl. Energy* **2016**, *161*, 583–590. [CrossRef]
3. Østergaard, P.A.; Sperling, K. Towards sustainable energy planning and management. *Int. J. Sustain. Energy Plan. Manag.* **2014**, *1*, 1.
4. Adam, A.D.; Apaydin, G. Grid connected solar photovoltaic system as a tool for greenhouse gas emission reduction in Turkey. *Renew. Sustain. Energy Rev.* **2016**, *53*, 1086–1091. [CrossRef]
5. Lang, T.; Ammann, D.; Girod, B. Profitability in absence of subsidies: A techno-economic analysis of rooftop photovoltaic self-consumption in residential and commercial buildings. *Renew. Energy* **2016**. [CrossRef]
6. Kausar, A.S.M.Z.; Reza, A.W.; Saleh, M.U.; Ramiah, H. Energizing wireless sensor networks by energy harvesting systems: Scopes, challenges and approaches. *Renew. Sustain. Energy Rev.* **2014**, *38*, 973–989. [CrossRef]
7. Heesen, H.t.; Herbort, V.; Rumpler, M. Performance of roof-top PV systems in Germany from 2012 to 2018. *Sol. Energy* **2019**, *194*, 128–135. [CrossRef]
8. Society and Environment—Housing. Statistisches Bundesamt. Available online: http://www.destatis.de/EN/Themes/Society-Environment/Housing/_no (accessed on 16 April 2020).
9. Fukushima Renewable Energy Institute. 2014. Available online: https://www.japantimes.co.jp/news/2018/03/11/national/fukushima-powers-toward-100-goal-renewables-grid-cost-woes-linger/ (accessed on 30 May 2020).
10. Xu, Q.; Li, H.; Hao, G.; Ding, Y. Study on Several Influencing Factors of Performance Evaluation Index of Photovoltaic System. *J. Clean Energy Technol.* **2016**, *4*, 424–429. [CrossRef]
11. Tazi, G.; Jbaihi, O.; Ghennioui, A.; Merrouni, A.A.; Bakkali, M. Estimating the Renewable Energy Potential in Morocco: Solar energy as a case study. *IOP Conf. Ser. Earth Environ. Sci.* **2018**, *161*, 1–8. [CrossRef]
12. ERC, Energy Regulatory Commission Biomass. 2019. Available online: https://www.epra.go.ke/ (accessed on 12 January 2020).
13. Karakaya, E.; Sriwannawit, P. Barriers to the adoption of photovoltaic systems: The state of the art. *Renew. Sustain. Energy Rev.* **2015**, *49*, 60–66. [CrossRef]

14. Marczinkowski, H.M.; Østergaard, P.A. Residential versus communal combination of photovoltaic and battery in smart energy systems. *Energy* **2018**, *152*, 466–475. [CrossRef]

15. Kwakwa, P.A.; Adu, G.; Osei-Fosu, A.K. A time series analysis of fossil fuel consumption in Sub-Saharan Africa: Evidence from Ghana, Kenya and South Africa. *Int. J. Sustain. Energy Plan. Manag.* **2018**, *17*, 4.

16. Ogola, P.F.A.; Davidsdottir, B.; Fridleifsson, I.B. Potential contribution of geothermal energy to climate change adaptation: A case study of the arid and semi-arid eastern Baringo lowlands, Kenya. *Renew. Sustain. Energy Rev.* **2012**, *16*, 4222–4246. [CrossRef]

17. Phillips, M.A. *Renewable Energy Incentives in Kenya: Feed-in-Tariffs and Rural Expansion*; Ministry of Energy: Nairobi, Kenya, 2016.

18. Lai, C.S.; McCulloch, M.D. Levelized cost of electricity for solar photovoltaic and electrical energy storage. *Appl. Energy* **2017**, *190*, 191–203. [CrossRef]

19. Ramirez, A. Indium Oxide as a Superior Catalyst for Methanol Synthesis by CO2 Hydrogenation. *Energy Access* **2016**. [CrossRef]

20. Oloo, F.; Olang, L.; Strobl, J. Spatial Modelling of Solar energy Potential in Kenya. *Int. J. Sustain. Energy Plan. Manag.* **2015**, *6*, 17–30.

21. Opiyo, N. Modelling PV-based communal grids potential for rural western Kenya. *Sustain. Energy Grids Netw.* **2015**, *4*, 54–61. [CrossRef]

22. Johannsen, R.M.; Østergaard, P.A.; Hanlin, R. Hybrid photovoltaic and wind mini-grids in Kenya: Techno-economic assessment and barriers to diffusion. *Energy Sustain. Dev.* **2020**, *54*, 111–126. [CrossRef]

23. Azimoh, C.L. Sustainability and Development Impacts of off-grid electrification in developing countries: An assessment of South Africa's rural electrification program. Ph.D. Thesis, Malardalen University, Vasteras, Sweden, 2016.

24. Sharma, P.; Walker, A.W.; Wheeldon, J.F.; Hinzer, K.; Schriemer, H. Enhanced efficiencies for high concentration, multijunction PV systems by optimizing grid spacing under non uniform illumination. *Int. J. Photoenergy* **2014**, *2014*, 582083. [CrossRef]

25. Child, M.; Haukkala, T.; Breyer, C. The Role of Solar Photovoltaics and Energy Storage Solutions in a 100 % Renewable Energy System for Finland in 2050. *Sustainability* **2017**, *9*, 1358. [CrossRef]

26. Hvelplund, F.; Möller, B.; Sperling, K. Local ownership, smart energy systems and better wind power economy. *Energy Strateg. Rev.* **2013**, *1*, 164–170. [CrossRef]

27. Hvelplund, F.; Djørup, S. Consumer ownership, natural monopolies and transition to 100% renewable energy systems. *Energy* **2019**, *181*, 440–449. [CrossRef]

28. Kirubi, C.; Jacobson, A.; Kammen, D.M.; Mills, A. Community-Based Electric Micro-Grids Can Contribute to Rural Development: Evidence from Kenya. *World Dev.* **2009**, *37*, 1208–1221. [CrossRef]

29. Simiyu, J.; Waita, S.; Musembi, R.; Ogacho, A.; Aduda, B. Promotion of PV uptake and sector growth in kenya through value added training in PV sizing, installation and maintenance. *Energy Procedia* **2014**, *57*, 817–825. [CrossRef]

30. Tigabu, A.; Kingiri, A.; Odongo, F.; Hanlin, R.; Andersen, M.H.; Lema, R. Capability development and collaboration for Kenya's solar and wind technologies: Analysis of major energy policy frameworks. *IREK Rep.* **2017**, *2*, 1–13.

31. Rolffs, P.; Ockwell, D.; Byrne, R. Beyond technology and finance: Pay-as-you-go sustainable energy access and theories of social change. *Environ. Plan. A* **2015**, *47*, 2609–2627. [CrossRef]

32. Munro, P.; van der Horst, G.; Willans, S.; Kemeny, P.; Christiansen, A.; Schiavone, N. Social enterprise development and renewable energy dissemination in Africa: The experience of the community charging station model in Sierra Leone. *Prog. Dev. Stud.* **2016**, *16*, 24–38. [CrossRef]

33. Werner, C.; Breyer, C. Analysis of mini-grid installations: An overview on system configurations. In Proceedings of the 27th European Photovoltaic Solar Energy Conference Exhibition, Frankfurt, Germany, 23–27 September 2012; pp. 3885–3892.

34. Mathiesen, B.V. Smart Energy Systems for coherent 100% renewable energy and transport solutions. *Appl. Energy* **2015**, *145*, 135–154. [CrossRef]

35. Energy, R. Renewable Energy Policies in a Time of Transition. Available online: http://energyaccess.org/wp-content/uploads/2018/04/IRENA_IEA_REN21_Policies_2018.pdf (accessed on 21 August 2020).

36. Da Silva, I.P.; Batte, G.; Ondraczek, J.; Ronoh, G.; Ouma, C.A. Diffusion of solar energy technologies in rural Africa: Trends in Kenya and the LUAV. In Proceedings of the 1st Africa Photovoltaic Solar Energy Conference and Exhibition, Durban, South Africa, 27–29 March 2014; Volume 1, pp. 27–29.

37. Ministry of Energy. *Republic of Kenya, Strategic Plan*; Ministry of Energy: Nairobi, Kenya, 2014.

38. Aris, A.M.; Shabani, B. Sustainable Power Supply Solutions for Off-Grid Base Stations. *Energies* **2015**, *8*, 10904–10941. [CrossRef]

39. Opiyo, N. A survey informed PV-based cost-effective electrification options for rural sub-Saharan Africa. *Energy Policy* **2016**, *91*, 1–11. [CrossRef]

40. Tian, Y.; Zhao, C.Y. A review of solar collectors and thermal energy storage in solar thermal applications. *Appl. Energy* **2013**, *104*, 538–553. [CrossRef]

41. Zhao, Z.; Venayagamoorthy, K.; Burg, T.; Groff, R.; Belotti, P. Optimal Energy Management for Micro grids. *Energies* **2018**, *11*, 1–22.

42. Elmer, U.; Brix, M. Review of Solar PV Market Development in East Africa. *UNEP Risø CentreUNEP Risø Cent. Work. Pap. Ser.* **2014**. [CrossRef]

43. Marigo, N.; Foxon, T.J.; Pearson, P.J. Comparing innovation systems for solar photovoltaics in the United Kingdom and in China. Available online: https://core.ac.uk/download/pdf/9715618.pdf (accessed on 21 August 2020).

44. Maclean, L.M.; Brass, J.N. Foreign Aid, NGOs and the Private Sector: New Forms of Hybridity in Renewable Energy Provision in Kenya and Uganda. *Afr. Today* **2015**, *62*, 57–82. [CrossRef]

45. Osman, J. Utilizing Solar Energy in King Faisal Specialist Hospital & Research Center Riyadh, Saudi Arabia. Master's Thesis, King Faisal Specialist Research Center, Riyadh, Saudi Arabia, 2012.

46. Attari, K.; Elyaakoubi, A.; Asselman, A. Performance analysis and investigation of a grid-connected photovoltaic installation in Morocco. *Energy Rep.* **2016**, *2*, 261–266. [CrossRef]

47. Nygaard, I.; Hansen, U.E.; Pedersen, M.B. Measures for diffusion of solar PV in selected African countries. *J. Clean. Prod.* **2015**, 1–15. [CrossRef]

48. Borah, R.R.; Palit, D.; Mahapatra, S. Comparative analysis of solar photovoltaic lighting systems in India. *Energy Procedia* **2014**, *54*, 680–689. [CrossRef]

49. Government of Kenya (Ministry of Energy and Petroleum). *Draft National Energy and Petroleum Policy*; Ministry of Energy: Nairobi, Kenya, 2015; pp. 1–130.

50. Parthasarathy, M.; Anandaraj, A.V. *Vision—2030*; Government of Kenya: Nairobi, Kenya, 2011.

51. Elmer, U.; Gregersen, C.; Lema, R.; Samoita, D.; Wandera, F. Technological shape and size: A disaggregated perspective on sectoral innovation systems in renewable electrification pathways. *Energy Res. Soc. Sci.* **2018**, *42*, 13–22.

52. Lund, H. Practical operation strategies for pumped hydroelectric energy storage (PHES) utilising electricity price arbitrage Practical operation strategies for pumped hydroelectric energy storage (PHES) utilising electricity price arbitrage. *Energy Policy* **2011**, *39*, 4189–4196.

53. Williams, B.K. No Batteries Required: Pumped Hydro for Solar Energy Storage. Available online: http://reneweconomy.com.au/2016/no-batteries-needed-pumped-hydro-for-energy-storage-79785 (accessed on 12 August 2020).

54. Schlecht, I.; Weigt, H. Linking Europe: The Role of the Swiss Electricity Transmission Grid until 2050. *Swiss J. Econ. Stat.* **2015**, *151*, 125–165. [CrossRef]

55. Maximov, S.A.; Harrison, G.P.; Friedrich, D. Long Term Impact of Grid Level Energy Storage on Chilean Electric System. *Energies* **2019**, *12*, 1070. [CrossRef]

56. Limpens, G.; Moret, S. The role of storage in the Swiss energy transition. In Proceedings of the 32nd International Conference On efficiency, Cost, Optimization, Simulation and Environmental Impact of Energy Systems, Wroclaw, Poland, 23–28 June 2019; Volume 1, pp. 761–774.

57. Generation, W.O. 100% Renewable Energy for Australia. *Swiss J. Econ. Stat.* **2016**. Available online: http://www.uts.edu.au/sites/default/files/article/downloads/ISF_100%25_Australian_Renewable_Energy_Report.pdf (accessed on 12 June 2020).

58. Canales, F.A.; Jurasz, J.K.; Beluco, A. *A Geothermal Hydro Wind Pv Hybrid System with Energy Storage in an Extinct Volcano for 100 % Renewable Supply in*; pp. 1–19. Available online: https://arxiv.org/pdf/1907.04357 (accessed on 10 July 2020).

59. Al-ghussain, L.; Samu, R.; Taylan, O. Techno-Economic Comparative Analysis of Renewable Energy Systems: Case Study in Zimbabwe. Available online: https://www.researchgate.net/publication/342707080_Techno-Economic_Comparative_Analysis_of_Renewable_Energy_Systems_Case_Study_in_Zimbabwe (accessed on 21 August 2020).

60. Profile, S. Zimbabwe. 2016. Available online: https://constructionreviewonline.com/2019/05/zimbabwe-to-construction-of-three-250mw-solar-power-plants (accessed on 6 August 2020).

61. Makonese, T. Renewable Energy in Zimbabwe Renewable Energy in Zimbabwe. In Proceedings of the Domestic Use of Energy (DUE) 2018 International Conference, Cape Town, South Africa, 3–5 April 2018; pp. 1–5.

62. Zeyringer, M.; Pachauri, S.; Schmid, E.; Schmidt, J.; Worrell, E.; Morawetz, U.B. Energy for Sustainable Development Analyzing grid extension and stand-alone photovoltaic systems for the cost-effective electrification of Kenya. *Energy Sustain. Dev.* **2020**, *25*, 75–86. [CrossRef]

63. Moner-Girona, M. Decentralized rural electrification in Kenya: Speeding up universal energy access. *Energy Sustain. Dev.* **2019**, *52*, 128–146. [CrossRef]

64. Chakir, A. Optimal energy management for a grid connected PV-battery system. *Energy Rep.* **2020**, *6*, 218–231. [CrossRef]

65. Hamrouni, N.; Younsi, S.; Jraidi, M. A flexible active and reactive power control strategy of a LV grid connected PV system. *Energy Procedia* **2019**, *162*, 325–338. [CrossRef]

66. Lund, H.; Østergaard, P.A. Electric grid and heat planning scenarios with centralised and distributed sources of conventional, CHP and wind generation. *Energy* **2000**, *25*, 299–312. [CrossRef]

67. Østergaard, P.A. Geographic aggregation and wind power output variance in Denmark. *Energy* **2008**, *33*, 1453–1460. [CrossRef]

68. Østergaard, P.A. Transmission-grid requirements with scattered and fluctuating renewable electricity-sources. *Appl. Energy* **2003**, *76*, 247–255. [CrossRef]

69. Østergaard, P.A. Regulation strategies of cogeneration of heat and power (CHP) plants and electricity transit in Denmark. *Energy* **2010**, *35*, 2194–2202. [CrossRef]

70. Østergaard, P.A. Modelling grid losses and the geographic distribution of electricity generation. *Renew. Energy* **2005**, *30*, 977–987. [CrossRef]

71. Østergaard, P.A. Reviewing EnergyPLAN simulations and performance indicator applications in EnergyPLAN simulations. *Appl. Energy* **2015**, *154*, 921–933. [CrossRef]

72. Lund, H.; Münster, E. Modelling of energy systems with a high percentage of CHP and wind power. *Renew. Energy* **2003**, *28*, 2179–2193. [CrossRef]

73. Danish Energy Agency. Technology data Energy Storage. 2018. Available online: https://ens.dk/sites/ens.dk/files/Analyser/technology_data_catalogue_for_energy_storage.pdf (accessed on 16 September 2019).

74. Lund, H. Energy Storage and Smart Energy Systems. *Int. J. Sustain. Energy Plan. Manag.* **2016**, *11*, 3–14.

75. Sorknæs, P.; Lund, H.; Andersen, A.N.; Ritter, P. Small-scale combined heat and power as a balancing reserve for wind—the case of participation in the German secondary control reserve. *Int. J. Sustain. Energy Plan. Manag.* **2014**, *4*. [CrossRef]

76. Andersen, A.N.; Lund, H. New CHP partnerships offering balancing of fluctuating renewable electricity productions. *J. Clean. Prod.* **2007**, *15*, 288–293. [CrossRef]

77. Lund, R.; Mathiesen, B.V. Large combined heat and power plants in sustainable energy systems. *Appl. Energy* **2015**, *142*, 389–395. [CrossRef]

78. Østergaard, P.A.; Andersen, A.N. Booster heat pumps and central heat pumps in district heating. *Appl. Energy* **2016**, *184*, 1374–1388. [CrossRef]

79. Østergaard, P.A.; Andersen, A.N. Economic feasibility of booster heat pumps in heat pump-based district heating systems. *Energy* **2018**, *155*, 921–929. [CrossRef]

80. Bacekovic, I.; Østergaard, P.A. Local smart energy systems and cross-system integration. *Energy* **2018**, *151*, 812–825. [CrossRef]

81. Prina, M.G. Smart energy systems applied at urban level: The case of the municipality of Bressanone-Brixen. *Int. J. Sustain. Energy Plan. Manag.* **2016**, *10*, 33–52.

82. Connolly, D. *Smart Energy Systems: Holistic and Integrated Energy Systems for the era of 100% Renewable Energy*; Aalborg University: Aalborg, Denmark, 2013.

83. Lund, H.; Mathiesen, B.V.; Connolly, D.; Østergaard, P.A. A smart energy systems approach to the choice and modelling of 100 % renewable solutions. *Renew. Energy Syst.* **2014**, *39*, 5–12.

84. Pfeifer, A.; Krajačić, G.; Ljubas, D.; Duić, N. Increasing the integration of solar photovoltaics in energy mix on the road to low emissions energy system—economic and environmental implications. *Renew. Energy* **2019**, *5*, 80. [CrossRef]

85. Lund, H.; Kempton, W. Integration of renewable energy into the transport and electricity sectors through V2G. *Energy Policy* **2008**, *36*, 3578–3587. [CrossRef]

86. Juul, N.; Pantuso, G.; Iversen, J.E.B.; Boomsma, T.K. Strategies for Charging Electric Vehicles in the Electricity Market. *Int. J. Sustain. Energy Plan. Manag.* **2015**, *7*, 67–74.

87. Samoita, D.; Remmen, A.; Nzila, C.; Østergaard, P.A. *Renewable Electrification in Kenya: Potentials and Barriers*; (IREK Working Paper No. 7). Copenhagen/Nairobi/Eldoret: AAU, ACTS and MU; 2019. Available online: https://www.irekproject.net/wp-content/uploads/IREK.Paper7_.pdf (accessed on 12 November 2019).

Publisher's Note: MDPI stays neutral with regard to jurisdictional claims in published maps and institutional affiliations.

Review

A Review on the Thermochemical Recycling of Waste Tyres to Oil for Automobile Engine Application

Mohammad I. Jahirul [1,2,*], Farhad M. Hossain [3,4], Mohammad G. Rasul [1,2] and Ashfaque Ahmed Chowdhury [1,2]

[1] School of Engineering and Technology, Central Queensland University, Rockhampton, QLD 4702, Australia;
m.rasul@cqu.edu.au (M.G.R.); a.chowdhury@cqu.edu.au (A.A.C.)

[2] Centre for Intelligent Systems, Clean Energy Academy, Central Queensland University,
Rockhampton, QLD 4702, Australia

[3] Biofuel Engine Research Facility, Queensland University of Technology (QUT), Brisbane, QLD 4000, Australia;
farhad.hossain@qut.edu.au or mfarhad03@yahoo.com

[4] Green Distillation Technologies Corporation Limited (GDTC), P.O. Box 4075, Richmond, VIC 3142, Australia

* Correspondence: md_jahirul@yahoo.com or m.j.islam@cqu.edu.au; Tel.: +614-1380-9227

Abstract: Utilising pyrolysis as a waste tyre processing technology has various economic and social advantages, along with the fact that it is an effective conversion method. Despite extensive research and a notable likelihood of success, this technology has not yet seen implementation in industrial and commercial settings. In this review, over 100 recent publications are reviewed and summarised to give attention to the current state of global tyre waste management, pyrolysis technology, and plastic waste conversion into liquid fuel. The study also investigated the suitability of pyrolysis oil for use in diesel engines and provided the results on diesel engine performance and emission characteristics. Most studies show that discarded tyres can yield 40–60% liquid oil with a calorific value of more than 40 MJ/kg, indicating that they are appropriate for direct use as boiler and furnace fuel. It has a low cetane index, as well as high viscosity, density, and aromatic content. According to diesel engine performance and emission studies, the power output and combustion efficiency of tyre pyrolysis oil are equivalent to diesel fuel, but engine emissions (NO_X, CO, CO, SO_X, and HC) are significantly greater in most circumstances. These findings indicate that tyre pyrolysis oil is not suitable for direct use in commercial automobile engines, but it can be utilised as a fuel additive or combined with other fuels.

Keywords: waste tyre; waste management; pyrolysis; automobile engine

Citation: Jahirul, M.I.; Hossain, F.M.; Rasul, M.G.; Chowdhury, A.A. A Review on the Thermochemical Recycling of Waste Tyres to Oil for Automobile Engine Application. *Energies* **2021**, *14*, 3837. https://doi.org/10.3390/en14133837

Academic Editor: Idiano D'Adamo

Received: 10 May 2021
Accepted: 21 June 2021
Published: 25 June 2021

Publisher's Note: MDPI stays neutral with regard to jurisdictional claims in published maps and institutional affiliations.

1. Introduction

The fast growth of industrialisation around the world has resulted in an expansion in vehicle production as a main mode of transportation to mobilise the population and expand economies. At the same time, oil consumption in the transportation sector is fast increasing, resulting in a rapid depletion of non-renewable petroleum-based fuel [1–3]. Alternative renewable and environmentally friendly sources of car fuel, such as biodiesel [4–8], oxygenated fuel [6,9,10], and blends with petroleum-based fuels [11,12], have received increased attention in recent decades. However, due to economic and environmental concerns, waste-to-fuel technology has received increased attention from researchers around the world in recent years [13]. Solid waste disposal in landfills is both expensive and damaging to the environment [14,15]. As a result, waste-to-fuel technology offers enormous potential to reduce global waste while also replacing petroleum-based gasoline.

The increasing use of transportation vehicles results in a global stockpile of waste tyres, which is one of the biggest sources of pollution [5,16–19]. Around 1.5 billion tyres are produced worldwide each year, which implies the same number of tyres end up as waste tyres, amounting to nearly 17 million tons [20–22]. About 15–20 per cent of tyres are considered for recycling or reuse once they have reached the end of their useful

53

life, while the remaining 70–80 per cent are disposed of in landfills and remain in the environment [23]. Every year, one billion WT are disposed of in landfills around the world, and one car per person is disposed of each year in industrialised countries [6]. Due to the high likelihood of hazardous fumes from fire, these landfills are a severe hazard for the environment and human health [24], and they provide ideal conditions for rats, snakes, and mosquito breeding.

Due to their highly complicated structure, the variable composition of the raw material, and the chemical structure of the rubber from which the tyres are formed, recycling waste tyres is exceedingly challenging [18]. Tyres are made up of 45–47% rubber, 21.5–22% carbon black, 16.5–25% steel belts, and 4.5–5.5% textile overlays, which give the tyre its ultimate form and practical features. In addition, depending on the production method and specification, numerous different materials can be added to the tyre [25,26]. The cross-linkages formed between the elastomer and various components throughout the production process produce a three-dimensional chemical network, resulting in excellent elasticity and strength. Tyres are difficult to break down due to their complicated chemical composition [25,27,28]. As a result, decomposition in the landfill will take more than a century [29]. Furthermore, landfilling ignores the enormous energy potential of waste tyres while also posing a fire risk, resulting in dangerous gas emissions as well as the poisoning of water and soil. Several investigations have been undertaken in the last few decades to create effective technology for converting used tyres to energy [30–32]. Pyrolysis [33–36], gasification [37,38], and hydrothermal liquefaction [39] are the most prevalent methods for turning waste tyres into energy in the form of fuels. Pyrolysis, in particular, has received a lot of interest for scraping tyre waste treatment because of its efficiency compared to other methods. Pyrolysis can be used to turn waste tyres into petrol and diesel, as well as fuel oil, without harming the environment. It is the mechanism of thermally degrading long-chain molecules into smaller molecules by heat and pressure in an oxygen-free environment, which results in the production of liquid hydrocarbons (oil), gases, and char [35,40,41]. During pyrolysis, the tyres are cracked in a medium temperature range between 400 and 700 °C, which produces char, tar, and gaseous fuels as well as steel [16]. This technique produces oil that can be utilised directly in industrial applications and diesel engines, or it can be refined further. In comparison to petroleum-derived fuel oils, the most essential feature of this oil is its low exhaust pollution. There has been a lot of research on the performance and emissions of diesel engines utilising tyre pyrolysis oil [21,42–44].

In recent decades, waste tyre pyrolysis technology has shown to be an effective waste-management strategy. This technology's ultimate goal is to manufacture high-quality fuels from scraping tyres that can compete with and eventually replace non-renewable fossil fuels. Despite extensive study and great advancements, waste tyre to vehicle fuel technology has not yet reached its full potential. This technology will need more development before it can be scaled up to an industrial level. However, in order to advance waste tyre to energy technology and upgrade the technology on an industrial scale, it is critical to thoroughly comprehend the current development stage. This paper review over 100 up-to-date papers from the literature and discussed the key findings, the current status, and the development of this technology. The information only considered from the peer-reviewed literature published in reputed international journals, conference proceedings, and reports. More emphasis was given to the recently published literature on the related topic. For the analysis, data were only taken from the literature where the experiments were carried out by the authors themselves in accordance with internationally recognised testing standards. Certain extreme information was removed from the database due to the unanticipated nature of the outcomes. The novelty of this article is to elaborate the way to utilise tyre pyrolysis oil as a substitution for conventional petroleum-based automobile fuel. Additionally, limitations of current waste tyre to automobile fuel technology have been identified and based on the observation of literature research; the future direction of research for commercialising the technology has been indicated. It has been expected that the findings of this literature review will serve as a basis on which the industrial production of waste tyre pyrolysis automobile engine oil will be possible.

2. Waste Tyre Management Practice

The goal of waste tyre management is to identify the most efficient approach to limit the waste's environmental impact. Reduction in consumption, reuse/recycling, and energy recovery are all strategies for solving the WT problem. The primary reason for developing those methods was the restrictions imposed by the government for collecting tyres for landfills. In recent years, the methods that are used for waste tyre management includes: reuse and rethreading, product recycling, and recovery of energy [45]. Figure 1 depicts the process of a typical waste tyre management system.

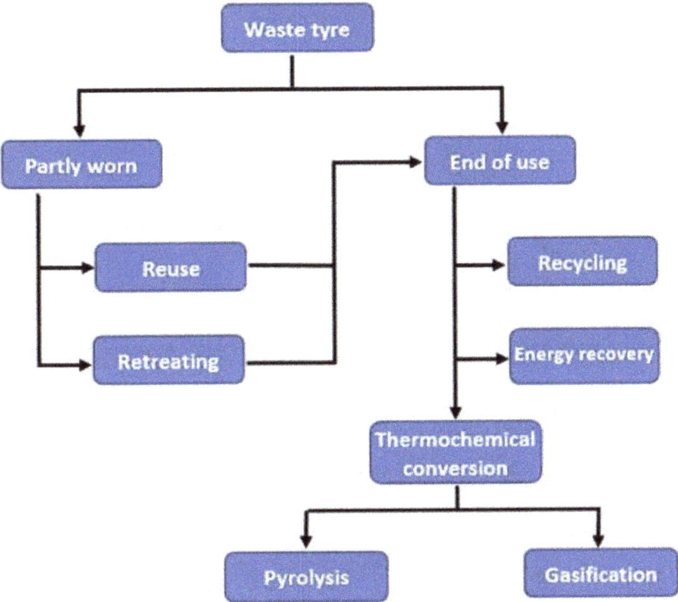

Figure 1. A typical waste tire management system.

2.1. Reuse and Rethreading

The rethreading process is used for the repair of limited worn-out tyres because of the initial wearing and certification requirement of the damaged areas that need to be rethreaded [46]. It is one of the most ideal strategies that require the limited deployment of additional energy and resources, mainly rubber, which accounts for 50% of the material involved. It involves 20% usage in the damaged situation, whereas the remaining part constitutes a tyre carcass [47]. Practically, rethreading requires about 25% of raw materials and 30% of energy and only 30% of the energy needed to produce new tyres, which make this process economically profitable [48]. However, because of the safety issue at high speed, rethreading tyres is not used in automobile application [49].

2.2. Recycling

Recycling is also one of the widely used methods for managing waste tyre. Sienkiewicz et al. [22] referred to recycling that does not include any treatment, be it physical or chemical in nature, and only disintegration is important. Moreover, the role of the shape of the tyre, their initial development in terms of size, durability, elasticity, vibration capacity, high damping, etc., makes this material very useful and important in works of many other purposes. There are many applications for the tyres recycled; some of the end products from recycled rubber include: manufacturing new tyres, tracks for athletics, insulation in buildings, matting surface, surfaces for playgrounds, marine non-slip surfaces, etc [50].

2.3. Energy Recovery

Various industrial processes rely on non-renewable energy sources, such as coal. The abundance of coal and the combustion process are sufficient to meet the demand. Recovery of waste tyres from other rubber products is one of the processes that can be used to deploy them as energy sources, such as fuel. Waste tyres can be used as an alternative fuel in cement kilns and for electricity generation, but their combustion and burning pollutes the environment. The cement industry uses tyres as a source of energy, making it a cost-effective way to meet their high-temperature requirements. However, compared to pollution levels in the air caused by the coal combustion process [22], this has a minor impact. Figure 2 shows that 53% of used tyres in the United States are capable of meeting fuel requirements [51]. Cement and paper mills use 68% of waste tyre for their energy needs, as shown in Figure 2.

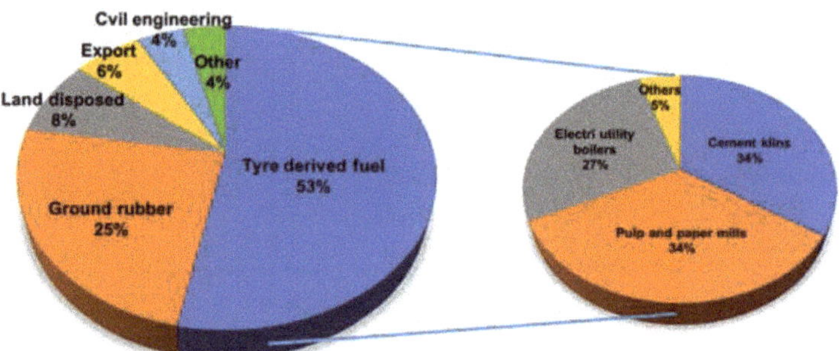

Figure 2. US scrap tyre disposition.

3. Waste Tyre to Fuel Using Thermochemical Conversion

Thermochemical conversion is conducted at high temperatures, with or without the presence of oxygen, to chemically degrade waste tyres. To produce bio-oil, syngas, and char, mostly pyrolysis and gasification conversion methods are used. In comparison to gasification, this study focused solely on pyrolysis because of its high liquid fuel recovery and low environmental impact [51–54].

3.1. Waste Tyre Conversion Using Pyrolysis

Due to its nature and fewer processing steps, the pyrolysis process, such as other similar thermochemical processes, is thought to be the most environmentally friendly [39]. The process involves the decomposition of the solid at a considerably inflated temperature of around 300–900 °C in an environment that is free of oxygen and as a result produces producing char, oil, and gas [54–58]. The important products of pyrolysis gas in most cases are H_2, CO, CO_2, CH_4, whereas the liquid consists of mainly CH_3OH, CH_3COOH, and H_2O. The rest of the solid products consist of carbon and ash [58–60]. The steps of the waste tyre pyrolysis process are depicted in Figure 3.

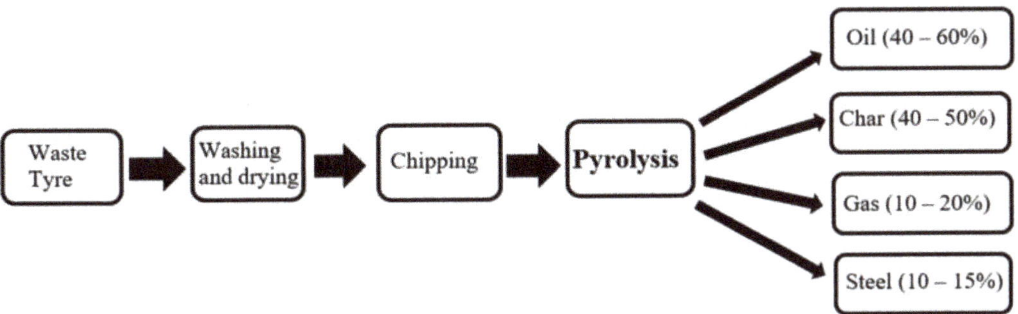

Figure 3. Waste tyre pyrolysis process.

Yield outcomes of the pyrolysis process depend on the different operating conditions and reactor settings [57,61]. There are three main kinds of pyrolysis: (i) slow pyrolysis process, (ii) fast pyrolysis process, and (iii) flash pyrolysis process. Slow pyrolysis, also known as conventional pyrolysis, is commonly used to produce wood charcoal from wood biomass. Fast and flash pyrolysis, on the other hand, is used to produce bio-fuel [59,62]. The main configurations of bio-fuel are esters, organic acids, phenols, alkenes, furfurals, and certain inorganic species. These products are easier to transport and store than solid biomass, which is converted into valuable biofuels and chemicals [63].

The pyrolysis processes depend on factors such as temperature, material size, residence period, etc [54,60]. Different types of pyrolysis reactors were developed and examined in recent years to produce waste tyre for oil, char, and gas. Pyrolysis production yields depend on the feedstock's preparation, reactor types, and pyrolysis reaction condition [57,61]. However, temperature is the main factor to control the configuration of the pyrolysis process [64].

3.1.1. Tyre Pyrolysis Reactors

The reactor is the main component of the pyrolysis process, in which tyre waste decomposes in the absence of oxygen. A substantial amount of research on the pyrolysis reactor has been conducted in order to improve the essential characteristics of heating methods and rate, waste tyre feeding, pyrolysis residence time, vapour condensation, and product collection. Fixed-bed, vacuum, fluidised bed, moving screw bed, rotary kiln, and other reactor types are investigated for waste tyre pyrolysis. Table 1 summarises the product yields of various pyrolysis reactors and their waste tyre product yields. The fixed bed reactor is the most commonly used reactor type for tyre pyrolysis, in which the processed waste tyres are heated externally using an electric furnace and an inert gas such as nitrogen is used as a carrier gas. The shredded tyre is continuously fed into the hot reactor in most other types of reactors, such as fluidised bed reactors. Usually, the decomposition of tyre materials starts at near 400 °C temperature, and therefore, most of the pyrolysis investigations shown in Table 1 have been conducted between the 450 and 600 °C temperature range [65].

Table 1. Various amount of product yield from waste tyre using pyrolysis reactor.

Reactor Type	Temp. (°C)	Maximum Oil Yield (wt%)			References
		Oil	Char	Gas	
Fixed bed, batch	500	40.26	47.88	11.86	[17]
Fixed bed, batch	450	63	30	7	[66]
Fixed bed, batch	500	54.12	20.22	26.40	[67]
Fixed bed, batch	400	38.8	34.0	27.2	[34]
Fixed bed, batch,	475	55	36	9	[68]
Fluidised bed	450	55	42.5	2.5	[69]
Fluidised bed	450	52	28	15	[33]
Moving screw bed	600	48.4	39.9	11.7	[70]
Two stages fixed bed	600	22	37	27.2	[71]
Microwave oven reactor	-	43	45	12	[46]
Rotary kiln	600	40.21	52.67	7.12	[72]
Rotary kiln	500	45.1	41.3	13.6	[73]
Rotary kiln	550	38.12	49.09	2.39	[73]
Vacuum	500	56.5	33.4	10.1	[74]
Vacuum	550	47.1	36.9	16	[74]

Fixed Bed Pyrolysis Reactor

Fixed bed pyrolysis reactors, as shown in Figure 4, are relatively simple to construct and efficient at producing clean fuel. These reactors are typically operated in batch mode. The waste tyre is fed into a fixed bed inside a cylindrical steel pyrolyser. Heat is supplied to the waste tyre via the pyrolyser wall by an electrical heater or furnace mounted around the reactor. By purging pressurised nitrogen (N_2) from the external cylinder, all the oxygen inside the pyrolyser is eliminated. When waste tyres decompose, solid char accumulates at the bottom of the pyrolyser, while vapour (both condensable and non-condensable) escapes to the top. The vapour is then cooled by a condenser, which condenses the condensable vapour into oil, which is then stored and collected in a liquid storage container. The non-condensable vapour remains gaseous and is collected as syngas [74]. The basic characteristics of fixed bed reactors are higher carbon conservation rate, lower velocity of gas, lower gas carryover rate, and a long residence period of solid. Generally, small-scale heat and power applications are considered for these reactors [75,76]. The removal of tar from fixed bed reactors is a major issue; however, recent advances in thermal and catalytic tar conversion have provided a possible solution to eliminate the problem [77–79]. In a 1.15-L, nitrogen fixed bed reactor in a temperature range of 400–700 °C, Aydin and Ilkilic [17] investigate the pyrolysis of waste tyres in stationary reactors, with removed fabric and steel. The oil output increase from 31% at 400 °C to 40% at 500 °C, and the return at higher temperatures increased slightly. Kar [67] investigated the effect of pyrolysis temperature ranging from 375 to 500 °C in a laboratory model fixed bed pyrolysis reactor. The highest oil output of 60.0 wt% oils was reported to be attained at 425 °C in this study. However, the oil output reduced to 54.12 wt% at higher pyrolysis temperatures, at 500 °C. The output of gas increased from 2.99% to 20.22%W while the output of char fell from 50.67% to 26.40%, with pyrolysis temperature increasing from 375 °C to 500 °C. In a similar pyrolysis reactor, Banar et al. [34] obtained 38.8 wt% of oil, 27.2 wt% of gas and 34 wt% of char at 400 °C pyrolysis temperature. The study further investigated the effect on waste tyre pyrolysis on the heating rate and found that the heating rate was influenced by a decrease in oil production to 35.1 wt% and gas yield to 33.8 wt% as the heating rate increased at a pyrolysis temperature of 400 °C from 5 °C min^{-1} to 35 °C min^{-1}.

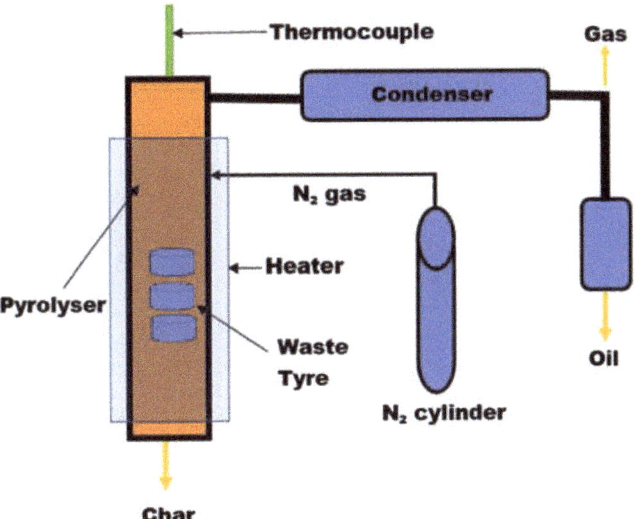

Figure 4. Diagram of a fixed bed pyrolysis reactor.

Fluidised Bed Reactor

In the oil and chemical industries, fluidised bed pyrolysis reactors are widely used. In contrast to fixed bed reactors, waste tyres in fluidised bed reactors can be processed continuously, making them very efficient and cost-effective, particularly in industrial plants [80]. This type of reactor can produce high-quality oil, with a liquid yield of 50–60% of the dry weight of the waste tyre. Waste plastics are shredded into smaller pieces and continuously fed into a pyrolyser's hot sand or other stable solid beds. To make the pyrolyser oxygen-free, the solid bed is fluidised with N_2 or other inert gases. The shredded tyre quickly heated up on the solid hotbed and decomposed into vapour, char, and aerosols [81–83]. Char is removed with a cyclone separator, and the remaining vapour is quickly cooled into bio-oil and stored with a quench cooler, as shown in Figure 5. The heat required to run this type of reactor can be created by burning a portion of the produced gas in the bed or by burning char produced in another chamber and transferring the heat to the solid bed [84]. One essential feature of the fluidising bed reactor is that the shredded tyre particle sizes of less than 2–3 mm are needed to achieve the desired heating rate [85,86]. The investigations were made in this types of reactor including an industrial scale with a throughput of 200 kg h^{-1}, a pilot scale with a throughput of 30 kg h^{-1}, and a laboratory scale with a throughput of 1 kg h^{-1} [65]. Williams et al. [69] pyrolysed tyre crumb (up to 1.4 mm size) in a laboratory-scale fluidised bed reactor ranging the temperature between 450 and 600 °C. The reactor dimensions were 100 cm in height and 7.5 cm in diameter, with a feedstock processing capacity of 220 gm kg h^{-1}. The quartz sand was utilised as a fluidised bed, with nitrogen serving as the fluidising air, and was pre-heated to 400 °C using an external electrical furnace. For pyrolysis vapour condensation, a series of water-cooled and dry ice/acetone-cooled condensers were used. This study found a maximum oil yield of 55% at 450 °C. However, the oil yield was reduced to 43.5% at 600 °C pyrolysis temperature and at the same time, the gas yield was increased from 2.5% to 14%.

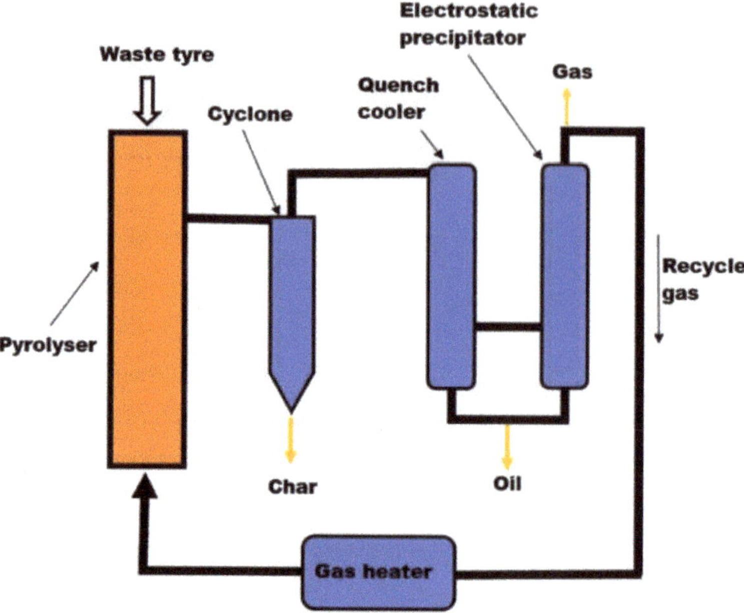

Figure 5. Diagram of a fluidised bed pyrolysis reactor.

Rotary Kiln Reactor

In a rotary kiln reactor shown in Figure 6, the waste tyres are fed into the front end of the reactor. The waste tyres inside the reactor are heated, moving down the cylinder, and pyrolytic gases are released. The reactor is slowly rotated and its inclined feature ensures the mixing and uniform heating of waste tyre [87]. This reactor offers various unique characteristics, such as the capacity to process heterogeneous feedstock, flexibility in residence time adjustment, proper and uniform waste tyre mixing, no need for waste tyre pre-treatment, and simple and easy maintenance. However, because of the sluggish heating rate, these reactors are mostly employed for slow pyrolysis applications. The reason for the low heating rates is that heat is transported to the feedstock from the outside reactor via the reactor wall only, and the particle size and contact area between the reactor wall surface and the feedstock are minimal. The heating rate can be achieve a maximum of 100 °C/min with a minimum residence time of 1 h [88]. Galvagno et al. [73] conducted an experiment on waste tyre pyrolysis in a pilot-scale rotary kiln reactor. The rotary reactor had a diameter of 0.4 m and rotated at 3 rpm, and was heated externally using 4.8 kg h^{-1} electrical furnaces. Condensed fractions of heavy and light pyrolysis oil were condensed in a condensation system, and the non-condensable gases were cleaned to remove acid gases before being combusted in a flare. The pyrolysis char (residue) was continuously released in a water-cooled tank. They obtained 45.1% of oil, 44.3% of char, and 13.6% of gas while conducting waste tyre pyrolysis at 500 °C.

%endparacol

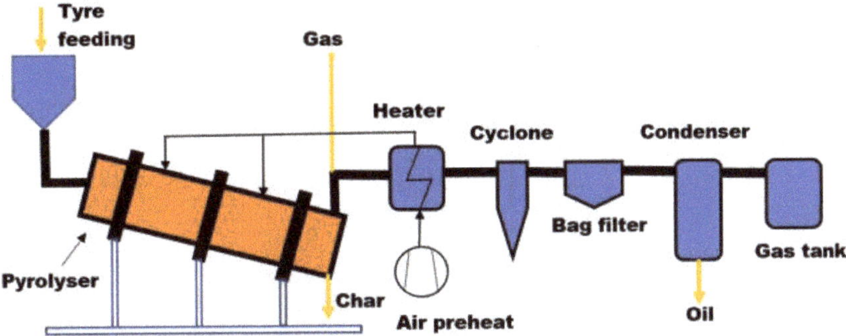

Figure 6. Diagram of a rotary kiln reactor pyrolysis reactor.

Screw Bed Reactor

The screw bed reactors consist of a rotating screw (Figure 7), are tubular in shape, and operate continuously. Rotation of the screw assists feedstock delivery into the reactor, while heat required for the pyrolysis process is conveyed across the tubular reactor wall. Thus, the screw serves two functions: first, mixing the feedstock and second, regulating the feedstock residence time in the reactor. A large hopper is used to feeds waste tyres into the screw bed. Inert gases are usually N_2. The hopper is supplied to the hopper to eliminate oxygen from the feedstock and also make the pyrolysis system oxygen-free. By creating a modest positive pressure at the screw bed, the inert gas also aids in the transmission of pyrolysis vapour [89]. Steel and ceramic pellets are normally packed with feedstock particles, which serves as a solid heat carrier for pyrolysis. This allows feedstock particles to interact more closely as they move through the screw bed. The oil is produced by drawing the vapours produced by the pyrolysis process into a condenser. An important advantage of a screw bed reactor is that it may be made to be very compact and even portable in some situations, allowing the reactor to be used at the feedstock generating site or anywhere there is lots of feedstock. On-site feedstock processing reduces operating expenses by lowering feedstock transportation expenses to the biorefinery [90]. However, if the reactor is not designed properly, there will be poor heat transfer and temperature control, resulting in the deposition of polymeric materials in the reactor's interior [91].

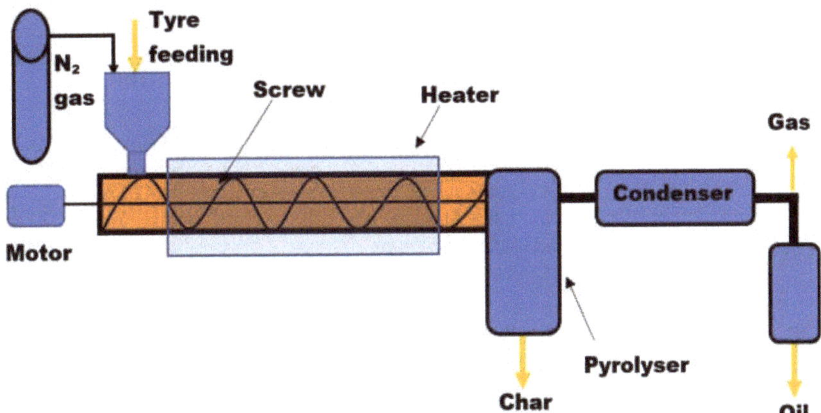

Figure 7. Diagram of a rotary screw bed pyrolysis reactor.

4. Waste Tyre to Oil, Carbon, and Steel

The recycling of waste tyres into useful products is of interest for both environmental and economic reasons. Many researchers have been working to solve the aforementioned issues and convert waste tyres into valuable products such as oil, carbon, and steel [69,92,93]. Waste-tyre oil could be used for heating by industry, refined further for use in diesel engines, or used directly as blended fuel in some stationary diesel engines. Carbon has a plethora of industrial uses, from toothpaste to electrodes and pharmaceutical goods, as well as being about 35% cleaner than coal and burning hotter, while steel can be sold as scrap metal or returned to tyre manufacturers for reuse.

4.1. Oil from Waste Tyres

The oil extracted from waste tyres via a thermochemical conversion process varies depending on the conversion process and the operating conditions. The colour of pyrolysis tyre oil (TPO) is generally black, and it has a distinct odour. Table 2 summarises the physicochemical properties of oil produced from waste tyre pyrolysis reported in the recent literature and compares the value to conventional diesel and biodiesel. For automobile engine application, one of the important quality parameters is the higher heating value (HHV) of the oil. The HHV of TPO is reported in the range of 38–42 MJ/kg, which is less than conventional diesel but is, however, higher than biodiesel. The Cetane index of PTO was found to be 28.6, which is much lower than both diesel (53.2) and biodiesel (58.6). Furthermore, TPO has a much higher density, kinematic viscosity, and aromatic content than both diesel and biodiesel. As a result of these findings, TPO may not be suitable for direct use as a fuel in commercial automobile engines; however, blending TPO with diesel and biodiesel can be utilised.

Table 2. Physicochemical properties of tyre oil, WCBD and diesel.

Properties	Units	Diesel	Biodiesel	TPO	References
Density	kg/L	0.83	0.88	0.91–0.96	[19,43,94]
HHV	MJ/kg	45.6	37.2	38–42	[19,43,94]
Water content	mg/kg	<30	–	118	[19,43]
Aromatic	% m/m	26.0	–	39.3	[19,43,95]
Kinematic V.	mm^2/s	2.66	4.73	3.22–6.3	[19,43,94]
Carbon (C)	wt%	87.0	76.9	79.61–88	[18,19,43,94]
Hydrogen (H)	wt%	13.0	12.2	9.4–11.73	[18,19,43,94]
Nitrogen (N)	wt%	–	–	0.40–1.05	[18,19,43,94]
Oxygen (O)	wt%	0	10.9	0.5–4.62	[19,43,94,96]
Ash content	wt%	0.01		0.02–0.31	[19,96]
Flash point	°C	50	130	20–65	[19,96,97]
Cetane index		53.2	58.6	28.6	[43,94]

4.2. Char from Waste Tyres

Another valuable byproduct of waste-tyre pyrolysis is char. It is reported that the char produced from tyre conversion ranges from 22 to 49% by weight [19]. There are many researchers investigating the characteristics of the char. Table 3 shows the properties of waste tyre char. The chars have a low heating value of 29.3–31.5 MJ/kg, compared to tyre oil, which has a range of 38–42 MJ/kg. Therefore, tyre char can be used in a variety of industries, including cement and fertiliser.

Table 3. Properties of waste-tyre chars.

Properties	Units	Tyre Pyrolysis Char			
		[98]	[87]	[73]	[99]
HHV	MJ/kg	30.8	31.5	30.7	29.3
Water content	mg/kg	0.37	2.35	3.57	
Carbon (C)	wt%	88.19	82.17	85.31	80.3
Hydrogen (H)	wt%	0.6	2.28	1.77	1.3
Nitrogen (N)	wt%	0.1	0.61	0.34	0.3
Sulphur (S)	wt%	1.9	2.32	2.13	2.7
Ash content	wt%	8.27	12.32	15.33	

4.3. Gas from Waste Tyre

The gas obtained from waste tyre pyrolysis is called pyrogas or syngas. It is normally a mixture of hydrogen (H_2) and olefins, and paraffins with carbon numbers ranging from one to six (C_2–C_6) with a low concentration of nitrogen (N_2) and sulphur (S) compounds [100,101]. The syngas' calorific value ranges from 50 to 70 MJ/m^3 [93–95]. It has been commonly reported that the syngas produces from waste tyre pyrolysis is sufficient for supplying the required heat for the process [18,73,102,103].

4.4. Steel from Waste Tyre

Waste tyres also produce steel when converted by a thermochemical process. It is reported that the amount of steel recovered from waste tyres typically ranges from 10 to 15% by weight of the waste tyre [19]. The recovered steel can be reused by the tyre manufacturer or diverted to steel re-rolling mills.

5. Diesel Engine Performance and Exhaust Emission Using Tyre Oil

According to various researchers [95,104], the properties of waste-tyre pyrolysis are similar to those of diesel and gasoline. In today's world, the diesel engine is the most widely used internal combustion engine. Increased demand for diesel fuel, combined with limited resources, has prompted a search for alternative fuels for diesel engines, such as alcohol, LPG, biodiesel, and compressed natural gas (CNG) [55]. The results of studies on engine testing with tyre oil in the literature vary due to the different properties of the test fuels and different test-engine technology [102]. In an engine-emissions analysis, many variables must be controlled, such as engine speed, fuel composition, and load condition. Tyre fuel has proven to be one of the most important and useful research outputs. However, funding for the use of tyre-derived pyrolytic fuel or diesel-blend fuel has been limited because the effects on overall engine performance and emissions have not been sufficiently confirmed. As a result, additional research focusing on diesel engine emissions using oil from waste tyres is expected to have a positive impact in alternative industries. Furthermore, it could be a promising option in the search for low-emission energy sources.

Several researchers have conducted tests on diesel engine performance with tyre oil in recent years. Table 4 summarises their findings. Vihar et al. [43] experimentally analysed the combustion characteristics and emission of tyre pyrolysis oil in a turbo-charged six-cylinder compression ignition engine using 100% TPO as fuel. They found a stable diesel running throughout the experiment with an almost similar thermal efficiency and specific fuel combustion. However, due to the higher density of TPO compared with diesel which has a direct link with fuel spray to the cylinder, the ignition delay (ID) of combustion and cylinder peak pressure (CPP) were found to be higher. Engine exhaust emission NO_X, CO, SO_2 and HC was found to be significantly higher (2–50%), whereas smoke emission was found slightly lower while running the engine with 100% TPO compared with diesel. Similar results were reported by Žvar Baškovič et. al. [105] when conducting an experiment in a 1.6-litre multi-cylinder common-rail diesel engine running with 100% pure TPO. Tudu et al. [42] examined the effect of diethyl ether in a diesel engine running on a tyre-derived fuel-diesel blend. They blended 40% tyre-pyrolysis oil with diesel and simultaneously 4%

diethyl ether to improve the CN of the blended fuel. It was reported that those blended fuels reduce the NO_X emission by approximately 25% with respect to diesel operation at full load [42]. Cumali and Huseyin [16] carried out an experimental investigation of fuel production from waste tyres using a catalytic pyrolysis process and tested it in a 0.75-litre single-cylinder diesel engine. This study ran the engine with blends of 5%, 10%, 15%, 25%, 35%, 50%, and 75% TPO with diesel and 100% TPO as fuel. It was reported that 50%, 75%, and 100% tyre-oil blends significantly increase CO, HC, SO_2, and smoke emissions compared to diesel emissions and are therefore not suitable for direct use in commercial diesel engines without engine modification. Hossain at al. [106] also reported a small changes in engine combustion performance running a 5.9-litre, six cylinder turbo-charged diesel engine with 10% and 20% of TPO. However, this study found a significant change in brake-specific emission of NO_X, CO_2, CO, and particle emission. The brake-specific NO_X reduced by 30%, whereas the CO emission increased by 10% with tyre oil blends, as shown in Figure 8.

Table 4. Diesel engine emission and performance with TPO.

Engine Specification	Test Condition	TPO Fraction	Power	Torque	BSFC	BTE	T_e	ID	CP	CO	Smoke	CO_2	NO_x	SO_2	HC	Ref.
6.8 L, 6C, 4S, CI, CR: 18:1 RP: 162 kW, RS: 2400 rpm	S: 1500–2400 rpm L: 80–140 Nm	100%						↑	↑	↑	↓		↑	↑	↑	[43]
5.9 L, 6C, 4S, CI, CR: 17.3:1 RP: 162 kW, RS: 2000 rpm	S: 1500 rpm L: 25–100%	10%, 20%	↓			↓			↑	↑		↓	↓			[106]
1.6 L, 4C, 4S, CI, CR: 18:1 RP: 66 kW, RS: 4000 rpm	S: 1500 rpm L: 80–140 Nm	100%			↑	↓		↑	↑	↑			↑		↑	[105]
0.66 L, 1C, 4S, CI, CR: 17.5:1 RP: 4.4 kW, RS: 1500 rpm	S: 1500 rpm L: 1.1–4.4 kW	40%			↓	↑							↓			[42]
0.40 L, 1C, 4S, CI, CR: 18:1 RP: 7.5 kW, RS: 3600 rpm	S: 1000–1500 rpm L: 1.1–4.4 kW	5–100%	↓						↑	↑	↑	↑	↑	↑	↑	[16]
1.6 L, 4C, 4S, CI, CR: 18:1 RP: 66 kW, RS: 4000 rpm	S: 1500–3000 rpm L: 60–100 Nm	50%						↑	↑				↓		↑	[107]
0.5 L, 1C, 4S, CI, CR: 17.5:1 RP: 3.7 kW, RS: 1500 rpm	S: 1500 rpm L: 80–140 Nm	10%				↑	↓	↑	↑	↑			↓	↑		[96]
0.5 L, 1C, 4S, CI, CR: 18.1:1 RP: 3.5 kW, RS: 1500 rpm	S: 1500 rpm L: 0.22–7.4 kg	10–90%	↑		↓	↑		↓	↑	↑	↑	↑	↑		↓	[108]
0.4 L, 1C, 4S, CI, CR: 18:1 RP: 5.4 kW, RS: 3000 rpm	S: 1500 rpm L: 3.75–15 Nm	10%				↑	↓	↑	↑			↓	↑	↑	↓	[1]
3.3 L, 4C, 4S, CI, CR: 18:1 RP: 60 kW, RS: 2200 rpm	S: 1000–2800 rpm	5–10%	↑	↑	↓		↓			↓			↑	↑		[109]
0.44 L, 1C, 4S, CI, CR: 20:1 RP: 7.1 kW, RS: 3600 rpm	S: 2000–3000 rpm L: 50–100%	20–80%			↑	↓		↑					↑			[110]

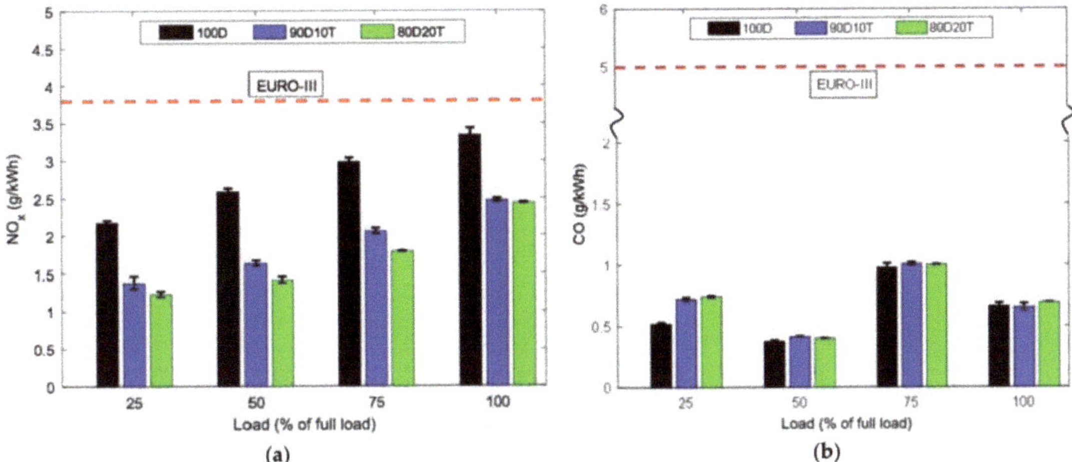

Figure 8. Brake-specific NOx (**a**) and CO (**b**) emission with TPO. (100D—100% diesel; 90D10T—90% diesel + 10% TPO; 80D20T—80% diesel + 20% TPO).

6. Discussion and Synthesis

The idea of waste tyre management is to find the best way to reduce the environmental impact produced by this waste. Waste tyre pyrolysis technology is proven as an efficient method in waste tyre management in recent decades. High-quality fuels from scrape tyre can be produced through pyrolysis, which will eventually replace non-renewable fossil fuels. Despite the fact that there has been a lot of research interest in waste tyre thermochemical conversion to fuel in recent decades, the commercialisation of TPO as an automotive engine fuel technology is still a long way off. It is necessary to fully recognise the current development stage as well as many technical and economical hurdles that need to be overcome for further development of waste tyre to energy technology and upgrade the technology on an industrial scale. There is minimal study regarding the industrial cost of tyre pyrolysis. It is essential that the financial and environmental benefits of the tyre pyrolysis have been thoroughly researched, and the cost has been decreased further for a large-scale commercial application to be viable in the long term. To realise the full potential of waste tyre pyrolysis technology, further research and development are needed, and some of the future challenges are described below:

- Conduct in-depth energy and economic studies of integrated waste tyre pyrolysis plants over their entire life cycle.
- Recognise the trade-offs between the scale of the waste tyre pyrolysis plant and feedstock, as well as the costs of transportation to a centralised upgrading facility.
- Development of the technology to overcome the limitations of the tyre pyrolysis reactor and process and improve the reliability.
- Identify TPO criteria and quality standards for manufacturers and end-user.
- Improve quality and consistency of TPO through the development of more effec -tive technologies.
- Develop catalyst for TPO upgrading in order to meet vehicle fuel-quality standards.
- Develop deoxygenated catalysts to extract oxygen-containing compounds for pyroly- sis processes for oil property improvement.
- Advocacy to develop relevant policy, regulation, and financial incentives for the tyre recyclers, refineries and start-ups who take up the challenges of recycling used tyres to oil.

7. Conclusions

The purpose of this study was to investigate the use of waste tyres as an alternative fuel to address the global problem of waste tyre management and environmental concerns. This paper summarises, describes, and presents research findings from recent publications on these topics. According to the literature, pyrolysis is the most common thermochemical method for addressing the waste tyre management issue due to its simplicity, high recovery of liquid and solid materials, and low environmental impact. As a result, the technological aspects of waste-tyre pyrolysis, including product yield, consistency, and applications, have received increased attention. The findings of this literature review lead to the following conclusions:

- ➢ The product yield and composition of waste tyre pyrolysis depends on reactor type and operating condition. The average production yield of oil, char, syngas, and steel from waste tyre pyrolysis is 40–60%, 40–50%, 10–20%, and 10–15% by weight.
- ➢ The pyrolysis temperature range of 450–500 °C is favourable for the high yield of liquid oil, whereas over 600 °C is for gas, and below 400 °C is for solid char production.
- ➢ Waste tyre pyrolysis products (oil, gas, and char) contain a high calorific value which is suitable to be directly used as a heat source in boiler, furnace, and other applications.
- ➢ The cetane index of WTO is much lower than that of petroleum diesel and biodiesel. The other important properties for automobile engine application, such as density, kinematic viscosity, and aromatic content of TPO, were higher than diesel and biodiesel.
- ➢ Many studies have attempted to run diesel with pure TPO and attempted to run diesel with pure TPO and blended with diesel to observe engine combustion performance and exhaust emissions. The summary of these research findings indicated similar power output of engines with TPO and diesel. However, ignition delay and cylinder pressure were found to be higher with TPO due to its high density and viscosity. Incorporated with high content of aromatic compound and combustion variations, TPO shows significantly higher NO_x, CO, CO, SO_x, and HC emission reported by most of the studies.
- ➢ Although pure TPO is not recommended to directly use in commercial automobile engine fuel due to engine durability and exhaust emission issues, a blending below 20% of TPO with diesel can be utilisd.
- ➢ The commercialisation and industrial production of waste tyre to commercial automobile liquid fuel requires more in-depth techno-economic assessment and quality improvement of WTO.

Author Contributions: Conceptualization, M.I.J., F.M.H.; Data collection and curation, M.I.J., F.M.H.; Formal analysis, M.I.J., F.M.H., M.G.R.; Writing original draft, M.I.J., F.M.H.; Writing—review and editing—M.G.R., A.A.C. All authors have read and agreed to the published version of the manuscript.

Funding: This research received no external funding.

Institutional Review Board Statement: Not applicable.

Informed Consent Statement: Not applicable.

Data Availability Statement: Not applicable as it is a review article.

Conflicts of Interest: The authors declare no conflict of interest.

References

1. Uyumaz, A.; Aydoğan, B.; Solmaz, H.; Yılmaz, E.; Yeşim Hopa, D.; Aksoy Bahtli, T.; Solmaz, Ö.; Aksoy, F. Production of waste tyre oil and experimental investigation on combustion, engine performance and exhaust emissions. *J. Energy Inst.* **2019**, *92*, 1406–1418. [CrossRef]
2. Murugan, S.; Ramaswamy, M.C.; Nagarajan, G. A comparative study on the performance, emission and combustion studies of a DI diesel engine using distilled tyre pyrolysis oil–diesel blends. *Fuel* **2008**, *87*, 2111–2121. [CrossRef]

3. Saidur, R.; Jahirul, M.I.; Moutushi, T.Z.; Imtiaz, H.; Masjuki, H.H. Effect of partial substitution of diesel fuel by natural gas on performance parameters of a four-cylinder diesel engine. *Proc. Inst. Mech. Eng. Part. A J. Power Energy* **2007**, *221*, 1–10. [CrossRef]

4. Jahirul, M.I.; Koh, W.; Brown, R.J.; Senadeera, W.; Hara, I.; Moghaddam, L. Biodiesel Production from Non-Edible Beauty Leaf (Calophyllum inophyllum) Oil: Process Optimization Using Response Surface Methodology (RSM). *Energies* **2014**, *7*, 5317–5331. [CrossRef]

5. Rahman, M.M.; Pourkhesalian, A.M.; Jahirul, M.I.; Stevanovic, S.; Pham, P.X.; Wang, H.; Masri, A.R.; Brown, R.J.; Ristovski, Z.D. Particle emissions from biodiesels with different physical properties and chemical composition. *Fuel* **2014**, *134*, 201–208. [CrossRef]

6. Jahirul, M.I.; Brown, R.J.; Senadeera, W.; Ashwath, N.; Rasul, M.G.; Rahman, M.M.; Hossain, F.M.; Moghaddam, L.; Islam, M.A.; O'Hara, I.M. Physio-chemical assessment of beauty leaf (Calophyllum inophyllum) as second-generation biodiesel feedstock. *Energy Rep.* **2015**, *1*, 204–215. [CrossRef]

7. Haseeb, A.S.M.A.; Fazal, M.A.; Jahirul, M.I.; Masjuki, H.H. Compatibility of automotive materials in biodiesel: A review. *Fuel* **2011**, *90*, 922–931. [CrossRef]

8. Jahirul, M.I.; Brown, R.J.; Senadeera, W.; O'Hara, I.M.; Ristovski, Z.D. The Use of Artificial Neural Networks for Identifying Sustainable Biodiesel Feedstocks. *Energies* **2013**, *6*, 3764–3806. [CrossRef]

9. Guan, C.; Cheung, C.S.; Li, X.; Huang, Z. Effects of oxygenated fuels on the particle-phase compounds emitted from a diesel engine. *Atmos. Pollut. Res.* **2017**, *8*, 209–220. [CrossRef]

10. Alptekin, E. Emission, injection and combustion characteristics of biodiesel and oxygenated fuel blends in a common rail diesel engine. *Energy* **2017**, *119*, 44–52. [CrossRef]

11. Feng, Z.; Zhan, C.; Tang, C.; Yang, K.; Huang, Z. Experimental investigation on spray and atomization characteristics of diesel/gasoline/ethanol blends in high pressure common rail injection system. *Energy* **2016**, *112*, 549–561. [CrossRef]

12. Valentino, G.; Corcione, F.E.; Iannuzzi, S.E.; Serra, S. Experimental study on performance and emissions of a high speed diesel engine fuelled with n-butanol diesel blends under premixed low temperature combustion. *Fuel* **2012**, *92*, 295–307. [CrossRef]

13. Wang, W.-C.; Bai, C.-J.; Lin, C.-T.; Prakash, S. Alternative fuel produced from thermal pyrolysis of waste tires and its use in a DI diesel engine. *Appl. Therm. Eng.* **2016**, *93*, 330–338. [CrossRef]

14. Mokhtar, N.M.; Omar, R.; Idris, A. Microwave Pyrolysis for Conversion of Materials to Energy: A Brief Review. *Energy Sources Part A Recovery Util. Environ. Eff.* **2012**, *34*, 2104–2122. [CrossRef]

15. Damodharan, D.; Rajesh Kumar, B.; Gopal, K.; de Poures, M.V.; Sethuramasamyraja, B. Utilization of waste plastic oil in diesel engines: A review. *Rev. Environ. Sci. BioTechnol.* **2019**, *18*, 681–697. [CrossRef]

16. İlkılıç, C.; Aydın, H. Fuel production from waste vehicle tires by catalytic pyrolysis and its application in a diesel engine. *Fuel Process. Technol.* **2011**, *92*, 1129–1135. [CrossRef]

17. Aydın, H.; İlkılıç, C. Optimization of fuel production from waste vehicle tires by pyrolysis and resembling to diesel fuel by various desulfurization methods. *Fuel* **2012**, *102*, 605–612. [CrossRef]

18. Martínez, J.D.; Puy, N.; Murillo, R.; García, T.; Navarro, M.V.; Mastral, A.M. Waste tyre pyrolysis—A review. *Renew. Sustain. Energy Rev.* **2013**, *23*, 179–213. [CrossRef]

19. Williams, P.T. Pyrolysis of waste tyres: A review. *Waste Manag.* **2013**, *33*, 1714–1728. [CrossRef]

20. ETRMA. European Tyre and Rubber Industry Statistics. European Tyre and Rubber Manufacturing Association. Available online: http://www.etrma.org (accessed on 20 December 2014).

21. Martínez, J.D.; Rodríguez-Fernández, J.; Sánchez-Valdepeñas, J.; Murillo, R.; García, T. Performance and emissions of an automotive diesel engine using a tire pyrolysis liquid blend. *Fuel* **2014**, *115*, 490–499. [CrossRef]

22. Sienkiewicz, M.; Kucinska-Lipka, J.; Janik, H.; Balas, A. Progress in used tyres management in the European Union: A review. *Waste Manag.* **2012**, *32*, 1742–1751. [CrossRef]

23. Parthasarathy, P.; Choi, H.S.; Park, H.C.; Hwang, J.G.; Yoo, H.S.; Lee, B.-K.; Upadhyay, M. Influence of process conditions on product yield of waste tyre pyrolysis—A review. *Korean J. Chem. Eng.* **2016**, *33*, 2268–2286. [CrossRef]

24. Verma, P.; Zare, A.; Jafari, M.; Bodisco, T.A.; Rainey, T.; Ristovski, Z.D.; Brown, R.J. Diesel engine performance and emissions with fuels derived from waste tyres. *Sci. Rep.* **2018**, *8*, 2457. [CrossRef]

25. Mastral, A.M.; Murillo, R.; Callén, M.S.; García, T. Application of coal conversion technology to tire processing. *Fuel Process. Technol.* **1999**, *60*, 231–242. [CrossRef]

26. Mastral, A.M.; Murillo, R.; Callén, M.S.; García, T.; Snape, C.E. Influence of Process Variables on Oils from Tire Pyrolysis and Hydropyrolysis in a Swept Fixed Bed Reactor. *Energy Fuels* **2000**, *14*, 739–744. [CrossRef]

27. Gent, A.N. Chapter 1—Rubber Elasticity: Basic Concepts and Behavior. In *The Science and Technology of Rubber*, 4th ed.; Mark, J.E., Erman, B., Roland, C.M., Eds.; Academic Press: Boston, MA, USA, 2013; pp. 1–26. [CrossRef]

28. Kyari, M.; Cunliffe, A.; Williams, P.T. Characterization of Oils, Gases, and Char in Relation to the Pyrolysis of Different Brands of Scrap Automotive Tires. *Energy Fuels* **2005**, *19*, 1165–1173. [CrossRef]

29. Leung, D.Y.C.; Wang, C.L. Kinetic study of scrap tyre pyrolysis and combustion. *J. Anal. Appl. Pyrolysis* **1998**, *45*, 153–169. [CrossRef]

30. Lorber, K.E.; Sarc, R.; Aldrian, A. Design and quality assurance for solid recovered fuel. *Waste Manag. Res.* **2012**, *30*, 370–380. [CrossRef] [PubMed]

31. Sarc, R.; Lorber, K.E. Production, quality and quality assurance of Refuse Derived Fuels (RDFs). *Waste Manag.* **2013**, *33*, 1825–1834. [CrossRef] [PubMed]

32. Sarc, R.; Lorber, K.E.; Pomberger, R.; Rogetzer, M.; Sipple, E.M. Design, quality, and quality assurance of solid recovered fuels for the substitution of fossil feedstock in the cement industry. *Waste Manag. Res.* **2014**, *32*, 565–585. [CrossRef]

33. Dai, X.; Yin, X.; Wu, C.; Zhang, W.; Chen, Y. Pyrolysis of waste tires in a circulating fluidized-bed reactor. *Energy* **2001**, *26*, 385–399. [CrossRef]

34. Banar, M.; Akyıldız, V.; Özkan, A.; Çokaygil, Z.; Onay, Ö. Characterization of pyrolytic oil obtained from pyrolysis of TDF (Tire Derived Fuel). *Energy Convers. Manag.* **2012**, *62*, 22–30. [CrossRef]

35. Jahirul, M.I.; Rasul, M.G.; Chowdhury, A.A.; Ashwath, N. Biofuels Production through Biomass Pyrolysis—A Technological Review. *Energies* **2012**, *5*, 4952–5001. [CrossRef]

36. Rasul, M.G.; Jahirul, M.I. *Recent Developments in Biomass Pyrolysis for Bio-Fuel Production: Its Potential for Commercial Applications*; Centre for Plant and Water Science, Faculty of Sciences, Engineering and Health, Central Queensland University: Norman Gardens, Australia, 2012; pp. 256–265.

37. Donatelli, A.; Iovane, P.; Molino, A. High energy syngas production by waste tyres steam gasification in a rotary kiln pilot plant. Experimental and numerical investigations. *Fuel* **2010**, *89*, 2721–2728. [CrossRef]

38. Janajreh, I.; Raza, S.S. Numerical simulation of waste tyres gasification. *Waste Manag. Res.* **2015**, *33*, 460–468. [CrossRef]

39. Zhang, L.; Zhou, B.; Duan, P.; Wang, F.; Xu, Y. Hydrothermal conversion of scrap tire to liquid fuel. *Chem. Eng. J.* **2016**, *285*, 157–163. [CrossRef]

40. Uddin, M.N.; Techato, K.; Taweekun, J.; Rahman, M.M.; Rasul, M.G.; Mahlia, T.M.I.; Ashrafur, S.M. An Overview of Recent Developments in Biomass Pyrolysis Technologies. *Energies* **2018**, *11*, 3115. [CrossRef]

41. Ward, J.; Rasul, M.G.; Bhuiya, M.M.K. Energy Recovery from Biomass by Fast Pyrolysis. *Procedia Eng.* **2014**, *90*, 669–674. [CrossRef]

42. Tudu, K.; Murugan, S.; Patel, S.K. Effect of diethyl ether in a DI diesel engine run on a tyre derived fuel-diesel blend. *J. Energy Inst.* **2016**, *89*, 525–535. [CrossRef]

43. Vihar, R.; Seljak, T.; Rodman Oprešnik, S.; Katrašnik, T. Combustion characteristics of tire pyrolysis oil in turbo charged compression ignition engine. *Fuel* **2015**, *150*, 226–235. [CrossRef]

44. Sharma, A.; Murugan, S. Potential for using a tyre pyrolysis oil-biodiesel blend in a diesel engine at different compression ratios. *Energy Convers. Manag.* **2015**, *93*, 289–297. [CrossRef]

45. Zebala, J.; Ciepka, P.; Reza, A.; Janczur, R. Influence of rubber compound and tread pattern of retreaded tyres on vehicle active safety. *Forensic Sci. Int.* **2007**, *167*, 173–180. [CrossRef]

46. Song, Z.; Yang, Y.; Zhao, X.; Sun, J.; Wang, W.; Mao, Y.; Ma, C. Microwave pyrolysis of tire powders: Evolution of yields and composition of products. *J. Anal. Appl. Pyrolysis* **2017**, *123*, 152–159. [CrossRef]

47. Ferrer, G. The economics of tire remanufacturing. *Resour. Conserv. Recycl.* **1997**, *19*, 221–255. [CrossRef]

48. Gieré, R.; Smith, K.; Blackford, M. Chemical composition of fuels and emissions from a coal+tire combustion experiment in a power station. *Fuel* **2006**, *85*, 2278–2285. [CrossRef]

49. Pipilikaki, P.; Katsioti, M.; Papageorgiou, D.; Fragoulis, D.; Chaniotakis, E. Use of tire derived fuel in clinker burning. *Cem. Concr. Compos.* **2005**, *27*, 843–847. [CrossRef]

50. Collins, K.J.; Jensen, A.C.; Mallinson, J.J.; Roenelle, V.; Smith, I.P. Environmental impact assessment of a scrap tyre artificial reef. *ICES J. Mar. Sci.* **2002**, *59*, S243–S249. [CrossRef]

51. Zhang, L.; Xu, C.; Champagne, P. Overview of recent advances in thermo-chemical conversion of biomass. *Energy Convers. Manag.* **2010**, *51*, 969–982. [CrossRef]

52. Bridgewater, A.V. *Thermal Conversion of Biomass and Waste*; Bio-Energy Research Group Aston University: Birmingham, UK, 2001.

53. Bridgewater, A.V.; Peacocke, G.V.C. Fast pyrolysis processes for biomass. *Renew. Sustain. Energy Rev.* **2000**, *4*, 1–73. [CrossRef]

54. Srirangan, K.; Akawi, L.; Moo-Young, M.; Chou, C.P. Towards sustainable production of clean energy carriers from biomass resources. *Appl. Energ.* **2012**, *100*, 172–186. [CrossRef]

55. Doğan, O.; Çelik, M.B.; Özdalyan, B. The effect of tire derived fuel/diesel fuel blends utilization on diesel engine performance and emissions. *Fuel* **2012**, *95*, 340–346. [CrossRef]

56. White, J.E.; Catall, W.J.; Legendre, B.L. Biomass pyrolysis kinetics: A comparative critical review with relevant agricultural residue case studies. *J. Anal. Appl. Pyrol.* **2011**, *91*, 1–33. [CrossRef]

57. Panwar, N.L.; Kothari, R.; Tyagi, V.V. Thermo chemical conversion of biomass—Eco friendly energy routes. *Renew. Sustain. Energy Rev.* **2012**, *16*, 1801–1816. [CrossRef]

58. Balat, M.; Balat, M.; Kırtay, E.; Balat, H. Main routes for the thermo-conversion of biomass into fuels and chemicals. Part 1: Pyrolysis systems. *Energy Convers. Manag.* **2009**, *50*, 3147–3157. [CrossRef]

59. Goyal, H.B.; Seal, D.; Saxena, R.C. Bio-fuels from thermochemical conversion of renewable resources: A review. *Renew. Sustain. Energy Rev.* **2008**, *12*, 504–517. [CrossRef]

60. Capareda, S.C. Biomass Energy Conversion. Texas A&M University, USA. Available online: https://www.intechopen.com/ (accessed on 22 November 2013).

61. Ceylan, R.; Bredenberg, J.B.s. Hydrogenolysis and hydrocracking of the carbon-oxygen bond. 2. Thermal cleavage of the carbon-oxygen bond in guaiacol. *Fuel* **1982**, *61*, 377–382. [CrossRef]

62. Mohan, D.; Pittman, C.U.; Steele, P.H. Pyrolysis of Wood/Biomass for Bio-oil: A Critical Review. *Energy Fuels* **2006**, *20*, 848–889. [CrossRef]
63. Zhang, Q.; Chang, J.; Wang, T.; Xu, Y. Review of biomass pyrolysis oil properties and upgrading research. *Energy Convers. Manag.* **2007**, *48*, 87–92. [CrossRef]
64. Sharma, A.; Pareek, V.; Zhang, D. Biomass pyrolysis—A review of modelling, process parameters and catalytic studies. *Renew. Sustain. Energy Rev.* **2015**, *50*, 1081–1096. [CrossRef]
65. Kumaravel, S.T.; Murugesan, A.; Kumaravel, A. Tyre pyrolysis oil as an alternative fuel for diesel engines—A review. *Renew. Sustain. Energy Rev.* **2016**, *60*, 1678–1685. [CrossRef]
66. Miranda, M.; Pinto, F.; Gulyurtlu, I.; Cabrita, I. Pyrolysis of rubber tyre wastes: A kinetic study. *Fuel* **2013**, *103*, 542–552. [CrossRef]
67. Kar, Y. Catalytic pyrolysis of car tire waste using expanded perlite. *Waste Manag.* **2011**, *31*, 1772–1782. [CrossRef]
68. Islam, M.R.; Joardder, M.U.H.; Kader, M.; Sarker, M. Valorization of solid tire wastes available in Bangladesh by thermal treatment. In Proceedings of the WasteSafe 2011—2nd International Conference on Solid Waste Management in the Developing Countries, Khulna, Bangladesh, 13–15 February 2011.
69. Williams, P.T.; Brindle, A.J. Catalytic pyrolysis of tyres: Influence of catalyst temperature. *Fuel* **2002**, *81*, 2425–2434. [CrossRef]
70. Aylón, E.; Fernández-Colino, A.; Murillo, R.; Navarro, M.; García, T.; Mastral, A. Valorisation of waste tyre by pyrolysis in a moving bed reactor. *Waste Manag.* **2010**, *30*, 1220–1224. [CrossRef] [PubMed]
71. Zhang, Y.; Williams, P.T. Carbon nanotubes and hydrogen production from the pyrolysis catalysis or catalytic-steam reforming of waste tyres. *J. Anal. Appl. Pyrolysis* **2016**, *122*, 490–501. [CrossRef]
72. Luo, S.; Feng, Y. The production of fuel oil and combustible gas by catalytic pyrolysis of waste tire using waste heat of blast-furnace slag. *Energy Convers. Manag.* **2017**, *136*, 27–35. [CrossRef]
73. Galvagno, S.; Casu, S.; Casabianca, T.; Calabrese, A.; Cornacchia, G. Pyrolysis process for the treatment of scrap tyres: Preliminary experimental results. *Waste Manag.* **2002**, *22*, 917–923. [CrossRef]
74. Roy, C.; Chaala, A.; Darmstadt, H. The vacuum pyrolysis of used tires: End-uses for oil and carbon black products. *J. Anal. Appl. Pyrolysis* **1999**, *51*, 201–221. [CrossRef]
75. Sangeeta, C.; Jain, A.K. A review of fixed bed gasification systems for biomass. *Agric. Eng. Int.* **2007**, *9*, 5.
76. Altafini, C.R.; Wander, P.R.; Barreto, R.M. Prediction of the working parameters of a wood waste gasifier through an equilibrium model. *Energy Convers. Manag.* **2003**, *44*, 2763–2777. [CrossRef]
77. Leung, D.Y.C.; Yin, X.L.; Wu, C.Z. A review on the development and commercialization of biomass gasification technologies in China. *Renew. Sustain. Energy Rev.* **2004**, *8*, 565–580. [CrossRef]
78. Hu, C.; Xiao, R.; Zhang, H. Ex-situ catalytic fast pyrolysis of biomass over HZSM-5 in a two-stage fluidized-bed/fixed-bed combination reactor. *Bioresour. Technol.* **2017**, *243*, 1133–1140. [CrossRef] [PubMed]
79. Wang, J.; Zhong, Z.; Ding, K.; Li, M.; Hao, N.; Meng, X.; Ruan, R.; Ragauskas, A.J. Catalytic fast co-pyrolysis of bamboo sawdust and waste tire using a tandem reactor with cascade bubbling fluidized bed and fixed bed system. *Energy Convers. Manag.* **2019**, *180*, 60–71. [CrossRef]
80. Yang, M.; Shao, J.; Yang, H.; Zeng, K.; Wu, Z.; Chen, Y.; Bai, X.; Chen, H. Enhancing the production of light olefins and aromatics from catalytic fast pyrolysis of cellulose in a dual-catalyst fixed bed reactor. *Bioresour. Technol.* **2019**, *273*, 77–85. [CrossRef] [PubMed]
81. Lewandowski, W.M.; Januszewicz, K.; Kosakowski, W. Efficiency and proportions of waste tyre pyrolysis products depending on the reactor type—A review. *J. Anal. Appl. Pyrolysis* **2019**, *140*, 25–53. [CrossRef]
82. Edwin Raj, R.; Robert Kennedy, Z.; Pillai, B.C. Optimization of process parameters in flash pyrolysis of waste tyres to liquid and gaseous fuel in a fluidized bed reactor. *Energy Convers. Manag.* **2013**, *67*, 145–151. [CrossRef]
83. Rodríguez, E.; Gutiérrez, A.; Palos, R.; Azkoiti, M.J.; Arandes, J.M.; Bilbao, J. Cracking of Scrap Tires Pyrolysis Oil in a Fluidized Bed Reactor under Catalytic Cracking Unit Conditions. Effects of Operating Conditions. *Energy Fuels* **2019**, *33*, 3133–3143. [CrossRef]
84. Kaminsky, W. 8—Fluidized bed pyrolysis of waste polymer composites for oil and gas recovery. In *Management, Recycling and Reuse of Waste Composites*; Goodship, V., Ed.; Woodhead Publishing: Sawston, UK, 2010; pp. 192–213. [CrossRef]
85. Hasan, M.M.; Rasul, M.G.; Khan, M.M.K.; Ashwath, N.; Jahirul, M.I. Energy recovery from municipal solid waste using pyrolysis technology: A review on current status and developments. *Renew. Sustain. Energy Rev.* **2021**, *145*, 111073. [CrossRef]
86. Heo, H.S.; Park, H.J.; Park, Y.-K.; Ryu, C.; Suh, D.J.; Suh, Y.-W.; Yim, J.-H.; Kim, S.-S. Bio-oil production from fast pyrolysis of waste furniture sawdust in a fluidized bed. *Bioresour. Technol.* **2010**, *101*, S91–S96. [CrossRef]
87. Li, S.Q.; Yao, Q.; Chi, Y.; Yan, J.H.; Cen, K.F. Pilot-scale pyrolysis of scrap tires in a continuous rotary kiln reactor. *Ind. Eng. Chem. Res.* **2004**, *43*, 5133–5145. [CrossRef]
88. Fantozzi, F.; Colantoni, S.; Bartocci, P.; Desideri, U. Rotary Kiln Slow Pyrolysis for Syngas and Char Production From Biomass and Waste—Part I: Working Envelope of the Reactor. *J. Eng. Gas. Turbines Power* **2007**, *129*, 901–907. [CrossRef]
89. Bortolamasi, M.F.J. Design and sizing of screw feeders. Proceedings of International Congress for Particle Technology (PARTEC), Nuremberg, Germany, 27–29 March 2001.
90. Badger, P.C.; Fransham, P. Use of mobile fast pyrolysis plants to densify biomass and reduce biomass handling costs—A preliminary assessment. *Biomass. Bioenergy* **2006**, *30*, 321–325. [CrossRef]

91. Qureshi, M.S.; Oasmaa, A.; Pihkola, H.; Deviatkin, I.; Tenhunen, A.; Mannila, J.; Minkkinen, H.; Pohjakallio, M.; Laine-Ylijoki, J. Pyrolysis of plastic waste: Opportunities and challenges. *J. Anal. Appl. Pyrolysis* **2020**, *152*, 104804. [CrossRef]
92. Kaminsky, W.; Mennerich, C.; Zhang, Z. Feedstock recycling of synthetic and natural rubber by pyrolysis in a fluidized bed. *J. Anal. Appl. Pyrolysis* **2009**, *85*, 334–337. [CrossRef]
93. Pantea, D.; Darmstadt, H.; Kaliaguine, S.; Roy, C. Heat-treatment of carbon blacks obtained by pyrolysis of used tires. Effect on the surface chemistry, porosity and electrical conductivity. *J. Anal. Appl. Pyrolysis* **2003**, *67*, 55–76. [CrossRef]
94. Islam, M.A.; Rahman, M.M.; Heimann, K.; Nabi, M.N.; Ristovski, Z.D.; Dowell, A.; Thomas, G.; Feng, B.; von Alvensleben, N.; Brown, R.J. Combustion analysis of microalgae methyl ester in a common rail direct injection diesel engine. *Fuel* **2015**, *143*, 351–360. [CrossRef]
95. Murugan, S.; Ramaswamy, M.C.; Nagarajan, G. The use of tyre pyrolysis oil in diesel engines. *Waste Manag.* **2008**, *28*, 2743–2749. [CrossRef]
96. Hariharan, S.; Murugan, S.; Nagarajan, G. Effect of diethyl ether on Tyre pyrolysis oil fueled diesel engine. *Fuel* **2013**, *104*, 109–115. [CrossRef]
97. Jahirul, M.I.; Brown, J.R.; Senadeera, W.; Ashwath, N.; Laing, C.; Leski-Taylor, J.; Rasul, M.G. Optimisation of Bio-Oil Extraction Process from Beauty Leaf (Calophyllum Inophyllum) Oil Seed as a Second Generation Biodiesel Source. *Procedia Eng.* **2013**, *56*, 619–624. [CrossRef]
98. Conesa, J.A.; Martín-Gullón, I.; Font, R.; Jauhiainen, J. Complete Study of the Pyrolysis and Gasification of Scrap Tires in a Pilot Plant Reactor. *Environ. Sci. Technol.* **2004**, *38*, 3189–3194. [CrossRef]
99. Olazar, M.; Aguado, R.; Arabiourrutia, M.; Lopez, G.; Barona, A.; Bilbao, J. Catalyst effect on the composition of tire pyrolysis products. *Energy Fuels* **2008**, *22*, 2909–2916. [CrossRef]
100. Zhang, X.; Wang, T.; Ma, L.; Chang, J. Vacuum pyrolysis of waste tires with basic additives. *Waste Manag.* **2008**, *28*, 2301–2310. [CrossRef]
101. Teng, H.; Serio, M.A.; Wojtowicz, M.A.; Bassilakis, R.; Solomon, P.R. Reprocessing of used tires into activated carbon and other products. *Ind. Eng. Chem. Res.* **1995**, *34*, 3102–3111. [CrossRef]
102. Aylón, E.; Murillo, R.; Fernández-Colino, A.; Aranda, A.; García, T.; Callén, M.S.; Mastral, A.M. Emissions from the combustion of gas-phase products at tyre pyrolysis. *J. Anal. Appl. Pyrolysis* **2007**, *79*, 210–214. [CrossRef]
103. Leung, D.Y.C.; Yin, X.L.; Zhao, Z.L.; Xu, B.Y.; Chen, Y. Pyrolysis of tire powder: Influence of operation variables on the composition and yields of gaseous product. *Fuel Process. Technol.* **2002**, *79*, 141–155. [CrossRef]
104. Quek, A.; Balasubramanian, R. Liquefaction of waste tires by pyrolysis for oil and chemicals—A review. *J. Anal. Appl. Pyrolysis* **2013**, *101*, 1–16. [CrossRef]
105. Žvar Baškovič, U.; Vihar, R.; Seljak, T.; Katrašnik, T. Feasibility analysis of 100% tire pyrolysis oil in a common rail Diesel engine. *Energy* **2017**, *137*, 980–990. [CrossRef]
106. Hossain, F.M.; Nabi, M.N.; Rainey, T.J.; Bodisco, T.; Bayley, T.; Randall, D.; Ristovski, Z.; Brown, R.J. Novel biofuels derived from waste tyres and their effects on reducing oxides of nitrogen and particulate matter emissions. *J. Clean. Prod.* **2020**, *242*, 118463. [CrossRef]
107. Vihar, R.; Žvar Baškovič, U.; Seljak, T.; Katrašnik, T. Combustion and emission formation phenomena of tire pyrolysis oil in a common rail Diesel engine. *Energy Convers. Manag.* **2017**, *149*, 706–721. [CrossRef]
108. Pote, R.N.; Patil, R.K. Combustion and emission characteristics analysis of waste tyre pyrolysis oil. *SN Appl. Sci.* **2019**, *1*, 294. [CrossRef]
109. Koc, A.B.; Abdullah, M. Performance of a 4-cylinder diesel engine running on tire oil–biodiesel–diesel blend. *Fuel Process. Technol.* **2014**, *118*, 264–269. [CrossRef]
110. Frigo, S.; Seggiani, M.; Puccini, M.; Vitolo, S. Liquid fuel production from waste tyre pyrolysis and its utilisation in a Diesel engine. *Fuel* **2014**, *116*, 399–408. [CrossRef]

Article

Catalyst Characteristics and Performance of Silica-Supported Zinc for Hydrodeoxygenation of Phenol

Hamed Pourzolfaghar [1], Faisal Abnisa [2], Wan Mohd Ashri Wan Daud [1],*,
Mohamed Kheireddine Aroua [3,4] and Teuku Meurah Indra Mahlia [5],*

[1] Department of Chemical Engineering, Faculty of Engineering, University of Malaya,
    Kuala Lumpur 50603, Malaysia; h_pourzolfaghar@yahoo.com
[2] Department of Chemical and Materials Engineering, Faculty of Engineering, King Abdulaziz University,
    Jeddah 21589, Saudi Arabia; fta@kau.edu.sa
[3] Centre for Carbon Dioxide Capture and Utilization (CCDCU), School of Science and Technology,
    Sunway University, Bandar Sunway, Petaling Jaya 47500, Malaysia; kheireddinea@sunway.edu.my
[4] Department of Engineering, Lancaster University, Lancaster LA1 4YW, UK
[5] School of Information, Systems and Modelling, Faculty of Engineering and Information Technology,
    University of Technology Sydney, Sydney, NSW 2007, Australia
* Correspondence: ashri@um.edu.my (W.M.A.W.D.); TMIndra.Mahlia@uts.edu.au (T.M.I.M.)

Received: 25 April 2020; Accepted: 23 May 2020; Published: 1 June 2020

Abstract: The present investigation aimed to study the physicochemical characteristics of supported catalysts comprising various percentages of zinc dispersed over SiO_2. The physiochemical properties of these catalysts were surveyed by N_2 physisorption (BET), thermogravimetry analysis (TGA), H_2 temperature-programmed reduction, field-emission scanning electron microscopy (FESEM), inductively coupled plasma-optical emission spectrometry (ICP-OES), and NH_3 temperature-programmed desorption (NH_3-TPD). In addition, to examine the activity and performance of the catalysts for the hydrodeoxygenation (HDO) of the bio-oil oxygenated compounds, the experimental reaction runs, as well as stability and durability tests, were performed using 3% Zn/SiO_2 as the catalyst. Characterization of silica-supported zinc catalysts revealed an even dispersion of the active site over the support in the various dopings of the zinc. The acidity of the calcinated catalysts elevated clearly up to 0.481 mmol/g. Moreover, characteristic outcomes indicate that elevating the doping of zinc metal led to interaction and substitution of proton sites on the SiO_2 surface that finally resulted in an increase in the desorption temperature peak. The experiments were performed at temperature 500 °C, pressure 1 atm; weight hourly space velocity (WHSV) 0.32 (h^{-1}); feed flow rate 0.5 (mL/min); and hydrogen flow rate 150 (mL/min). Based on the results, it was revealed that among all the prepared catalysts, that with 3% of zinc had the highest conversion efficiency up to 80%. However, the selectivity of the major products, analyzed by gas chromatography flame-ionization detection (GC-FID), was not influenced by the variation in the active site doping.

Keywords: heterogeneous catalyst; hydrodeoxygenation (HDO); zinc; phenol; bio-oil

---

1. Introduction

Due to global warming, many countries have moved from fossil fuel-based energy toward renewable energy. There are several forms of renewable energy that have been effectively executed out there. Many countries have successfully implemented renewable energy generated from solar [1,2]. Where some others have successfully utilized hydro, wave, geothermal, and wind as a secondary energy source [1,3], Malaysia and Indonesia have successfully utilized biofuel as one of the alternative energy sources [4–6]. Biofuel has been proven to reduce greenhouse gas emissions significantly in these countries [7]. Therefore, among all alternative sources of energy to replace the restricted natural fossil fuel reservoirs, biofuel represents the sole available sustainable energy source that could properly replace petroleum [8,9]. Nevertheless, due to its component's complexity, which results in its instability, corrosiveness, and low heating value, the pyrolyzed biomass, known as bio-oil, must be upgraded before utilization [10–12]. Hydrodeoxygenation is a catalytic reaction that applies hydrogen to eliminate oxygen from the oxygenated compounds. These reactions are an efficient alternative method to attain bio-oils from biomass-derived oxygenated compounds [13]. Nevertheless, several issues must first be addressed before the hydrodeoxygenation (HDO) process can be entirely commercialized. The development of efficient, robust, cost-effective, and selective catalysts enable the advancement of this method to partly prevail these disadvantages [14].

Various kinds of active sites have been tested in the last decade for the HDO of the oxygenated compounds, which have had promising results including transition metal oxides such as MoO_3 [15,16], Mo_2C [17], Pt/(Al_2O_3, SiO_2, H-Beta zeolie, activated carbon) [18–20], Fe/(SiO_2, activated carbon) [21,22], Ni/(Al_2O_3, SiO_2, HZSM-5 zeolite) [23], and Ga/(HBETA, SiO_2, ZSM-5) [24]; precious metals such as Ru/TiO_2 [25], and W/carbon [26]; phosphides such as Ni_2P/SiO_2, Fe_2P/SiO_2, MoP/SiO_2, Co_2P/SiO_2, and WP/SiO_2 [27]; and bifunctional such as $NiMo/Al_2O_3$ [28] and $Pd-FeOX/SiO_2$ [29].

Zinc has been reported as an active site for the hydroprocessing of bio-oil oxygenated compounds, resulting in aromatic compounds and/or hydrocarbons [30–38]. The interdependent effect of nickel and zinc metals on Al_2O_3 support has been investigated by Cheng et al. [39]. They found that the pine sawdust bio-oil could efficiently be converted by 15%Ni.5%Zn/Al_2O_3 at 44.64 wt%. According to their report, the highest hydrocarbon content of 50.12% could be achieved by applying this bifunctional catalyst. Bifunctional catalysts possess two types of active sites and catalyze two dissimilar types of reactions. In another study by Parsell et al. [40], Zn/Pd was used for the HDO of lignin. Based on the author's claims, 80%–90% of conversion yield could be achieved using Pd/Zn as the HDO catalyst. Besides, they reported that the bifunctional Pd/Zn/C catalyst was more effective in the HDO of lignin molecules (β-O-4) and aromatic compounds, in comparison to the Pd/C sample. Additionally, they expressed that the bifunctional Pd/Zn/C sample is recyclable and robust, and no further addition of the zinc metal is required after each reaction cycle. The efficiency upsurge of the Kapuk seed oil hydrocracking process using the zinc metal has been reported recently by Mirzayanti et al. [41]. They stated that using Zn-Mo/HZSM-5 with a loading of 2.99 wt% for Zn and 7.55 wt% for Mo, the highest performance of the catalyst for the hydrocracking process could be reached.

Due to the complexity of the bio-oils causing complicated reactions during the HDO process, model compounds such as phenol, lignin, and others have been applied. Using the model compounds, one can attain enough data regarding the mechanism and reaction networks of the process. Phenol and its derivatives are known as the most refractory compound in the bio-oil and, hence, it has been selected as the model compound in this investigation [42].

The current research aims to study the physicochemical characteristics of the Zn/SiO_2 catalyst as a potential and promising catalyst for the HDO of bio-oil oxygenated compounds using various analytical equipment. Furthermore, using the phenol as a model compound for the oxygenated bio-oil compound, the reactivity, performance, stability, and reusability of the selected catalyst have been examined in a continuous fixed-bed reactor.

2. Materials and Methods

2.1. Materials

Zinc powder (used to synthesize Zn/SiO_2), anhydrous benzene (99.8%), phenol (≥89%), HPLC grade cyclohexane (99.9%), acetone (HPLC grade), and analytical tetrahydrobenzene or cyclohexene (99%) were purchased from Sigma Aldrich (St. Louis, MO, USA). Synthesis grade n-decane (≥95%) and SiO_2 (Pure silica) have been purchased from Aldrich and R & M chemicals, respectively.

2.2. Catalyst Synthesis

The incipient wetness impregnation method was used in this investigation to prepare the catalyst samples with various doping percentages of the zinc metal onto the silica surface. Silica-supported 3% zinc was prepared using 10 g of silica (catalyst support) and 0.31 g of the powdered Zn metal and some water. The compounds were mixed in a stirrer, roughly 2 h. Afterward, it was dried overnight in an electric oven at 105 °C. The dried sample was then taken to a furnace for the calcination step for 5 h under 500 °C and air atmosphere. Other impregnated samples were prepared with the same procedure.

2.3. Characterization of the Catalysts

The prepared calcinated catalysts and the bare support were analyzed by XRD (PANalytical X'Pert with Cu-Kα radiation). Scherrer's formula was applied to estimate the crystallite size of silica-supported zinc catalysts. In Equation (1), λ represents the X-ray wavelength, d_{xrd} denotes the mean size of the crystallite, β_{hkl} stands for the full-width at half-maximum (FWHM), k is the shape factor, and θ denotes the Bragg angle. Data were collected at ambient conditions at a scanning range from 5° to 80° with a step size of 0.0260°. The slit and the time step were set to 0.1 mm and 0.1 s, respectively.

$$d_{xrd} = \frac{k\lambda}{\beta_{hkl}cos\theta} \tag{1}$$

N_2-adsorption analysis was performed using a Micromeritics TrisStar II 3020 V1.04 to detect the BET surface area, pore size, and pore volume of the calcinated catalysts in this study. The measurement temperature was set at −196 °C and the degassing of the catalysts were performed for one hour at 90 °C and, subsequently, 4 h at 200 °C, prior to the test and under vacuum atmosphere.

Thermogravimetric (TGA) analysis was employed for the bare SiO_2 and freshly calcinated catalysts to examine the extent of water and organic compounds on the samples. A TGA Q500 V20.10 (TA instruments) instrument was applied in this survey. In a typical run, 80 mg of a sample was placed in the sample holder and was heated with a heating rate of 20 °C/min up to 800 °C.

H_2 temperature-programmed reduction (H_2-TPR) was conducted by a Micromeritics Chemisorb 2720 apparatus. Prior to hydrogen adsorption, using the pure nitrogen at 500 °C, the samples were outgassed for 30 min. Subsequently, the catalysts were reduced, using 5% hydrogen diluted in argon gas, for one hour and 500 °C. In the following, the catalysts were cooled and, to remove the physically adsorbed molecules, flushed with N_2 gas at a flow rate of 20 mL/min. Finally, the catalysts were heated up to 900 °C in the nitrogen atmosphere. A thermal conductivity detector (TCD) monitored and recorded the H_2 consumption during the reduction process.

A QUANTA 450 FEG instrument was applied for the FESEM analysis. The morphology of two calcinated catalysts (1% Zn/SiO$_2$ and 3% Zn/SiO$_2$) and the bare silica was examined.

Inductively coupled plasma optical emission spectrometry (ICP–OES) was applied to determine the metal content of the prepared catalysts. In this analysis, 100 mg of each catalyst was digested with HNO_3 (10 mL). Zinc metal solutions for the calibration standard preparation were made using a certain quantity of Zn metal to 7 mL of nitric acid and 14 mL of distilled water, for each standard sample.

The total acidity of the prepared catalysts was tested using the NH_3-TPD method. Prior to outgassing the catalyst (50 mg) in the quartz U-tube, it was heated from room temperature up to 300 °C. Then, under a flow rate of 20 mL/min of helium, the sample was degassed for 30 min at 300 °C. Subsequently, the catalyst was cooled down to room temperature and was then ready for the ammonia chemisorption step. A 5% NH_3/He gas mixture was selected for this step and the ammonia chemisorbed on the sample for 30 min. To remove the physisorbed molecules, the sample was purged using the helium gas at 100 °C (30 min). Finally, applying the heating rate of 40 °C/min, the sample was heated up to 900 °C in the helium atmosphere (20 mL/min). For this analysis, a Micromeritics apparatus was used (Model: Chemisorb 2720).

2.4. Reactivity Studies

The reactivity and selectivity analysis of the prepared catalysts was performed by passing the phenol gas through the catalyst in a stainless-steel fixed-bed reactor. An accurate reaction temperature was observed using an external thermocouple (Type K) located inside the reactor and next to the sample. In a typical run, 1 g of the sample was placed in the middle of the reactor. Prior to the main experiments at 500 °C, the samples were purged for one hour using the nitrogen gas with a 150 mL/min flow rate. Subsequently, the catalyst was activated for 2 h by hydrogen gas at 500 °C and 100 mL/min of flow rate. The main experiment began by injecting the feed to the reactor using an HPLC pump (Model: PU-980, JASCO) under N_2 gas atmosphere. For analysis of the reaction products, the gas-phase products were collected using a condenser connected to a chiller. To obtain the run time of each product including cyclohexane, cyclohexene, and benzene, as well as n-hexane, phenol, and acetone, the standard curves have been created using the standard solutions.

2.5. Catalyst Stability and Reusability Study

To perform the stability test of the prepared catalysts in this study, a freshly prepared 3% Zn/SiO_2 catalyst was selected. The collected data from the HDO reaction at various times on stream (TOSs) were used for calculating the conversion efficiency of the phenol. The obtained results will reveal the variation in activity and selectivity of the catalyst applied with TOS. A reusability study was done in the fixed-bed reactor and the sample applied for the stability test was regenerated at 500 °C under the air atmosphere (for 1 h in an electrical furnace). The liquid products were collected and further analyzed using GC-FID analysis.

3. Results

3.1. Characterization of the Catalyst

3.1.1. XRD

Figure 1 represents the XRD diffractograms of the silica-supported zinc catalysts at various loadings and the bare SiO_2. The XRD patterns of the samples confirmed the existence of Zn metal in all zinc-doped catalysts (b, c, d, and e). According to the literature, the observed peaks at certain 2θ values (32°, 34°, 36°, 47°, 56°, 63°, and 68°) represent hexagonal zinc [43]. Table 1 shows the calculated average size of zinc (calculated from 2θ = 32°, 34°, and 36°) crystallite, which was elevated with the increment in Zn metal doping. This phenomenon occurs due to the agglomeration of the zinc particles outside the pores [44]. The diffraction peaks of some planes including (112), (200), and (110) have not been found, due to the good dispersion of the small crystallites formed during the impregnation or forming in negligible quantity. Comparable degrees of crystallinity and slight peak broadening of all XRD diffractograms indicate that the addition of 0.5–3 wt% Zn metal insignificantly transformed the crystallinity of the samples. The peaks of Zn metal crystallites (Figure 1) were not distinguished, probably due to the decent diffusion of Zn particles in the surface of the support (SiO_2) or the low quantity of the zinc.

Table 1. Physicochemical characteristics results of the bare SiO_2 and Zn-doped catalysts.

Catalyst (Nominal Loading)	S_{BET} (m^2/g)	d_{XRD} (nm)	Pore Volume (cm^3/g)	Pore Size (Å)	Real Doped Zn (wt%)	Acidity of Catalyst (mmol/g)
Bare SiO_2	185.5	-	1.300	280.220	-	0.108
0.5% Zn/SiO_2	185.2	48	1.300	280.220	0.46	0.357
1% Zn/SiO_2	184.7	49	1.300	280.940	0.91	0.370
2% Zn/SiO_2	181	51	1.274	281.906	1.8	0.422
3% Zn/SiO_2	173.3	60	1.284	283.819	2.53	0.481

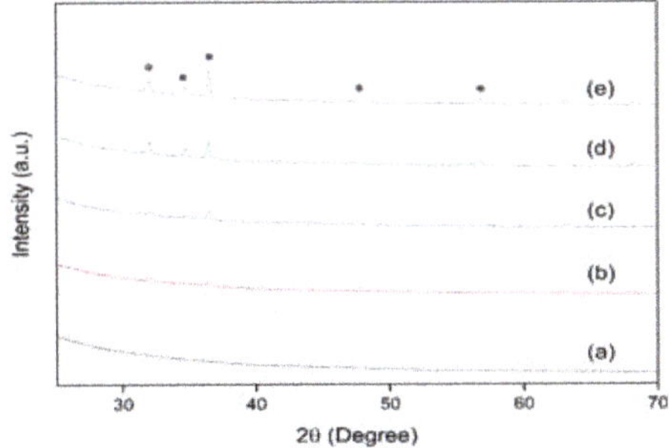

Figure 1. XRD patterns of bare SiO_2 and doped catalysts applied in this study [bare SiO_2 (a), 0.5% Zn/SiO_2 (b), 1% Zn/SiO_2 (c), 2% Zn/SiO_2 (d), and 3% Zn/SiO_2 (e)].

3.1.2. BET

The surface area, pore size, and pore volume of the bare silica and the freshly calcinated sample values have been detailed in Table 1. The analysis results specified that the BET surface area of the calcinated impregnated samples was slightly lower than the bare SiO_2. Nie [45] has also reported the same range of surface area for the silica. The surface area and pore volume of the impregnated samples, as expected and documented previously in the literature, decreased by increasing the zinc load on the support surface [46,47]. In a study by Clerk et al., they declared that various loadings of molybdenum and the surface result in a decrease in the BET surface area [48]. Whilst the loading of the active site resulted in an increase in the BET surface area and pore volume of the samples, it caused decreases in the pore sizes of the samples. This expected trend is due to the selective closing of the small pores with the loading of the zinc metal of the surface of the SiO_2 [49]. According to the results, all samples offered type IV isotherms, which indicate a mesoporous texture with capillary condensation.

3.1.3. Thermogravimetric Analysis (TGA)

Thermogravimetry analysis was performed to detect the content of the water and organic compounds of all calcinated samples, and their weight loss was recorded as a function of temperature. TGA results of the samples are presented in Figure 2. In all cases, the weight loss occurred up to 100 °C, which is mainly due to the humidity. As it can be observed from Figure 2, calcinated bare SiO_2 has around a 2% weight loss whilst the calcinated metalized samples have a minor one (less than 1%). This specifies that by loading the support by zinc metal, less humidity can be absorbed, thus having lower weight loss during the HDO experiments. No high-temperature weight loss was monitored in TGA analysis, which indicates that the samples had no organic carbon impurites.

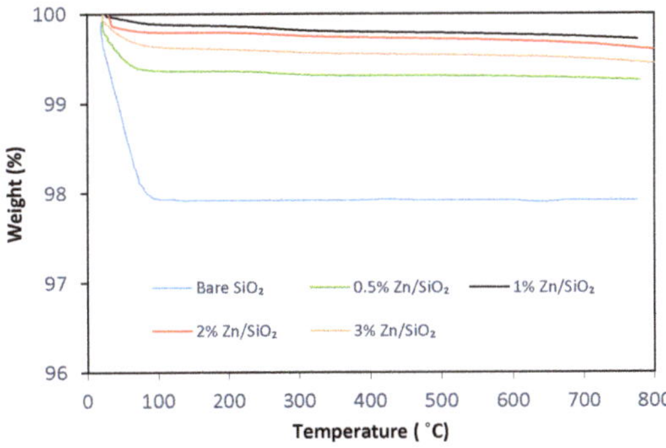

Figure 2. Thermogravimetric analysis for the samples as a function of temperature.

3.1.4. H$_2$-TPR

Figure 3 represents the results for the reducibility of the bare silica and the calcinated samples, which were investigated by H$_2$-TPR. Some preliminary tests of the SiO$_2$ support showed that it did not consume any hydrogen (between 300 °C to 700 °C). Accordingly, any usage of hydrogen gas in the impregnated catalysts could be credited to the Zn metal doped on the silica support. The catalyst with 0.5% Zn loading showed the lowest maximum reduction temperature (T_{max}) at 448 °C. The T_{max} of the impregnated samples gradually raised with the elevation in Zn doping probably due to the elevation in the crystallite size of the active metal on the surface of the support. This outcome agrees with the abovementioned XRD findings. Therefore, the 3% Zn/SiO$_2$ sample offered the uppermost maximum reduction in temperature among other catalysts (498 °C).

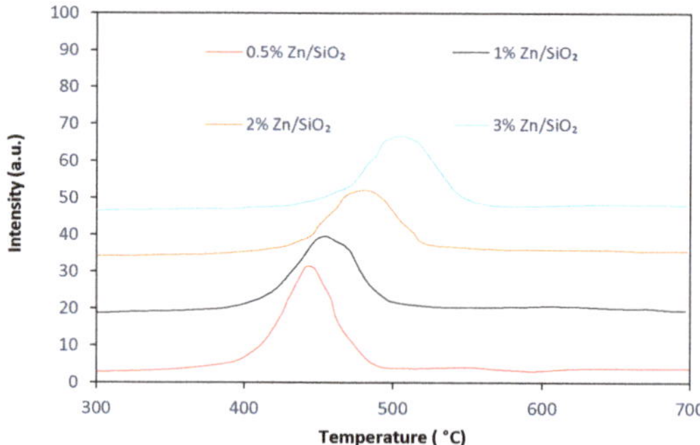

Figure 3. Temperature-programmed reduction profiles of the impregnated catalysts.

3.1.5. FESEM

Micrographs of the bare silica and the selected impregnated catalysts are presented in Figure 4 (1% Zn/SiO$_2$ and 3% Zn/SiO$_2$). The image of the silica represents a big slab formed surface with clusters of silica nanoparticles. From the micrographs of the impregnated samples, it can be perceived that the bright spots, which are the zinc metal, become more densely colonized as the Zn loading was elevated from 1% to 3%. However, no agglomeration of the zinc metal particles has been detected in the sample with the highest loading of the active site (3% Zn/SiO$_2$). Previous studies have mentioned that by overloading the catalyst surface by an active site, some of the particles might cause agglomeration and might, therefore, result in a reduction in the catalyst performances [44].

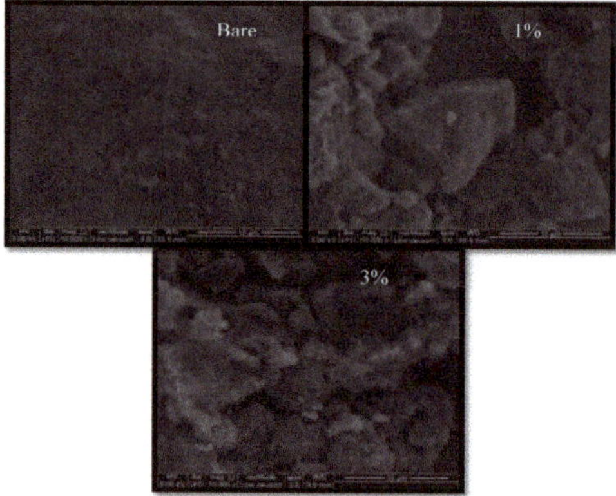

Figure 4. Field-emission scanning electron microscopy (FESEM) micrographs of the selected samples.

3.1.6. ICP-OES

The metal content of each catalyst was detected using ICP-OES analysis (Perkin Elmer, Optima 5300DV). Table 1 indicates the analysis result for the catalysts. In all samples, as shown in the Table, the real metal contents were slightly lower than the calculated extent. The variant between the calculated and real extent of zinc metal on the silica support was between 8%–15%. The calculated variant was mainly because of the impurity of the Zn as the error percentages elevated with the increase in the doped percentages of the impregnated catalysts.

3.1.7. NH_3-TPD

Figure 5 represents the NH3-TPD profile of the catalysts. It can be observed that all samples absorb ammonia in a broad range of temperatures (100 °C–800 °C). From Figure 5 and Table 1, it can be observed that after the 0.5%–3% zinc metal was doped on SiO_2, the total acidity of the impregnated samples evidently raised and an elevation in the area of the ammonia desorption peak could be perceived. Furthermore, considering the desorption peak temperature of the samples, the position of the NH_3 desorption peak temperatures increased by elevating the zinc doping to the samples. It could be concluded that doping the zinc metal onto the silica surface results in the interaction and exchange of proton sites on the support surface. This might be due to the interaction and exchange of proton spots on the silica surface with the loading of zinc metal, initiating the renewal of new proton spots on the samples. As Table 1 indicates, the total acidity of the catalysts was expressively raised from 0.108 to 0.481 mmol/g within the upsurge in zinc doping from 0% to 3%. According to Kernajanakon et al. [50], loading the optimum amount of active sites on the support is crucial and ranges from 1% to 2.5%. By increasing the loading amount such as 5% and 10%, the acidity of the catalysts decreases due to the concealing of acid spots by the produced ZnO cluster. Furthermore, the authors implied that the dispersion efficiency of the metal on the surface of the support might be influenced by the amount of the loaded metal.

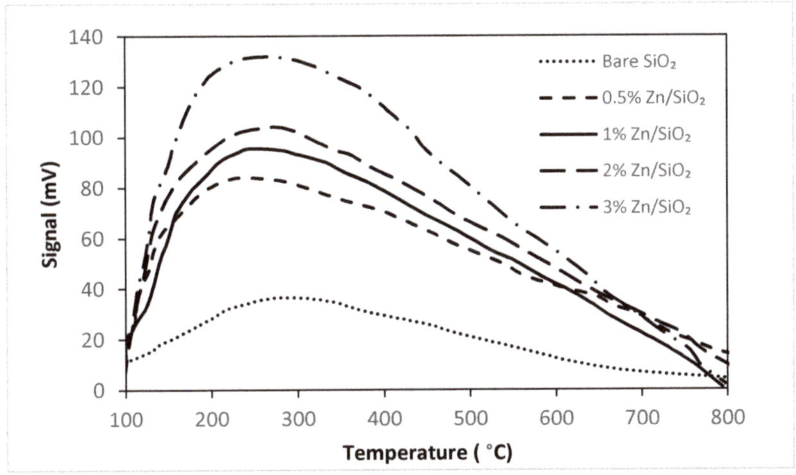

Figure 5. NH_3-TPD profile of the impregnated catalysts and the bare SiO_2.

3.2. Reactivity

Products selectivity and phenol conversion over xZn/SiO_2 catalysts in the hydrodeoxygenation of phenol at various loadings of the zinc metal are shown in Figure 6. From Figure 6, it can be perceived that all impregnated samples are active and able to convert the phenol. Table 2 details the data attained by GC-FID analysis. Phenol conversion efficiency using various loads of Zn metal, 0.5–3%, varied from 15% to 80%, respectively. By elevating the zinc metal loading of the catalysts, the total conversion increased, and the highest conversion efficiency was achieved with 3% doping of the active site. In a separate study [42], we observed that a loading of 4% of the active metal (Zn) on the support surface (SiO$_2$) resulted in a slight reduction in the HDO conversion efficiency. The main reason was found to be the occupation of the support surface pores by zinc metal. However, the zinc metal loading (0.5–3%) had a slight effect (up to 13%) on the selectivity of the products including cyclohexane, cyclohexene, and phenol. The selectivity of the cyclohexane represents the highest value (71.62%) using the 2% Zn/SiO$_2$. However, around a 3.14% decrease has been observed in its selectivity, by elevating the loading of the active site from 2% to 3%. This minor reduction in selectivity might be referred to the occupying of the surface porosities by the zinc metal nanoparticles.

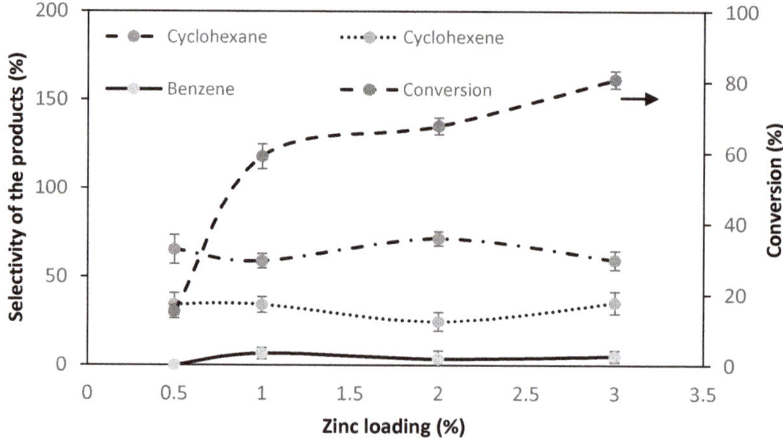

Figure 6. Conversion efficiency and selectivity of the products at various dopings of Zn metal.

Table 2. Results of selectivity and performance of the catalysts.

Catalyst (Nominal Loading)	Conversion of Phenol (%)	Selectivity (%)		
		Benzene	Cyclohexene	Cyclohexane
0.5% Zn/SiO$_2$	15	0	34.57	65.43
1% Zn/SiO$_2$	59	6.62	34.40	58.98
2% Zn/SiO$_2$	68	3.60	24.78	71.62
3% Zn/SiO$_2$	80	6.10	25.42	68.48

Note: The experimental conditions for all samples include temperature 500 °C; pressure 1 atm; WHSV (h^{-1}) 0.32; feed flow rate (mL/min) 0.5; and hydrogen flow rate (mL/min) 150.

3.3. Catalyst Stability and Reusability

The catalyst with the highest reactivity, 3% Zn/SiO$_2$, was selected for the stability studies of phenol HDO. To identify the activity alteration of the catalyst with the time on stream, the conversion of the phenol was assessed using the extracted experimental data at various TOS (Figure 7). As can be illustrated from the Figure, the catalyst was highly active up to 240 min with a minor decrease in inactivity. Following that, the activity of the catalyst steadily diminished with elevating TOS. After around 340 min of TOS, the alteration in the catalyst activity became insignificant and was around 43%. Additionally, the variation in the product selectivity was likewise insignificant after 420 min of the TOS. In terms of the selectivity of the products, after a TOS of 420 min, the selectivity of the cyclohexane slightly decreased with a minor increase in the selectivity on cyclohexene. Henceforth, from the stability results, it could be concluded that the silica-supported zinc with 3% loading is highly active until 340 min and, after that, must be replaced or regenerated.

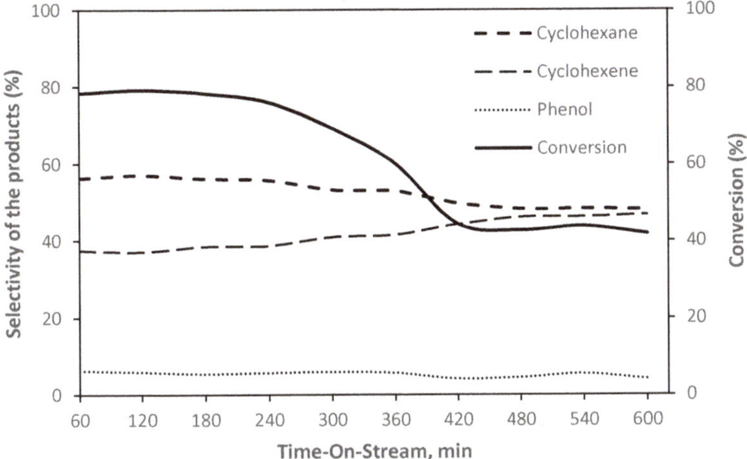

Figure 7. Alteration of the products selectivity and conversion efficiency of phenol with time on stream (TOS). Catalyst 3% Zn/SiO$_2$; temperature 500 °C; pressure 1 atm; weight hourly space velocity (WHSV) (h^{-1}) 0.32; feed flow rate (mL/min) 0.5; and hydrogen flow rate (mL/min) 150.

The applied catalyst for the stability test has been used for the reusability study. The spent catalyst was regenerated by calcination at 500 °C for one hour under the air atmosphere after each cycle. The freshly calcinated sample was then examined by the HDO of phenol. This procedure was continued until four cycles of regeneration. Figure 8 represents the results of the regeneration study. As can be perceived from the Figure, the catalyst recovered fully and had almost the same results as the fresh catalyst in its first and second regeneration steps. In the third and fourth cycles of regeneration, the conversion efficiency of the catalyst was eliminated. The main reason for eliminating the phenol conversion efficiency after the third regeneration cycle might be because of coking and occupying the pores of the support.

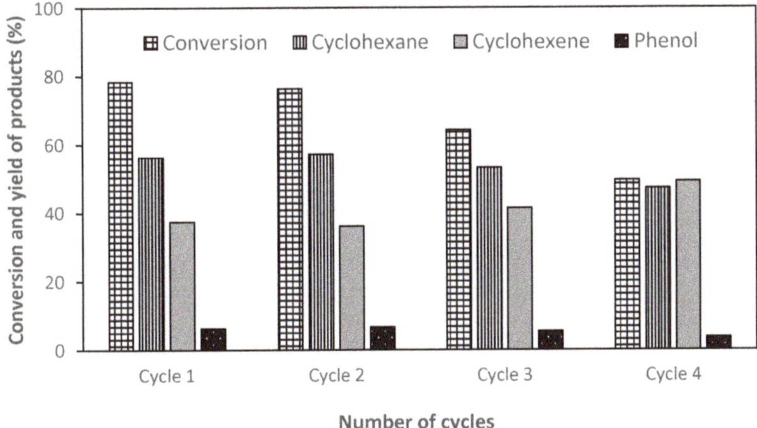

Figure 8. Catalyst reusability survey displaying the conversion and yield of products for hydrodeoxygenation of phenol over 3% Zn/SiO$_2$. Temperature 500 °C; pressure 1 atm; WHSV (h^{-1}) 0.32; feed flow rate (mL/min) 0.5; and H$_2$ volumetric flow rate (mL/min) 150.

4. Discussions

A catalyst characterization study using various analytical techniques including XRD, BET, TGA, H_2-TPR, NH_3-TPD, and ICP-OES along with topography analysis of the samples using the FESEM method revealed that the zinc metal is a promising active site to be applied for hydrodeoxygenation of oxygenated compounds such as phenol. Zinc is an abundant element in the earth's crust, which makes it a promising candidate as a cost-effective active site. Thermogravimetry analysis revealed that the silica-supported zinc catalysts are highly stable at high temperatures up to 800 °C, which is a crucial factor for a catalyst to be applied for HDO reactions. High surface area and large porosities of the samples, analyzed by N_2-adsorption analysis, make the surface reactions possible during the HDO process. Significant H_2 consumption of the samples in the range of 400–550 °C explored by H_2-TPR proved that the catalysts have decent reducibility due to the zinc metal. Furthermore, as it can be perceived from Figure 3, all the samples have one broad peak, which is representative of the one-step reduction of the active metal. Based on the literature [50], the surface acidity of a catalyst plays an important role in the reactivity and selectivity of the HDO reactions. In other words, higher acidity results in a higher conversion efficiency and selectivity of the products. NH_3-TPD analysis proved that the total acidities of the samples were in the range of 0.108–0.481 mmol/g. Elevating the active site doping resulted in a higher value for the total acidity, and the sample with 3% of Zn represented the highest surface acidity of 0.408 mmol/g. The topography analysis micrographs (Figure 4) of the selected samples showed no sign of agglomeration of the zinc metals on the silica surface. Agglomeration is an imperative reason for a reduction in conversion efficiency [51].

Along with the characterization analysis, reactivity analysis revealed that the highest conversion efficiency could be achieved using the sample with 3% metal loading (80%), which was predictable. The selectivity of the HDO reaction products was analyzed using the GC-FID method and revealed that the products of the reactions were cyclohexane, cyclohexene, and benzene. According to the literature [52–54], hydrodeoxygenation of phenol progresses through two dissimilar reaction mechanisms. The first one is breaking the C-O bonds of phenol through direct hydrogenolysis. The final main product of this mechanism is cyclohexane. The second mechanism proceeds through hydrogenation of the aromatic ring, which results in different products in comparison to the first mechanism such as cyclohexanone, cyclohexanol, cyclohexene, and cyclohexane. Accordingly, regarding the selectivity of the products, the second mechanism is the main route in this survey.

5. Conclusions

In summary, the catalyst characterization study revealed that the physicochemical properties of zinc-supported catalysts depended on their active site content. The incorporation of zinc as an active site over SiO_2 mostly involved the filling of pores with zinc, which occupied up to 6.58% of the surface area. The accumulation of zinc on silica also caused an upsurge in acid site concentration due to the regeneration of new proton sites on the catalyst. The effect of metal loading amount on the catalyst performance was also studied in a continuous fixed-bed reactor. It was found that 3%Zn/SiO_2 represents the highest phenol conversion efficiency among other samples. However, the results indicate that the selectivity of the products including cyclohexane, cyclohexene, and phenol remained almost steady by varying the doping amount of the active site. The time-on-stream investigation showed that the silica-supported zinc was highly active up to 240 min of phenol HDO, with a conversion efficiency up to 80%, and after 420 min of TOS, the activity decreased to around a conversion rate of 43%. Deposition of the impurities and coke on the surface of the catalysts is responsible for the deactivation of the catalysts. Reusability tests revealed that the catalyst displayed outstanding reusability and could be regenerated fully after several reusing rounds. Further investigation on catalyst design, reaction mechanism, and regeneration should facilitate the progress of bio-oil oxygenated compounds using silica-supported zinc catalysts.

Author Contributions: Conceptualization, H.P., W.M.A.W.D. and F.A.; Methodology, H.P. and F.A.; Software, H.P. and M.K.A.; Validation, H.P., F.A. and W.M.A.W.D.; Formal analysis, H.P. and F.A.; investigation, H.P.; resources, F.A., W.M.A.W.D., M.K.A. and T.M.I.M.; Data curation, H.P. and F.A.; Writing—original draft preparation, H.P.; Writing—review and editing, F.A. and T.M.I.M.; Visualization, H.P.; Supervision, W.M.A.W.D., M.K.A. and F.A.; Project administration, W.M.A.W.D. and F.A.; Funding acquisition, W.M.A.W.D., F.A. and T.M.I.M. All authors have read and agreed to the published version of the manuscript.

Funding: The authors gratefully acknowledge the financial support from the University of Malaya, Malaysia, and the Deanship of Scientific Research (DSR) at King Abdulaziz University, Jeddah.

Acknowledgments: The authors gratefully acknowledge the financial support from the University of Malaya, Malaysia, and the Deanship of Scientific Research (DSR) at King Abdulaziz University, Jeddah.

Conflicts of Interest: The authors declare no conflict of interest.

References

1. Ismail, M.; Moghavvemi, M.; Mahlia, T.M.I. Characterization of PV panel and global optimization of its model parameters using genetic algorithm. *Energy Convers. Manag.* **2013**, *73*, 10–25. [CrossRef]

2. Alinejad, M.; Arof, A.K. Effect of Extraction Time and Concentration of Crocus sativus on Behavior of Dye-Sensitized Solar Cells. *Ionics* **2017**, *23*, 1579–1584. [CrossRef]

3. Mason, I.; Page, S.; Williamson, A. A 100% renewable electricity generation system for New Zealand utilising hydro, wind, geothermal and biomass resources. *Energy Policy* **2010**, *38*, 3973–3984. [CrossRef]

4. Silitonga, A.; Shamsuddin, A.; Mahlia, T.M.I.; Milano, J.; Kusumo, F.; Siswantoro, J.; Dharma, S.; Sebayang, A.; Masjuki, H.; Ong, H.C. Biodiesel synthesis from Ceiba pentandra oil by microwave irradiation-assisted transesterification: ELM modeling and optimization. *Renew. Energy* **2020**, *146*, 1278–1291. [CrossRef]

5. Ong, H.C.; Milano, J.; Silitonga, A.S.; Hassan, M.H.; Shamsuddin, A.H.; Wang, C.-T.; Mahlia, T.M.I.; Siswantoro, J.; Kusumo, F.; Sutrisno, J. Biodiesel production from Calophyllum inophyllum-Ceiba pentandra oil mixture: Optimization and characterization. *J. Clean. Prod.* **2019**, *219*, 183–198. [CrossRef]

6. Mahlia, T.; Syazmi, Z.; Mofijur, M.; Abas, A.P.; Bilad, M.; Ong, H.C.; Silitonga, A. Patent landscape review on biodiesel production: Technology updates. *Renew. Sustain. Energy Rev.* **2020**, *118*, 109526. [CrossRef]

7. Ong, H.C.; Hassan, M.; Mahlia, T.M.I.; Silitonga, A.S.; Chong, W.T.; Yusaf, T. Engine performance and emissions using Jatropha curcas, Ceiba pentandra and Calophyllum inophyllum biodiesel in a CI diesel engine. *Energy* **2014**, *69*, 427–445. [CrossRef]

8. Silitonga, A.S.; Masjuki, H.; Mahlia, T.M.I.; Ong, H.; Chong, W.T.; Boosroh, M. Overview properties of biodiesel diesel blends from edible and non-edible feedstock. *Renew. Sustain. Energy Rev.* **2013**, *22*, 346–360. [CrossRef]

9. Bhatt, A.H.; Zhang, Y.M.; Heath, G.A. *Air Pollutant Emissions and Regulatory Implications of Co-Processing Raw Bio-Oil in US Petroleum Refineries*; National Renewable Energy Lab (NREL): Golden, CO, USA, 2020.

10. Pourzolfaghar, H.; Abnisa, F.; Daud, W.M.A.W.; Aroua, M.K. Atmospheric hydrodeoxygenation of bio-oil oxygenated model compounds: A review. *J. Anal. Appl. Pyrolysis* **2018**, *133*, 117–127. [CrossRef]

11. Lee, X.J.; Ong, H.; Gan, Y.Y.; Chen, W.-H.; Mahlia, T.M.I. State of art review on conventional and advanced pyrolysis of macroalgae and microalgae for biochar, bio-oil and bio-syngas production. *Energy Convers. Manag.* **2020**, *210*, 112707. [CrossRef]

12. Goh, B.H.H.; Ong, H.; Cheah, M.Y.; Chen, W.-H.; Yu, K.L.; Mahlia, T.M.I. Sustainability of direct biodiesel synthesis from microalgae biomass: A critical review. *Renew. Sustain. Energy Rev.* **2019**, *107*, 59–74. [CrossRef]

13. Kim, S.; Kwon, E.E.; Kim, Y.T.; Jung, S.; Kim, H.J.; Huber, G.W.; Lee, J. Recent advances in hydrodeoxygenation of biomass-derived oxygenates over heterogeneous catalysts. *Green Chem.* **2019**, *21*, 3715–3743. [CrossRef]

14. Boullosa-Eiras, S.; Lødeng, R.; Bergem, H.; Stocker, M.; Hannevold, L.; Blekkan, A. Catalytic hydrodeoxygenation (HDO) of phenol over supported molybdenum carbide, nitride, phosphide and oxide catalysts. *Catal. Today* **2014**, *223*, 44–53. [CrossRef]

15. Prasomsri, T.; Nimmanwudipong, T.; Román-Leshkov, Y. Effective hydrodeoxygenation of biomass-derived oxygenates into unsaturated hydrocarbons by MoO3 using low H2 pressures. *Energy Environ. Sci.* **2013**, *6*, 1732. [CrossRef]

16. Prasomsri, T.; Shetty, M.; Murugappan, K.; Román-Leshkov, Y. Insights into the catalytic activity and surface modification of MoO 3 during the hydrodeoxygenation of lignin-derived model compounds into aromatic hydrocarbons under low hydrogen pressures. *Energy Environ. Sci.* **2014**, *7*, 2660–2669. [CrossRef]

17. Chen, C.-J.; Lee, W.-S.; Bhan, A. Mo2C catalyzed vapor phase hydrodeoxygenation of lignin-derived phenolic compound mixtures to aromatics under ambient pressure. *Appl. Catal. A Gen.* **2016**, *510*, 42–48. [CrossRef]

18. Nimmanwudipong, T.; Runnebaum, R.C.; Block, D.E.; Gates, B.C. Catalytic Conversion of Guaiacol Catalyzed by Platinum Supported on Alumina: Reaction Network Including Hydrodeoxygenation Reactions. *Energy Fuels* **2011**, *25*, 3417–3427. [CrossRef]

19. Zhu, X.; Lobban, L.L.; Mallinson, R.G.; Resasco, D.E. Bifunctional transalkylation and hydrodeoxygenation of anisole over a Pt/HBeta catalyst. *J. Catal.* **2011**, *281*, 21–29. [CrossRef]

20. Zanuttini, M.; Costa, B.D.; Querini, C.; Peralta, M. Hydrodeoxygenation of m-cresol with Pt supported over mild acid materials. *Appl. Catal. A Gen.* **2014**, *482*, 352–361. [CrossRef]

21. Olcese, R.N.; Francois, J.; Bettahar, M.M.; Petitjean, D.; Dufour, A. Hydrodeoxygenation of Guaiacol, A Surrogate of Lignin Pyrolysis Vapors, Over Iron Based Catalysts: Kinetics and Modeling of the Lignin to Aromatics Integrated Process. *Energy Fuels* **2013**, *27*, 975–984. [CrossRef]

22. Olcese, R.; Bettahar, M.; Malaman, B.; Ghanbaja, J.; Tibavizco, L.; Petitjean, D.; Dufour, A. Gas-phase hydrodeoxygenation of guaiacol over iron-based catalysts. Effect of gases composition, iron load and supports (silica and activated carbon). *Appl. Catal. B Environ.* **2013**, *129*, 528–538. [CrossRef]

23. Palla, V.C.S.; Shee, D.; Maity, S.K. Kinetics of hydrodeoxygenation of octanol over supported nickel catalysts: A mechanistic study. *RSC Adv.* **2014**, *4*, 41612–41621. [CrossRef]

24. Ausavasukhi, A.; Huang, Y.; To, A.T.; Sooknoi, T.; Resasco, D.E. Hydrodeoxygenation of m-cresol over gallium-modified beta zeolite catalysts. *J. Catal.* **2012**, *290*, 90–100. [CrossRef]

25. Wan, S.; Pham, T.; Zhang, S.; Lobban, L.; Resasco, D.; Mallinson, R.G. Direct catalytic upgrading of biomass pyrolysis vapors by a dual function Ru/TiO$_2$ catalyst. *AIChE J.* **2013**, *59*, 2275–2285. [CrossRef]

26. Ren, H.; Chen, Y.; Huang, Y.; Deng, W.; Vlachos, D.G.; Chen, J.G. Tungsten carbides as selective deoxygenation catalysts: Experimental and computational studies of converting C3 oxygenates to propene. *Green Chem.* **2014**, *16*, 761–769. [CrossRef]

27. Zhao, H.; Li, D.; Bui, P.; Oyama, S.T. Hydrodeoxygenation of guaiacol as model compound for pyrolysis oil on transition metal phosphide hydroprocessing catalysts. *Appl. Catal. A Gen.* **2011**, *391*, 305–310. [CrossRef]

28. Joshi, N.; Lawal, A. Hydrodeoxygenation of acetic acid in a microreactor. *Chem. Eng. Sci.* **2012**, *84*, 761–771. [CrossRef]

29. Yang, J.; Li, S.; Zhang, L.; Liu, X.; Wang, J.; Pan, X.; Li, N.; Wang, A.; Cong, Y.; Wang, X.; et al. Hydrodeoxygenation of furans over Pd-FeOx/SiO2 catalyst under atmospheric pressure. *Appl. Catal. B Environ.* **2017**, *201*, 266–277. [CrossRef]

30. Zhao, X.; Wei, L.; Cheng, S.; Cao, Y.; Julson, J.; Gua, Z. Catalytic cracking of carinata oil for hydrocarbon biofuel over fresh and regenerated Zn/Na-ZSM-5. *Appl. Catal. A Gen.* **2015**, *507*, 44–55. [CrossRef]

31. Viswanadham, N.; Pradhan, A.; Ray, N.; Vishnoi, S.; Shanker, U.; Rao, T.P. Reaction pathways for the aromatization of paraffins in the presence of H-ZSM-5 and Zn/H-ZSM-5. *Appl. Catal. A Gen.* **1996**, *137*, 225–233. [CrossRef]

32. Smiekova, A. Aromatization of light alkanes over ZSM-5 catalystsInfluence of the particle properties of the zeolite. *Appl. Catal. A Gen.* **2004**, *268*, 235–240. [CrossRef]

33. Li, Y.; Liu, S.; Xie, S.; Xu, L. Promoted metal utilization capacity of alkali-treated zeolite: Preparation of Zn/ZSM-5 and its application in 1-hexene aromatization. *Appl. Catal. A Gen.* **2009**, *360*, 8–16. [CrossRef]

34. Fanchiang, W.-L.; Lin, Y.-C. Catalytic fast pyrolysis of furfural over H-ZSM-5 and Zn/H-ZSM-5 catalysts. *Appl. Catal. A Gen.* **2012**, *419*, 102–110. [CrossRef]

35. Luzgin, M.V.; Rogov, V.A.; Arzumanov, S.S.; Toktarev, A.V.; Stepanov, A.G.; Parmon, V.N. Methane aromatization on Zn-modified zeolite in the presence of a co-reactant higher alkane: How does it occur? *Catal. Today* **2009**, *144*, 265–272. [CrossRef]

36. Ni, Y.; Sun, A.; Wu, X.; Hai, G.; Hu, J.; Li, T.; Li, G. The preparation of nano-sized H[Zn, Al]ZSM-5 zeolite and its application in the aromatization of methanol. *Microporous Mesoporous Mater.* **2011**, *143*, 435–442. [CrossRef]

37. Fang, Y.; Tang, J.; Huang, X.; Shen, W.; Song, Y.; Sun, C. Aromatization of Dimethyl Ether over Zn/H-ZSM-5 Catalyst. *Chin. J. Catal.* **2010**, *31*, 264–266. [CrossRef]

38. Hu, Y.; Song, T.-Y.; Wang, Y.-J.; Hu, G.; Xie, G.; Luo, M. Gas Phase Dehydrochlorination of 1, 1, 2-Trichloroethane over Zn/SiO2 Catalysts: Acidity and Deactivation. *Acta Phys.-Chim. Sin.* **2017**, *33*, 1–11.

39. Klein, I.; Marcum, C.; Kenttamaa, H.I.; Abu-Omar, M.M. Mechanistic investigation of the Zn/Pd/C catalyzed cleavage and hydrodeoxygenation of lignin. *Green Chem.* **2016**, *18*, 2399–2405. [CrossRef]

40. Parsell, T.H.; Jarrell, T.M.; Haupert, L.J.; Amundson, L.M.; Abu-Omar, M.M.; Owen, B.J.; Klein, I.; Marcum, C.; Kenttamaa, H.I.; Ribeiro, F.H.; et al. Cleavage and hydrodeoxygenation (HDO) of C–O bonds relevant to lignin conversion using Pd/Zn synergistic catalysis. *Chem. Sci.* **2013**, *4*, 806–813. [CrossRef]

41. Mirzayanti, Y.W.; Prajitno, D.H.; Roesyadi, A. Catalytic hydrocracking of Kapuk seed oil (Ceiba pentandra) to produce biofuel using Zn-Mo supported HZSM-5 catalyst. *IOP Conf. Ser. Earth Environ. Sci.* **2017**, *67*, 12023. [CrossRef]

42. Pourzolfaghar, H.; Abnisa, F.; Daud, W.M.A.W.; Aroua, M.K. Gas-phase hydrodeoxygenation of phenol over Zn/SiO2 catalysts: Effects of zinc load, temperature, weight hourly space velocity, and H2 volumetric flow rate. *Biomass Bioenergy* **2020**, *138*, 105556. [CrossRef]

43. Rusu, D.; Rusu, G.; Luca, D. Structural Characteristics and Optical Properties of Thermally Oxidized Zinc Films. *Acta Phys. Pol. A* **2011**, *119*, 850–856. [CrossRef]

44. Sacco, O.; Vaiano, V.; Sannino, D.; Picca, R.; Cioffi, N. Ag modified ZnS for photocatalytic water pollutants degradation: Influence of metal loading and preparation method. *J. Colloid Interface Sci.* **2019**, *537*, 671–681. [CrossRef]

45. Nie, L. Catalytic Hydrodeoxygenation of Phenolic Compounds of Importance in Bio-oil Upgrading. 2014. Available online: https://hdl.handle.net/11244/10369 (accessed on 21 March 2020).

46. Oyama, S.T.; Clark, P.; Da Silva, V.T.; Lede, E.J.; Requejo, F.G. XAFS Characterization of Highly Active Alumina-Supported Molybdenum Phosphide Catalysts (MoP/Al2O3) for Hydrotreating. *J. Phys. Chem. B* **2001**, *105*, 4961–4966. [CrossRef]

47. Phillips, D.C.; Sawhill, S.J.; Self, R.; Bussell, M.E. Synthesis, Characterization, and Hydrodesulfurization Properties of Silica-Supported Molybdenum Phosphide Catalysts. *J. Catal.* **2002**, *207*, 266–273. [CrossRef]

48. Clark, P.; A Clark, P.; Oyama, S.T. Alumina-supported molybdenum phosphide hydroprocessing catalysts. *J. Catal.* **2003**, *218*, 78–87. [CrossRef]

49. Asphaug, S. Catalytic Hydrodeoxygenation of Bio-oils with Supported MoP-Catalysts. Master's Thesis, Norwegian University of Science and Technology, Trondheim, Norway, 2013.

50. Karnjanakom, S.; Bayu, A.; Hao, X.; Kongparakul, S.; Samart, C.; Abudula, A.; Guan, G.-Q. Selectively catalytic upgrading of bio-oil to aromatic hydrocarbons over Zn, Ce or Ni-doped mesoporous rod-like alumina catalysts. *J. Mol. Catal. A Chem.* **2016**, *421*, 235–244. [CrossRef]

51. Huynh, T.M.; Armbruster, U.; Kreyenschulte, C.R.; Nguyen, L.H.; Phan, B.M.Q.; Nguyen, D.A.; Martin, A. Understanding the Performance and Stability of Supported Ni-Co-Based Catalysts in Phenol HDO. *Catalysts* **2016**, *6*, 176. [CrossRef]

52. Bu, Q.; Lei, H.; Zacher, A.H.; Wang, L.; Ren, S.; Liang, J.; Wei, Y.; Liu, Y.; Tang, J.; Zhang, Q.; et al. A review of catalytic hydrodeoxygenation of lignin-derived phenols from biomass pyrolysis. *Bioresour. Technol.* **2012**, *124*, 470–477. [CrossRef]

53. Echeandia, S.; Arias, P.L.; Barrio, V.; Pawelec, B.; Fierro, J.L.G. Synergy effect in the HDO of phenol over Ni–W catalysts supported on active carbon: Effect of tungsten precursors. *Appl. Catal. B Environ.* **2010**, *101*, 1–12. [CrossRef]

54. Zhao, C.; He, J.; Lemonidou, A.A.; Li, X.; Lercher, J.A. Aqueous-phase hydrodeoxygenation of bio-derived phenols to cycloalkanes. *J. Catal.* **2011**, *280*, 8–16. [CrossRef]

 *energies*

Article

Modeling and Optimization of Microwave-Based Bio-Jet Fuel from Coconut Oil: Investigation of Response Surface Methodology (RSM) and Artificial Neural Network Methodology (ANN)

Mei Yin Ong [1,2], Saifuddin Nomanbhay [1,2,*], Fitranto Kusumo [1,2], Raja Mohamad Hafriz Raja Shahruzzaman [1,2] and Abd Halim Shamsuddin [2]

[1] Institute of Sustainable Energy, Universiti Tenaga Nasional (UNITEN), Kajang 43000, Selangor, Malaysia; me089475@hotmail.com (M.Y.O.); fitrantokusumo@yahoo.co.id (F.K.); raja.hafriz@uniten.edu.my (R.M.H.R.S.)

[2] AAIBE Chair of Renewable Energy, Universiti Tenaga Nasional (UNITEN), Kajang 43000, Selangor, Malaysia; AbdHalim@uniten.edu.my

* Correspondence: saifuddin@uniten.edu.my; Tel.: +603-8921-7285

Abstract: In this study, coconut oils have been transesterified with ethanol using microwave technology. The product obtained (biodiesel and FAEE) was then fractional distillated under vacuum to collect bio-kerosene or bio-jet fuel, which is a renewable fuel to operate a gas turbine engine. This process was modeled using RSM and ANN for optimization purposes. The developed models were proved to be reliable and accurate through different statistical tests and the results showed that ANN modeling was better than RSM. Based on the study, the optimum bio-jet fuel production yield of 74.45 wt% could be achieved with an ethanol–oil molar ratio of 9.25:1 under microwave irradiation with a power of 163.69 W for 12.66 min. This predicted value was obtained from the ANN model that has been optimized with ACO. Besides that, the sensitivity analysis indicated that microwave power offers a dominant impact on the results, followed by the reaction time and lastly ethanol–oil molar ratio. The properties of the bio-jet fuel obtained in this work was also measured and compared with American Society for Testing and Materials (ASTM) D1655 standard.

Keywords: bio-jet fuel; microwave-assisted transesterification; RSM; ANN; optimization; coconut oil

Citation: Ong, M.Y.; Nomanbhay, S.; Kusumo, F.; Raja Shahruzzaman, R.M.H.; Shamsuddin, A.H. Modeling and Optimization of Microwave-Based Bio-Jet Fuel from Coconut Oil: Investigation of Response Surface Methodology (RSM) and Artificial Neural Network Methodology (ANN). *Energies* **2021**, *14*, 295. https://doi.org/10.3390/en14020295

Received: 15 December 2020
Accepted: 5 January 2021
Published: 7 January 2021

Publisher's Note: MDPI stays neutral with regard to jurisdictional claims in published maps and institutional affiliations.

1. Introduction

Amongst renewable energy technologies the current trend is technology which aims to optimally utilize clean and sustainable energy sources, based on current and future economic and societal needs. Social and economic development is always followed with an increase in energy demand. Currently, most of the energy demand is fulfilled by non-renewable fossil fuels, including natural gas, coal, and petroleum. The formation of fossil fuels took hundreds of million years. It is predicted that with the current consumption rate, fossil fuel will deplete in the near future [1,2]. Furthermore, the burning of fossil fuels results in negative environmental impacts, such as global warming, acid rain, climate change, and others. Hence, it is crucial to look for alternative energy resources that are clean and sustainable to replace the non-renewable fossil fuel.

There are a series of studies that have been done in exploring renewable energy technologies, such as solar energy, wind energy, tidal energy, and biofuels [3–8]. Among the available renewable energy, liquid biofuel from biomass resources has received a lot of attention [9–11], especially in the transportation sector, as it offers an opportunity to replace petroleum in operating the combustion engine with little to no modification [12,13]. In fact, over 90% of transportation nowadays is still dependent on non-renewable fossil fuel [14]. Therefore, many works have been done to enhance and improve the biofuel production

85

technologies, especially biodiesel, a renewable replacement of diesel fuel [15–18]. Nevertheless, there is only a little information about the production of bio-kerosene (bio-jet fuel). Bio-jet fuel is a promising alternative fuel for a jet engine. Other than the aviation sector, bio-jet fuel also can be used in the power generation sector to operate the gas turbine engine. Hence, the bio-jet fuel production technologies should be further explored and investigated in obtaining a comparable renewable replacement for petroleum-based kerosene at sufficient volume.

Based on the statistical report, flights produced 859 Mt of carbon dioxide (CO_2) globally in 2017. In the other words, approximately 2% of the CO_2 emitted through human activities is contributed by the global aviation sector. In addition, this 859 Mt of CO_2 is responsible for 12% of CO_2 emission from all transports sources [19,20]. CO_2 is one of the greenhouse gases that will trap heat within the Earth's atmosphere, causing global warming and climate change. In order to mitigate the CO_2 emissions from air transport and, at the same time, to address the global challenge of climate change, the International Air Transport Association (IATA) has adopted a set of targets and approaches. One of the ambitious targets is to achieve 50% reduction in net aviation carbon emissions by 2050 as compared to 2005, through the deployment of sustainable low-carbon fuels, such as bio-jet fuel [19].

Generally, aviation liquid fuel can be produced from different biomass feedstock with different methods [21–23]. Currently, transesterification and hydrotreating are the main production method of bio-jet fuel [24]. Unlike the hydrotreating process, transesterification requires an upgrading process to separate bio-jet fuel from the product of transesterification (biodiesel). Although an additional downstream process is required, however, the transesterification process operates under milder condition as compared to the hydrotreating process. Hence, its operating cost are relatively lower. The transesterification process involves three consecutive and reversible reactions (refer to Equations (1)–(3)) that react triglyceride with alcohol in the presence or absence of a catalyst and produce a mixture of fatty acid alkyl ester (FAAE or biodiesel) and glycerol as a by-product. In overall, transesterification requires 1 mole of triglyceride and 3 moles of alcohol to produce 3 moles of biodiesel, as shown in Equation (4). However, practically, the excess amount of alcohol is usually used to shift the equilibrium to the product side and allow the phase separation of biodiesel from the glycerol [25].

$$Triglyceride\ (oil\ or\ fat) + ROH\ (alcohol) \leftrightarrow Diglyceride + RICOOR \tag{1}$$

$$Diglyceride + ROH \leftrightarrow Monoglyceride + RICOOR \tag{2}$$

$$Monoglyceride + ROH \leftrightarrow Glycerol + RICOOR \tag{3}$$

$$Triglyceride\ (oil\ or\ fat) + 3ROH\ (alcohol) \leftrightarrow Glycerol + 3R'COOR\ (FAAE) \tag{4}$$

Commonly, methanol is used for biodiesel production via transesterification reactions. This is because methanol is cheaper, allows phase separation to be conducted more easily, and permits the transesterification process to be conducted under milder conditions [26]. However, methanol is very toxic to humans, as in the body methanol is metabolized into formaldehyde and then formic acid. Therefore, there is the scope of using ethanol for producing biodiesel. Ethanol has several advantages to methanol, such as offering better solvent properties and low toxicity relative to methanol [27,28] Furthermore, there is a study reported that the biodiesel formed using ethanol, the fatty acid ethyl ester (FAEE) present a higher cetane number, calorific value, oxidation stability, lubricant characteristics, lower cloud and pour points, and also have lower tailpipe emissions in comparison to the product collected using methanol, fatty acid methyl ester (FAME) [29,30].

In addition, microwave technology is deployed to replace conventional heating in this study. Microwave technology is a green processing method that offers several advantages, such as by being more environmental-friendly, in terms of lower energy consumption. Furthermore, the volumetric heating mechanism of microwave heating also allows rapid heating, enhances chemical reaction rate and selectivity, and improves the production quality and yield [25]. Microwave heating has received a lot of attention since the 1970s, especially in the chemical research. Conventional heating transfers heat into the reactant through the reactant vessel via conduction and convection (wall heating). However, microwave heating is highly dependent on the dielectric properties of the reactant [25,31]. It allows direct heating of the material without heat-up of the reactant vessel [32]. Hence, microwave heating allows selective and rapid heating, as mentioned previously.

Numerous studies have been done on the microwave-assisted transesterification process for biodiesel production [25,33,34]. Compared to other approaches, transesterification is the simplest and a widely-accepted method to reduce oil viscosity because it is cost-effective. Therefore, modeling the process and optimizing the process input variables involved are important in order to save time, achieve high product yield and reduce the overall cost to produce biodiesel. Response surface methodology (RSM) and artificial neural network (ANN) are one of the mathematical methods for modeling transesterification processes [35–38]. Some of the advantages that RSM has are the durability under optimal setting conditions and the ability to minimize the number of trials required to provide sufficient evidence for statistically acceptable results [37].

Artificial neural network (ANN) is an information processing system that has characteristics such as biological neural networks that imitates the behaviour and learning process of the human brain. An interesting characteristic of this ANN is its ability to learn (learning and training). The training process at ANN aims to find convergent weights between layers so that the weights obtained to produce the desired output. ANNs are universal approximators and their predictions are based on prior available data, have shown great ability in solving complex nonlinear systems [39].

Ant colony optimization (ACO) is a swarm intelligence technique which is inspired by natural metaphors, namely communication and cooperation between ants to find the shortest path from the nest ants to the food. It is desirable to integrate ACO with ANN model since ACO is capable of optimizing complex process parameters [15,40]. Coupling ACO with the ANN is desirable since the ACO algorithm is capable of optimizing complex process parameters [41].

This study proposed the production of bio-jet fuel through microwave-assisted catalytic transesterification from coconut oil. Coconut oil was selected as the raw feedstock as it consists of a high percentage of medium-chain triglycerides, which made it suitable to be used for bio-jet fuel production [22,42]. An optimization study was conducted in this work based on three parameters, including oil to ethanol molar ratio, reaction time and microwave power, using response surface methodology (RSM) and artificial neural network (ANN) coupled with ant colony optimization (ACO) algorithm. Additionally, the relevant characterizations of coconut oil through the application of gas chromatographic-mass spectrometry and Fourier transform infrared spectrometry analyses of the coconut oil, FAEE, and bio-jet fuel were conducted and reported. The physicochemical properties of the bio-jet fuel collected and their comparison with the ASTM standard are also reported in this paper.

2. Materials and Methods

2.1. Materials and Equipment

Coconut oil was purchased from the local market in Selangor, Malaysia. The coconut oil was then analyzed using gas chromatography, GC (Agilent, 7890A, Wilmington, DE, USA) to obtain its fatty acid profile, as reported in Table 1. Ethanol (C_2H_5OH) and potassium hydroxide (KOH) were purchased from Sigma–Aldrich (St. Louis, MO, USA).

Table 1. Fatty acid composition of coconut oil.

Fatty Acid	Composition (wt%)
Caprylic acid, C8:0	6.0
Capric acid, C10:0	5.1
Lauric acid, C12:0	48.5
Myristic acid, C14:0	16.9
Palmitic acid, C16:0	9.6
Stearic acid, C18:0	2.3
Oleic acid, C18:1	8.2
Linoleic acid, C18:2	3.4

2.2. Microwave-Assisted Catalytic Transesterification

The experiments of this work involve the catalytic transesterification of coconut oil under microwave irradiation. Firstly, coconut oil and ethanol at a specific molar ratio with 0.5 wt% of potassium hydroxide (KOH) catalyst concentration was prepared. Then, the mixture (reactant) was poured into a 250 mL round bottom flask with a magnetic stirrer bar and put into the microwave reactor. Next, the reflux system was attached, and the reactant was then subjected to microwave irradiation with different power settings under different reaction times at a stirring speed of 200 rpm. At the same time, the temperature of the reflux system was maintained at $-4\,°C$ by using a chiller to condense back the vaporized reactant. After completion, the product of this transesterification process was cooled to room temperature. The equipment used in producing bio-jet fuel via transesterification was a modified microwave oven equipped with a reflux system, as shown in Figure 1.

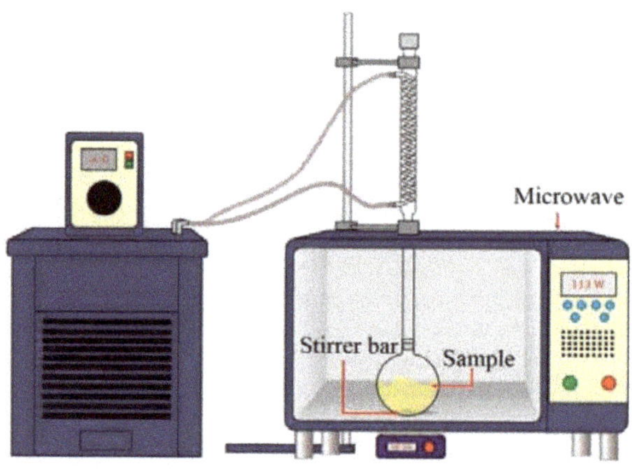

Figure 1. Experimental setup for microwave-assisted catalytic transesterification.

Then, the product was left in a separating funnel to split the liquid phases. The upper layer, which is the fatty acid ethyl ester (FAEE), was washed using warm water three times and dried at 100 °C for 1 h in the conventional oven to remove the moisture.

2.3. Distillation

The bio-jet fuel fraction of the FAEE was obtained by fractional distillation process, which was carried out as shown in Figure 2. The bio-jet fuel with a lower boiling point (C8-C14 of FAEE) was separated from the FAEE by using a rotary evaporator at 165 °C and

25 mbar for 120 min. The bio-jet fuel was collected, and the percentage yield of bio-jet fuel was calculated by using the equation below:

$$\%yield\ of\ biojet\ fuel = \frac{amount\ of\ biojet\ fuel\ obtained\ (g)}{amount\ of\ coconut\ oil\ feedstock\ (g)} \times 100\% \tag{5}$$

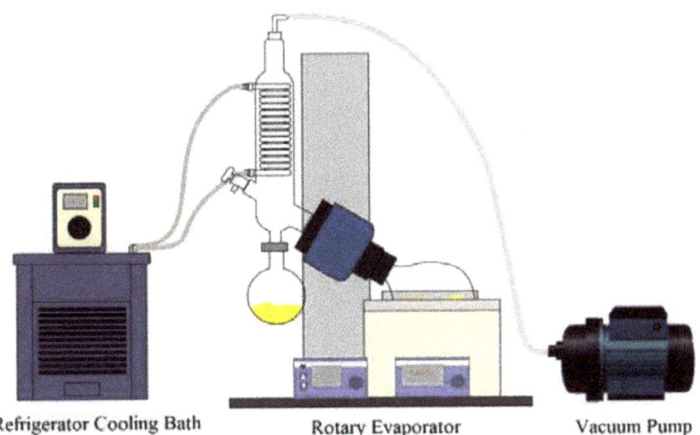

Figure 2. Experimental setup for distillation.

2.4. Response Surface Methodology (RSM)

To investigate the optimal condition for bio-jet fuel production through microwave-assisted catalytic transesterification, response surface methodology (RSM) was applied by using Design Expert 11 software (Stat-Ease, Minneapolis, MN, USA). RSM was normally used to optimize an experiment based on the selected variables [43–45]. Traditionally, the one-factor-at-a-time (OFAT) methodology, which varies one variable at a time while maintaining the others as constant, is time-consuming and expensive because many experimental runs are required to evaluate the relationship between the studied variables. Hence, RSM is recommended as the optimization can be achieved with less experimental runs as compared to OFAT by varying different parameters at a time. Moreover, a mathematical model can be generated to describe the experiments by using a polynomial function that fitted by the least square method:

$$Y = \beta_0 + \sum \beta_i X_i + \sum \beta_{ij} X_i X_j + \sum \beta_i \beta_i X_i X_i + e \tag{6}$$

where X_i indicates the studied variable, while Y symbolize the results to be optimized. β_0, β_i, and β_{ij}, however, are the regression coefficient. Finally, e represents the random error.

In this project, the Box–Behnken design was selected as it able to estimate the regression coefficient of a second-degree quadratic equation with less experimental runs in comparison to central composite design. In this optimization study, three parameters were considered, namely, coconut oil to ethanol molar ratio, reaction time, and microwave power. The respective levels of different parameters are summarized in Table 2 and note that the center point was repeated three times to determine the experimental errors. The 3-levels-3-parameters Box–Behnken Design was implemented and showed that a total 15 experimental runs with different reaction conditions (refer to Table 3) are required to conduct the optimization study. From there, bio-jet fuel was produced, and the result obtained was inserted into software for further analysis. Analysis of variance (ANOVA) was conducted to generate a mathematical model for the studied response, in this case, the biokerosene yield.

Table 2. Experimental level of studied parameters.

Experimental Level	Coconut Oil to Ethanol Molar Ratio, F_1	Reaction Time, F_2 (min)	Microwave Power, F_3 (W)
Low level, L (-1)	1:6	5	100
Medium level, M (0)	1:9	10	300
High level, H ($+1$)	1:12	15	500

Table 3. Fit summary table.

Source	Sequential p-Value	Lack of Fit p-Value	Adjusted R-Squared	Predicted R-Squared	Remarks
Linear	0.0418	0.0211	0.3787	0.2289	
2 Factors Interaction	0.9898	0.0143	0.1574	-0.4111	
Quadratic	<0.0001	0.4277	0.9792	0.9130	Suggested
Cubic	0.4277		0.9839		Aliased

2.5. Artificial Neural Network (ANN)

MATLAB's Neural Network in MATLAB R2011b (MathWorks Inc., Natick, MA, USA) was used to train the backpropagation ANN developed in this study. The hyperbolic *tangent sigmoid* (Equation (7)) and the *purelin* (Equation (8)) transfer function was used for the input layer to the hidden layer and the hidden layer to the output layer, respectively.

$$tangent\ sigmoid\ (x) = \frac{2}{(1 + e^{-2x})} - 1 \tag{7}$$

$$A = purelin\ (x) = x \tag{8}$$

The proposed ANN has an input layer with three neurons (coconut oil to ethanol molar ratio, reaction time and microwave power), a hidden layer and an output layer with one neuron (bio-jet fuel yield). The ANN model was trained until the mean square error (MSE) was minimized and the average correlation coefficient was close or equal to 1. The optimum number of hidden neurons was selected by heuristic procedure. The dataset containing the bio-jet fuel yield and the process input variables were divided into three subsets: training (70%), validating (15%), and testing (15%).

2.6. Ant Colony Optimization (ACO)

Ant colony optimization (ACO) is also known as a swarm intelligence technique, which is inspired by the foraging pattern of ant colonies. Ants can find the shortest path from a food source to their nest, without having to see it directly. The ants have a unique and very advanced solution, namely using a pheromone trail on a path to communicate and building a solution, the more pheromone traces are left, the other ants will follow that path. These pheromones too relate to the previous good element solutions formed by the ants. Equation (9) describes the probability of an ant move from one node (i) to another (j):

$$P_{i,j} = \frac{\left(\tau_{i,j}^{\alpha}\right)\left(n_{i,j}^{\beta}\right)}{\sum\left(\tau_{i,j}^{\alpha}\right)\left(n_{i,j}^{\beta}\right)} \tag{9}$$

where $\tau_{i,j}$ indicates the amount of pheromone on edge i,j (refer to Equation (10)) and the α symbolizes the factors selected to regulate the impact of $\tau_{i,j}$. $\tau_{i,j}$. Meanwhile, $n_{i,j}$ is the desirability of edge i,j (commonly $1/d_{i,j}$) and the β implies the factors selected to regulate the impact of $n_{i,j}$.

$$\tau_{i,j} = 1 - \rho\tau_{i,j} + \Delta\tau_{i,j} \tag{10}$$

where ρ is the rate of pheromone evaporation and $\Delta\tau_{i,j}$ is the amount of pheromone deposited.

If ant k travels on edge i,j, the amount of pheromone deposited is given by Equation (11):

$$\Delta\tau_{i,j}^{k} = \begin{cases} \frac{1}{L_k}, & \text{if ant k travels on edge } i,j \\ 0, & \text{Otherwise} \end{cases} \tag{11}$$

where L_k is the cost of the kth ant's tour (typically length).

2.7. Statistical Evaluation of the Developed Models

Different statistical measures, including correlation coefficient (R), coefficient of determination (R^2), root mean absolute error (RMSE), standard error of prediction (SEP), mean absolute error (MAE) and Chi-square were used to test the developed models, as described in Equations (12)–(17) [39]:

$$R = \frac{\sum_{i=1}^{n} \left(M_p - M_{p,avg}\right) \times \left(M_e - M_{e,avg}\right)}{\sqrt{\left[\sum_{i=1}^{n}\left(M_p - M_{p.avg}\right)^2\right]\left[\sum_{i=1}^{n}\left(M_e - M_{e.avg}\right)^2\right]}} \tag{12}$$

$$R^2 = 1 - \frac{\sum_{i=1}^{n}\left(M_e - M_p\right)^2}{\sum_{i=1}^{n}\left(M_{e,avg} - M_p\right)^2} \tag{13}$$

$$RMSE = \sqrt{\frac{1}{n}\sum_{i=1}^{n}\left(M_e - M_p\right)^2} \tag{14}$$

$$SEP = \frac{\sqrt{\frac{1}{n}\sum_{i=1}^{n}\left(M_e - M_p\right)^2}}{M_{e,avg}} \times 100 \tag{15}$$

$$MAE = \frac{1}{n}\left(\sum_{i=1}^{n}\left|\left(M_e - M_p\right)\right|\right) \tag{16}$$

$$Chi-square = \sum_{i=1}^{n}\frac{\left(M_e - M_p\right)^2}{M_e} \tag{17}$$

where n shows the number of points, and M_p, M_e, $M_{p,avg}$ and $M_{e,avg}$ are the predicted value, experimental value and the average of the predicted and experimental values, respectively.

2.8. Sensitivity Analysis

The importance of the studied variables (coconut oil to ethanol molar ratio, reaction time, and microwave power) was explored through conducting the sensitivity analysis. In this analysis, Equation (18) is used with the "sum of squares" values obtained from the ANOVA table generated from response surface methodology (RSM).

$$sensitivity\% = \frac{S_x}{S_y} \times 100 \tag{18}$$

where S_x and S_y indicate the sum of square of the individual variable and total sum of squares of all the variables, correspondingly.

The significance input variables for ANN, were calculated based on Equation (19) [37]:

$$F_k = \frac{\sum_{j=1}^{j=M_o}\left(\left(\frac{\left|W_{kj}^{ag}\right|}{\sum_{h=1}^{My}\left|W_{hj}^{ac}\right|}\right) \times \left|W_{jm}^{gl}\right|\right)}{\sum_{h=1}^{h=M_y}\left\{\sum_{j=1}^{j=M_o}\left(\left(\left|W_{qr}^{an}\right|\frac{\left|W_{hj}^{ag}\right|}{\sum_{d=1}^{Mp}\left|W_{hj}^{ag}\right|}\right) \times \left|W_{jm}^{gl}\right|\right)\right\}} \tag{19}$$

where, F_k is the relative significance of the kth input variable on the output variable. M_o is the number of input neurons and M_y is the number of hidden neurons. W is the connection weight. The superscript a, g, and l represent the input, output, and hidden layer, respectively, whereas the subscript h, j, and m represent the input, output, and hidden neuron, respectively.

2.9. Optimization of Transesterification Process Variables

RSM and ANN-ACO were used to evaluate the optimal value of the three respective studied parameters in order to obtain the highest bio-jet fuel yield. The bio-jet fuel yield and the studied parameters were set at "maximum" and "in the range" individually in the case of RSM. ACO was used to determine the optimal values with the maximum bio-jet fuel yield for ANN. By conducting triplicate experiments, the optimal values determined by each approach were validated and the average values obtained were compared with the expected values.

2.10. Bio-Jet Fuel Properties

The bio-jet fuel isolated from biodiesel (FAEE) was then analyzed and compared with the standard requirements for aviation turbine fuels (ASTM D1655) of the American Society for Testing and Materials. In this work, according to the respective ASTM test process, physicochemical properties including density at 15 °C, kinetic viscosity at −20 °C, flash point, and freezing point were calculated. Using a bomb calorimeter, the lower heating factor, also known as calorific value, was calculated as well.

3. Results

3.1. RSM

Based on this result, a fit summary table was generated to evaluate the suitable model for the optimization study. The fit summary table is shown in Table 3, and the outcomes indicate that the quadratic model is adequate to model the studied response (bio-jet fuel yield). This finding is based on the sequential p-value of the quadratic model, which is less than 0.05. In other words, the quadratic model consists of more than 95% confidence in modeling the experiments. Besides that, inclusion of cubic model terms might cause the model to be aliased and hence, the quadratic model is the best with the highest polynomial order.

The model used in this quadratic equation using a notation such as F_1 for the coconut oil ethanol molar ratio, F_2 for the reaction time, and F_3 for microwave power. The experimental and predicted results in this work are summarized in Table 4 and the uncoded quadratic equation is shown in the following equation:

$$\text{Bio} - \text{jet fuel yield (\%)} = 21.84F_1 + 2.419F_2 + 0.0416F_3 - 0.00717F_1F_2 + 0.00227F_1F_3 - 0.000165F_2F_3 - 1.18644F_1^2 - 0.082917F_2^2 - 0.00016F_3^2 - 50.84552 \tag{20}$$

Table 4. Experimental runs in RSM design with the different operating conditions and their respective experimental and predicted bio-jet fuel yield.

Experimental Run	Coconut Oil to Ethanol Molar Ratio, F_1	Reaction Time, F_2 (min)	Microwave Power, F_3 (W)	Experimental Bio-Jet Fuel Yield (%)	Predicted Bio-Jet Fuel Yield (%) RSM	Predicted Bio-Jet Fuel Yield (%) ANN
1	1:6	10	500	40.92	40.07	41.10
2	1:9	15	100	70.86	71.12	70.73
3	1:12	5	300	56.05	56.11	56.18
4	1:6	5	300	48.21	49.32	47.97
5	1:12	15	300	63.88	62.77	64.13
6	1:9	15	500	55.73	56.64	55.69
7	1:9	5	100	64.83	63.92	64.84
8	1:9	5	500	50.36	50.10	50.38

Table 4. *Cont.*

Experimental Run	Coconut Oil to Ethanol Molar Ratio, F_1	Reaction Time, F_2 (min)	Microwave Power, F_3 (W)	Experimental Bio-Jet Fuel Yield (%)	Predicted Bio-Jet Fuel Yield (%) RSM	Predicted Bio-Jet Fuel Yield (%) ANN
9	1:6	10	100	57.14	56.94	56.64
10	1:12	10	500	49.16	49.36	49.19
11	1:12	10	100	59.94	60.79	59.96
12 *	1:9	10	300	69.82	68.85	69.81
13	1:6	15	300	56.47	56.41	55.96
14 *	1:9	10	300	69.15	68.85	69.81
15 *	1:9	10	300	67.59	68.85	69.81

* The center points of the experiments are replicated for 3 times.

The effect of the studied parameters as linear, quadratic, and interaction coefficients on the studied responses was determined for their significance through analysis of variance (ANOVA). The ANOVA results of this work are reported in Table 5. The studied parameters coconut oil to ethanol molar ratio, reaction time and microwave power represent as F_1, F_2, and F_3 terms particularly. It is noted that the model used was statistically significant with a confidence level of 95%. From the table, it observed that parameters F_1, F_2, and F_3 have a significant influence on bio-jet fuel production (p-values < 0.0500). Furthermore, note that the "lack of fit" of the model consists of a p-value of 0.4227 (not significant as >0.05). In the other words, the model is fit to be used for further analysis.

Table 5. Analysis of Variance for the modeling of bio-jet fuel production.

Source	Sum of Squares	df	Mean Square	F-Value *	p-Value	Remarks
Model	1127.25	9	125.25	74.3	<0.0001	significant
F_1-Ethanol	86.4	1	86.4	51.25	0.0008	
F_2-Time	94.46	1	94.46	56.04	0.0007	
F_3-Power	400.45	1	400.45	237.55	<0.0001	
F_1F_2	0.0462	1	0.0462	0.0274	0.875	
F_1F_3	7.4	1	7.4	4.39	0.0903	
F_2F_3	0.1089	1	0.1089	0.0646	0.8095	
F_1^2	420.99	1	420.99	249.74	<0.0001	
F_2^2	15.11	1	15.11	8.96	0.0303	
F_3^2	150.55	1	150.55	89.31	0.0002	
Residual	8.43	5	1.69			
Lack of Fit	5.81	3	1.94	1.48	0.4277	not significant
Pure Error	2.62	2	1.31			
Cor Total	1135.68	14				

* The F value for a term shows the test in order to compare the variance of that particular term with the residual variance.

Effect of the Parameter

Effect of the Ethanol to Oil Molar Ratio

Figure 3a shows the effect of ethanol to oil molar ratio and reaction time for the bio-jet fuel production. It is shown, from the ANOVA analysis (Table 5), that ethanol to oil molar ratio has the positive effect for bio-jet fuel production (p value < 0.05). Increasing the ethanol to oil molar ratio from 1:6 to 1:9.36, the bio-jet fuel yield increases until an optimal point and then decreases. Similar trends have been reported in several studies [46]. As stated before, the stoichiometry of the transesterification requires three moles of alcohol to react with one mole of triglyceride. However, since it involves a reversible reaction, so, an excessive amount of alcohol is usually needed to shift the reaction equilibrium toward the production of biodiesel. In the work done by Encinar et al. [47], the conventional alkali-catalyzed transesterification was reported to be incomplete for the molar ratio of methanol to Cynara oil that was less than 4.05. As expected, the methyl esters yield increases with the methanol molar ratio and achieve an optimal yield at a molar ratio of 5.67. However, for the methanol molar ratio higher than 5.67, the methyl esters yield drops. This result tallies with current work. This phenomenon is observed because the separation and recovery process of glycerol has been interfered by the high alcohol molar ratio [48]. The high amount of ethanol, in this case, has increased the solubility of the glycerol in the ester phase and shifts the reaction equilibrium towards the reactant side. As the result, biodiesel yield decreases as well as bio-jet fuel.

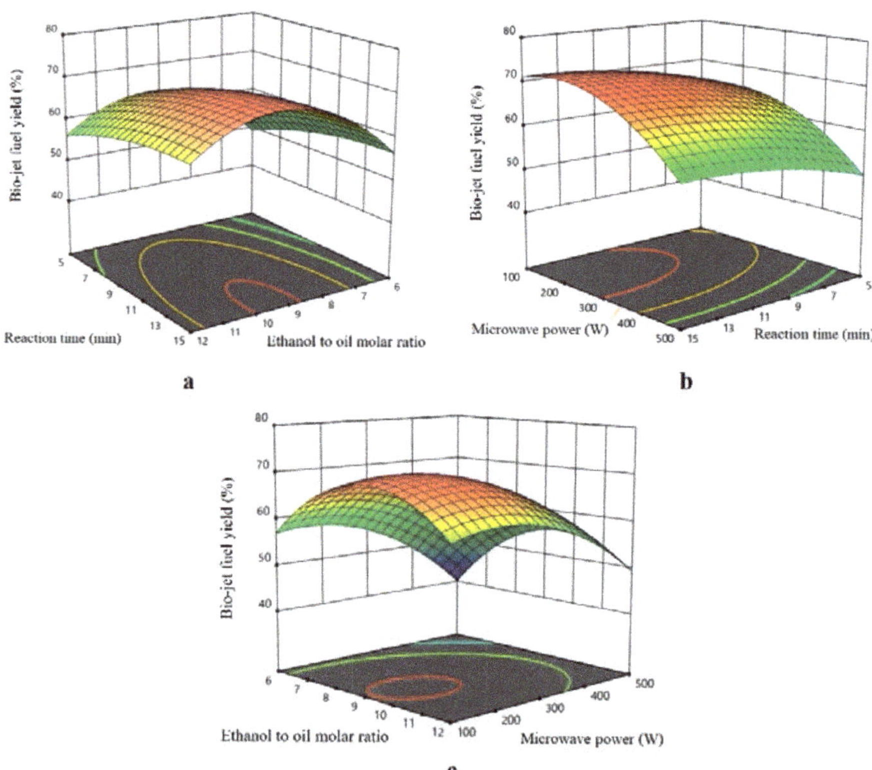

Figure 3. Three-dimensional response for the interaction on bio-jet fuel yield between (**a**) reaction time and ethanol to oil molar ratio, (**b**) microwave power and reaction time and, (**c**) ethanol to oil molar ratio and microwave power.

Effect of the Reaction Time

The relationship between reaction time and microwave power is shown in Figure 3b. The p-value = 0.0007 of the reaction time parameter suggested that there was a reasonably large impact of reaction time on the production of bio-jet fuel yield (Table 5). With the increase in reaction time from 5 to 14.34 min followed by the increase in microwave power from 100–189.53 W, bio-jet fuel yield improved, which may be due to the explanation that more reaction time causes a more productive reaction time. It can be noticed that the bio-jet fuel yield increases as the reaction time increases, similar to the studies done by other researchers [42].

Effect of the Microwave Power

It is observed that the interaction of the microwave power and ethanol molar ratio have positively approached the bio-jet fuel yield. From Figure 3c, it observed that increasing the microwave power from 100 to 189.53 W, and the ethanol to oil molar ratio from 1:6 to 1:9.36, will increase the bio-jet fuel into maximum point and then it will decrease. The bio-jet fuel yield was decreasing when the microwave power reached above 189.53 W, with an ethanol molar ratio of more than 1:9.36. This might be due to the possibility of high microwave power causes the destruction of triglycerides [49]. Hence, less triglycerides were reacted with alcohol, and resulting in lower yield.

3.2. Analysis of the Developed ANN Model

The design of the ANN architecture chosen for this work was based on the minimum mean square error (MSE) and the highest coefficient correlation value (R). The ANN network modeling output plots (predicted values) versus target (actual values) for training, validation, testing and entire datasets with R values of 0.997, 1, 1, and 0.99785, respectively, are shown in Figure 4. In this study, the best topology was found to be 3-8-1, (Figure 5). The architecture consists of a three-neuron input layer.

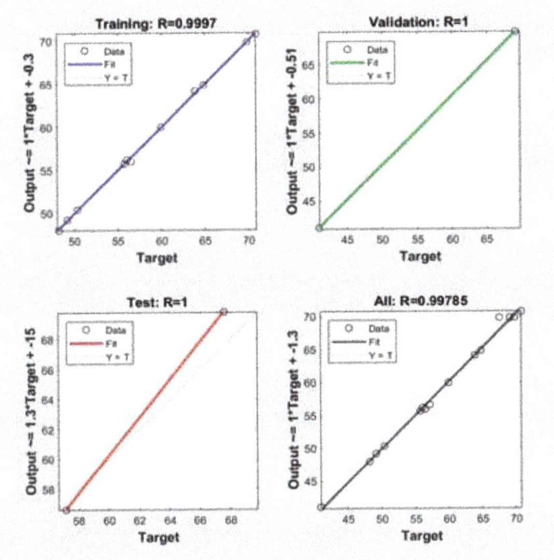

Figure 4. Artificial neural network modeling.

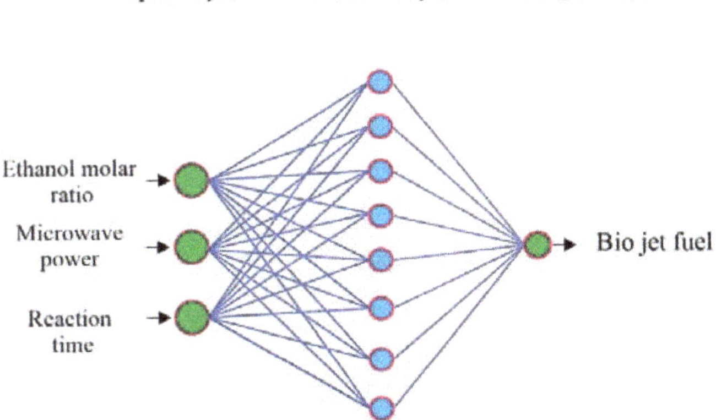

Figure 5. Artificial neural network architecture.

3.3. Comparison of the Predictive Capability of Models

Table 6 displays the findings obtained from different statistical metrics used to test the built models. For both developed models (RSM and ANN), the high R values (Table 6) indicate that there is a strong link between the real and the predicted yields of bio-jet fuel. For both RSM and ANN, the R and R^2, were 0.9963, 0.9926, and 0.9979,0.9957, respectively.

Table 6. Statistical analysis.

Statistical Analysis	RSM	ANN
R	0.9963	0.9979
R^2	0.9926	0.9957
RSME	0.5619	0.4048
SEP (%)	0.9563	0.6889
MAE	0.0414	0.0220
Chi-square	0.0096	0.0061

The good fit of the models is representative of these high values. RMSE is a test of the dataset adherence to the regression line. For both RSM and ANN models, the values of RMSE that were obtained were all low, confirming the models' good fit. To calculate the residuals (deviation from actual objective) of the built models, SEP (percent) and MAE were used. The ANN model had less divergence from experimental values of SEP (0.6889 percent) and MAE (0.022) than the RSM model, as seen in Table 6. In comparison, lower chi-square (0.782) values confirmed that the most reliable was the ANN model with the lowest error term values and highest R and R^2 values. It was noted in this analysis that ANN was found to be superior than RSM.

3.4. Sensitivity Analysis Results of the Input Variables on the Developed Models

The sensitivity analysis findings for both RSM and ANN are shown in Figure 6. The result for both models have the same pattern in that the most influential response input variable (bio-jet fuel yield) was seen to be microwave power, followed by reaction time and finally by ethanol to oil molar ratio. The degrees of significance distributions, however, differed between the methods of modeling. The significance level of 68.89% microwave power, 16.25% reaction time and 14.86% ethanol to the molar ratio for RSM. While for ANN, the important microwave power rating was 42.29%, the reaction time was 29.03% and the ethanol oil molar ratio was 28.68%.

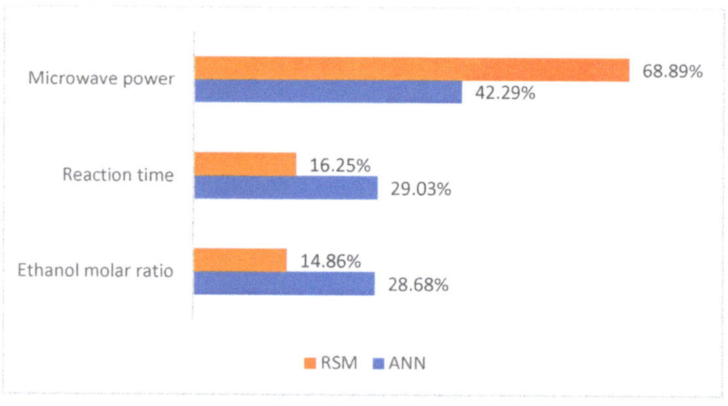

Figure 6. Sensitivity analysis for bio-jet fuel production.

3.5. Optimization of the Process Variables for Bio-Jet Fuel

To obtain the highest value of bio-jet fuel yield, the optimization of the studied variables was carried out using RSM and ANN-ACO. In the case of RSM, the in-build optimization method in the Design Expert software was used for the optimisation study. On the other hand, the ANN models acted as the fitness functions when combined with ACO in the case of optimization using the ANN-ACO method. Table 7 shows the optimum values expected by each system. For each model bio-jet fuel validated by triplicate experiments in the laboratory, the expected optimal condition and the average bio-jet fuel yields were reported (Table 7). From the result, it was observed that RSM predicted 72.49% of bio-fuel jet yield. While for ANN-ACO, the predicted yield was 74.45% with process variables (9.25:1 ethanol oil molar ratio, 12.66 min reaction time and 163.69% microwave power) was higher than most RSM values.

Table 7. Optimization of RSM and ANN-ACO.

Modeling Method	Ethanol	Time	Power	Predicted	Observed
RSM	9.35	14.34	189.53	72.49	73.02
ANN	9.25	12.66	163.69	74.8	74.45

3.6. FTIR

The results obtained from the FTIR study of the spectrum of coconut oil, FAEE and bio-jet fuel are given in Figure 7. The peaks shown in between 3000 to 2800 cm^{-1} indicate the stretching vibration of = C-H (alkene) and C-H (alkane) functional groups [50], which contributed by the esters chain of the samples. The characteristics peak at wavenumber of 1743 cm^{-1}, 1742 cm^{-1}, and 1740 cm^{-1} in the spectrum of coconut oil, FAEE and bio-jet fuel, respectively, show the occurrence of C=O stretching. Although this peak is shorter in the bio-jet fuel as compared to FAEE, however, the presence of this peak showing that there is a need for deoxygenation process to remove the oxygen content in the jet fuel. Besides, blending of the bio-jet fuel with the conventional fossil-based jet fuel is also recommended to balance the bio-jet fuel property deficiency. Furthermore, it is noted that an extra peak at 1437 cm^{-1} in the spectrum of FAEE and bio-jet fuel in comparison to the spectrum of coconut oil. This peak can be explained by the bending vibration of C-H groups in the samples. In addition, it confirms the transesterification process was successfully conducted and converted triglycerides in the coconut oil into ethyl esters in biodiesel (FAEE) and bio-jet fuel [44,51,52].

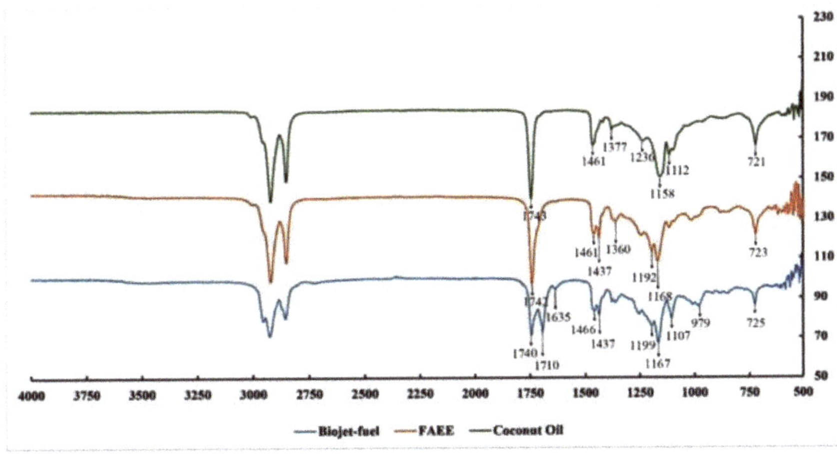

Figure 7. FTIR spectrum for coconut oil, FAEE, and bio-jet fuel.

3.7. Bio-Jet Fuel Properties

To evaluate the suitability of the bio-jet fuel produced in this work as alternative jet fuel, some of the major properties of bio-jet fuel were analyzed and compared with American Society for Testing and Materials (ASTM) D1655, the standard specification for aviation turbine fuels. The test results and the standard limit are summarized in Table 8.

Table 8. Comparison of properties between bio-jet fuel of the current work with ASTM D1655 standard.

Properties	Unit	ASTM D1655	This Work	Soybean [53]	Palm [42]
Density at 15 °C	kg/m^3	775–840	788	776	866.3
Kinetic viscosity at −20 °C	cSt	<8	6.52	3.30	-
Flash point	°C	>38	55	48.5	105
Freezing point	°C	<−47	−16	-	−50
Lower heating value	MJ/kg	>42.8	43.5	-	-

It can be seen that the kinematic viscosities and densities of the bio-jet fuel is lower than the maximum value provided by the ASTM D1655 standard (<8 cSt and 775–840 kg/m^3, respectively). The measurement of kinematic viscosity and density of fuel are important and should be within the acceptable limits established in the standard to avoid clogging in the fuel injectors and to achieve good fuel atomization. From the table, it is noticed that the density and kinematic viscosity of the bio-jet fuel in this study is higher compared to soybean, but lower than palm. The lower heating value of the bio-jet fuel in this study is 43.5 MJ/kg, which is well below the limit given in the ASTM D1655 standard, slightly lower than that for diesel (>42.8 MJ/kg). The flashpoint of the bio-jet fuel from FAEE coconut oil is 55 °C, which is significantly higher than the limit given in the ASTM D1655 (>38 °C). The flashpoint is higher than soybean, and much lower than the palm. As shown, bio-jet fuel obtained in the current study does not meet the criteria for the freezing point. A similar result has been reported by [42]. This showed that freezing point is the biggest challenge for bio-jet fuel technologies. This might be because bio-jet fuel consists of nearly no aromatics or cycloalkanes that are required for jet fuel [54]. Freezing point is one of the properties that affects the operability of bio-jet fuel at low temperature. Hence, further upgrading or blending is needed to bring down the freezing point and improve its jet fuel characteristics. Cheng and Brewer [55] suggested that blending of bio-jet fuel with alkyl-benzens might be the best, as it has a low molecular weight and emit less soot combustion as compared to other aromatics compounds. However, the effect of different aromatic compounds on the freezing point of bio-jet fuel has not been explored. As this objective is beyond the scope of this paper, it is suggested as a future work.

4. Conclusions

Other than the depletion issue of non-renewable fossil fuel, the increasing greenhouse gas emission has also driven the aviation industry toward sustainable development, such as exploration and commercialization of alternative renewable aviation fuels. This paper has investigated the effect of three parameters, including coconut oil to ethanol molar ratio, reaction time, and microwave power, on bio-jet fuel production. Bio-jet fuel production data was modeled using RSM and ANN. Statistical analysis proved that the ANN modeling was better than RSM. The optimal parameters predicted using ANN- ACO were, ethanol to oil molar ratio: 1:9.25, reaction time: 12.66 min and microwave power 74.8 W. Optimal yield of 74.45% were obtained experimentally in this optimal parameter conditions. Lastly, it was found that the bio-jet fuel collected in this work had comparable properties with the ASTM D1655, except freezing point. Hence, the aromatic additive is suggested to be added for properties enhancement. Besides, future work on evaluating the effect of catalyst and deoxygenation method on the improvement of bio-jet fuel properties will be conducted.

Author Contributions: Conceptualization, S.N. and M.Y.O.; methodology, M.Y.O., F.K. and R.M.H.R.S.; software, M.Y.O. and F.K.; validation, R.M.H.R.S. and F.K.; resources, F.K.; data curation, R.M.H.R.S.; writing—original draft preparation, M.Y.O.; writing—review and editing, M.Y.O., F.K. and S.N.; supervision, S.N.; funding acquisition, A.H.S. All authors have read and agreed to the published version of the manuscript.

Funding: This research was funded by AAIBE Chair of Renewable Energy (Project code: 202006KET-THA) and BOLD 2020 research grant (grant number: RJO010517844/116) under iRMC, UNITEN.

Institutional Review Board Statement: Not applicable.

Informed Consent Statement: Not applicable.

Data Availability Statement: The data presented in this study are available in Ong, M.Y.; Nomanbhay, S.; Kusumo, F.; Shahruzzaman, R.M.H.R.; Shamsuddin, A.H. Modeling and Optimization of Microwave-Based Bio-Jet Fuel from Coconut Oil: Investigation of Response Surface Methodology (RSM) and Artificial Neural Network Methodology (ANN). *Energies* **2021**, *14*, x. https://doi.org/10.3390/xxxxx.

Acknowledgments: The authors would like to acknowledge UNITEN for the research facilities. M.Y.O would also like to thank UNITEN for the UNITEN Postgraduate Excellence Scholarship 2019 (YCU-COGS).

Conflicts of Interest: The authors declare no conflict of interest.

References

1. Malaysiakini Ten Years Before Known Oil, Gas Reserves Run Dry 2019. Available online: https://www.malaysiakini.com/news/467781 (accessed on 24 June 2020).
2. Silitonga, A.S.S.; Masjuki, H.H.H.; Ong, H.C.; Sebayang, A.H.H.; Dharma, S.; Kusumo, F.; Siswantoro, J.; Milano, J.; Daud, K.; Mahlia, T.M.I.M.I.; et al. Evaluation of the engine performance and exhaust emissions of biodiesel-bioethanol-diesel blends using kernel-based extreme learning machine. *Energy* **2018**, *159*, 1075–1087. [CrossRef]
3. Holtsmark, N.; Agheb, E.; Molinas, M.; Høidalen, H.K. High frequency wind energy conversion system for offshore DC collection grid—Part II: Efficiency improvements. *Sustain. Energy Grids Netw.* **2016**, *5*, 177–185. [CrossRef]
4. Heriche, H.; Rouabah, Z.; Bouarissa, N. High-efficiency CIGS solar cells with optimization of layers thickness and doping. *Opt. Int. J. Light Electron. Opt.* **2016**, *127*, 11751–11757. [CrossRef]
5. Saifuddin, N.; Ong, M.Y.; Priatharsini, P. Optimization of photosynthetic hydrogen gas production by green alga in sulfur deprived condition. *Indian J. Sci. Technol.* **2016**, *9*, 1–13. [CrossRef]
6. Abdul Latif, N.I.S.; Ong, M.Y.; Nomanbhay, S. Hydrothermal liquefaction of Malaysia's algal biomass for high-quality bio-oil production. *Eng. Life Sci.* **2019**, *19*, 246–269. [CrossRef]
7. Mazaheri, H.; Ong, H.C.; Masjuki, H.H.; Amini, Z.; Harrison, M.D.; Wang, C.-T.; Kusumo, F.; Alwi, A. Rice bran oil based biodiesel production using calcium oxide catalyst derived from Chicoreus brunneus shell. *Energy* **2018**, *144*, 10–19. [CrossRef]
8. Mahlia, T.M.I.; Syazmi, Z.A.H.S.; Mofijur, M.; Abas, A.E.P.; Bilad, M.R.; Ong, H.C.; Silitonga, A.S. Patent landscape review on biodiesel production: Technology updates. *Renew. Sustain. Energy Rev.* **2020**, *118*, 109526. [CrossRef]
9. Salman, B.; Ong, M.Y.; Nomanbhay, S.; Salema, A.A.; Sankaran, R.; Show, P.L. Thermal analysis of nigerian oil palm biomass with sachet-water plasticwastes for sustainable production of biofuel. *Processes* **2019**, *7*, 475. [CrossRef]
10. Mohd Mokhta, Z.; Ong, M.Y.; Salman, B.; Nomanbhay, S.; Salleh, S.F.; Chew, K.W.; Show, P.L.; Chen, W.H. Simulation studies on microwave-assisted pyrolysis of biomass for bioenergy production with special attention on waveguide number and location. *Energy* **2020**, *190*, 116474. [CrossRef]
11. Rahman, S.M.A.; Fattah, I.M.R.; Maitra, S.; Mahlia, T.M.I. A ranking scheme for biodiesel underpinned by critical physicochemical properties. *Energy Convers. Manag.* **2021**, *229*, 113742. [CrossRef]
12. Ong, H.C.; Masjuki, H.H.; Mahlia, T.M.I.; Silitonga, A.S.; Chong, W.T.; Leong, K.Y. Optimization of biodiesel production and engine performance from high free fatty acid Calophyllum inophyllum oil in CI diesel engine. *Energy Convers. Manag.* **2014**, *81*, 30–40. [CrossRef]
13. Silitonga, A.S.; Hassan, M.H.; Ong, H.C.; Kusumo, F. Analysis of the performance, emission and combustion characteristics of a turbocharged diesel engine fuelled with Jatropha curcas biodiesel-diesel blends using kernel-based extreme learning machine. *Environ. Sci. Pollut. Res.* **2017**, *24*, 25383–25405. [CrossRef] [PubMed]
14. IEA. *Electricity Information 2019*; OECD: Paris, France, 2019.
15. Ong, H.C.; Milano, J.; Silitonga, A.S.; Hassan, M.H.; Shamsuddin, A.H.; Wang, C.T.; Indra Mahlia, T.M.; Siswantoro, J.; Kusumo, F.; Sutrisno, J. Biodiesel production from Calophyllum inophyllum-Ceiba pentandra oil mixture: Optimization and characterization. *J. Clean. Prod.* **2019**, *219*, 183–198. [CrossRef]

16. Silitonga, A.S.; Shamsuddin, A.H.; Mahlia, T.M.I.; Milano, J.; Kusumo, F.; Siswantoro, J.; Dharma, S.; Sebayang, A.H.; Masjuki, H.H.; Ong, H.C. Biodiesel synthesis from Ceiba pentandra oil by microwave irradiation-assisted transesterification: ELM modeling and optimization. *Renew. Energy* **2020**, *146*, 1278–1291. [CrossRef]

17. Ong, H.C.; Masjuki, H.H.; Mahlia, T.M.I.; Silitonga, A.S.; Chong, W.T.; Yusaf, T. Engine performance and emissions using Jatropha curcas, Ceiba pentandra and Calophyllum inophyllum biodiesel in a CI diesel engine. *Energy* **2014**, *69*, 427–445. [CrossRef]

18. Gholami, A.; Pourfayaz, F.; Maleki, A. Recent Advances of Biodiesel Production Using Ionic Liquids Supported on Nanoporous Materials as Catalysts: A Review. *Front. Energy Res.* **2020**, *8*, 144. [CrossRef]

19. ATAG. *Aviation Benefits beyond Borders*; Air Transport Action Group: Geneva, Switzerland, 2018.

20. Jeff, O. *Fact Sheet: The Growth in Greenhouse Gas Emissions from Commercial Aviation*; Environmental and Energy Study Institute: Washington, DC, USA, 2019.

21. Gutiérrez-Antonio, C.; Gómez-Castro, F.I.; de Lira-Flores, J.A.; Hernández, S. A review on the production processes of renewable jet fuel. *Renew. Sustain. Energy Rev.* **2017**, *79*, 709–729. [CrossRef]

22. Llamas, A.; García-Martínez, M.J.; Al-Lal, A.M.; Canoira, L.; Lapuerta, M. Biokerosene from coconut and palm kernel oils: Production and properties of their blends with fossil kerosene. *Fuel* **2012**, *102*, 483–490. [CrossRef]

23. Wang, W.C.; Tao, L. Bio-jet fuel conversion technologies. *Renew. Sustain. Energy Rev.* **2016**, *53*, 801–822. [CrossRef]

24. Wang, M.; Chen, M.; Fang, Y.; Tan, T. Highly efficient conversion of plant oil to bio-aviation fuel and valuable chemicals by combination of enzymatic transesterification, olefin cross-metathesis, and hydrotreating. *Biotechnol. Biofuels* **2018**, *11*, 30. [CrossRef]

25. Nomanbhay, S.; Ong, M.Y. A Review of Microwave-Assisted Reactions for Biodiesel Production. *Bioengineering* **2017**, *4*, 57. [CrossRef]

26. Yusoff, M.F.M.; Xu, X.; Guo, Z. Comparison of fatty acid methyl and ethyl esters as biodiesel base stock: A review on processing and production requirements. *JAOCS J. Am. Oil Chem. Soc.* **2014**, *91*, 525–531. [CrossRef]

27. Mendow, G.; Veizaga, N.S.; Sánchez, B.S.; Querini, C.A. Biodiesel production by two-stage transesterification with ethanol. *Bioresour. Technol.* **2011**, *102*, 10407–10413. [CrossRef]

28. Kishi, A.; Shironita, S.; Umeda, M. H_2O_2 detection analysis of oxygen reduction reaction on cathode and anode catalysts for polymer electrolyte fuel cells. *J. Power Sources* **2012**, *197*, 88–92. [CrossRef]

29. Encinar, J.M.; González, J.F.; Rodríguez, J.J.; Tejedor, A. Biodiesel fuels from vegetable oils: Transesterification of Cynara cardunculus L. Oils with ethanol. *Energy Fuels* **2002**, *16*, 443–450. [CrossRef]

30. Verma, P.; Sharma, M.P. Comparative analysis of effect of methanol and ethanol on Karanja biodiesel production and its optimisation. *Fuel* **2016**, *180*, 164–174. [CrossRef]

31. Dev, S.R.S.; Gariepy, Y.; Orsat, V.; Raghavan, G.S.V. FDTD modeling and simulation of microwave heating of in-shell eggs. *Prog. Electromagn. Res. M* **2010**, *13*, 229–243. [CrossRef]

32. Ong, M.Y.; Nomanbhay, S. Design and Modeling of an Enhanced Microwave Reactor for Biodiesel Production. *Int. J. Sci. Res. Publ.* **2018**, *8*, 527–534. [CrossRef]

33. Wahidin, S.; Idris, A.; Yusof, N.M.; Kamis, N.H.H.; Shaleh, S.R.M. Optimization of the ionic liquid-microwave assisted one-step biodiesel production process from wet microalgal biomass. *Energy Convers. Manag.* **2018**, *171*, 1397–1404. [CrossRef]

34. Patil, P.D.; Reddy, H.; Muppaneni, T.; Ponnusamy, S.; Cooke, P.; Schuab, T.; Deng, S. Microwave-mediated non-catalytic transesterification of algal biomass under supercritical ethanol conditions. *J. Supercrit. Fluids* **2013**, *79*, 67–72. [CrossRef]

35. Silitonga, A.; Mahlia, T.; Shamsuddin, A.; Ong, H.; Milano, J.; Kusumo, F.; Sebayang, A.; Dharma, S.; Ibrahim, H.; Husin, H.; et al. Optimization of Cerbera manghas Biodiesel Production Using Artificial Neural Networks Integrated with Ant Colony Optimization. *Energies* **2019**, *12*, 3811. [CrossRef]

36. Oloko-Oba, M.I.; Taiwo, A.E.; Ajala, S.O.; Solomon, B.O.; Betiku, E. Performance evaluation of three different-shaped bio-digesters for biogas production and optimization by artificial neural network integrated with genetic algorithm. *Sustain. Energy Technol. Assess.* **2018**, *26*, 116–124. [CrossRef]

37. Ishola, N.B.; Okeleye, A.A.; Osunleke, A.S.; Betiku, E. Process modeling and optimization of sorrel biodiesel synthesis using barium hydroxide as a base heterogeneous catalyst: Appraisal of response surface methodology, neural network and neuro-fuzzy system. *Neural Comput. Appl.* **2019**, *31*, 4929–4943. [CrossRef]

38. Dharma, S.; Masjuki, H.H.; Ong, H.C.; Sebayang, A.H.; Silitonga, A.S.; Kusumo, F.; Mahlia, T.M.I. Optimization of biodiesel production process for mixed Jatropha curcas-Ceiba pentandra biodiesel using response surface methodology. *Energy Convers. Manag.* **2016**, *115*, 178–190. [CrossRef]

39. Kusumo, F.; Mahlia, T.M.I.; Shamsuddin, A.H.; Ong, H.C.; Ahmad, A.R.; Ismail, Z.; Ong, Z.C.; Silitonga, A.S. The effect of multi-walled carbon nanotubes-additive in physicochemical property of rice brand methyl ester: Optimization analysis. *Energies* **2019**, *12*, 3291. [CrossRef]

40. Sebayang, A.H.; Masjuki, H.H.; Ong, H.C.; Dharma, S.; Silitonga, A.S.; Kusumo, F.; Milano, J. Optimization of bioethanol production from sorghum grains using artificial neural networks integrated with ant colony. *Ind. Crops Prod.* **2017**, *97*, 146–155. [CrossRef]

41. Chandra Mohan, B.; Baskaran, R. A survey: Ant Colony Optimization based recent research and implementation on several engineering domain. *Expert Syst. Appl.* **2012**, *39*, 4618–4627. [CrossRef]

42. Truong, N.T.T.; Boontawan, A. Development of Bio-Jet Fuel Production Using Palm Kernel Oil and Ethanol. *Int. J. Chem. Eng. Appl.* **2017**, *8*, 153–161. [CrossRef]

43. Milano, J.; Ong, H.C.; Masjuki, H.H.; Silitonga, A.S.; Kusumo, F.; Dharma, S.; Sebayang, A.H.; Cheah, M.Y.; Wang, C.T. Physicochemical property enhancement of biodiesel synthesis from hybrid feedstocks of waste cooking vegetable oil and Beauty leaf oil through optimized alkaline-catalysed transesterification. *Waste Manag.* **2018**, *80*, 435–449. [CrossRef]

44. Ong, M.Y.; Chew, K.W.; Show, P.L.; Nomanbhay, S. Optimization and kinetic study of non-catalytic transesterification of palm oil under subcritical condition using microwave technology. *Energy Convers. Manag.* **2019**, *196*, 1126–1137. [CrossRef]

45. Silitonga, A.S.; Mahlia, T.M.I.; Kusumo, F.; Dharma, S.; Sebayang, A.H.; Sembiring, R.W.; Shamsuddin, A.H. Intensification of Reutealis trisperma biodiesel production using infrared radiation: Simulation, optimisation and validation. *Renew. Energy* **2019**, *133*, 520–527. [CrossRef]

46. Binnal, P.; Amruth, A.; Basawaraj, M.P.; Chethan, T.S.; Murthy, K.R.S.; Rajashekhara, S. Microwave-assisted esterification and transesterification of dairy scum oil for biodiesel production: Kinetics and optimisation studies. *Indian Chem. Eng.* **2020**. [CrossRef]

47. Encinar, J.M.; González, J.F.; Sabio, E.; Ramiro, M.J. Preparation and properties of biodiesel from Cynara cardunculus L. oil. *Ind. Eng. Chem. Res.* **1999**, *38*, 2927–2931. [CrossRef]

48. Lee, J.S.; Saka, S. Biodiesel production by heterogeneous catalysts and supercritical technologies. *Bioresour. Technol.* **2010**, *101*, 7191–7200. [CrossRef]

49. Saifuddin, A.; Chua, K.H. Production of Ethyl Ester (Biodiesel) from Used Frying Oil: Optimization of Transesterification Process Using Microwave Irradiation. *MJ. Chem.* **2004**, *6*, 77–82.

50. Oyerinde, A.; Bello, E. Use of Fourier Transformation Infrared (FTIR) Spectroscopy for Analysis of Functional Groups in Peanut Oil Biodiesel and Its Blends. *Br. J. Appl. Sci. Technol.* **2016**, *13*, 1–14. [CrossRef]

51. Jimoh, A.; Abdulkareem, A.S.; Afolabi, A.S.; Odigure, J.O.; Odili, U.C. Production and Characterization of Biofuel from Refined Groundnut Oil. In *Energy Conservation*; InTech: London, UK, 2012; pp. 10–12.

52. Saifuddin, N.; Refal, H. Spectroscopic Analysis of Structural Transformation in Biodiesel Degradation. *Res. J. Appl. Sci. Eng. Technol.* **2014**, *8*, 1149–1159. [CrossRef]

53. Rabaev, M.; Landau, M.V.; Vidruk-Nehemya, R.; Koukouliev, V.; Zarchin, R.; Herskowitz, M. Conversion of vegetable oils on Pt/Al2O3/SAPO-11 to diesel and jet fuels containing aromatics. *Fuel* **2015**, *161*, 287–294. [CrossRef]

54. Amanda, B. *The Future of Advanced Bio-Jet Fuel*; Linköping University: Linköping, Sweden, 2017.

55. Cheng, F.; Brewer, C.E. Producing jet fuel from biomass lignin: Potential pathways to alkyl-benzenes and cycloalkanes. *Renew. Sustain. Energy Rev.* **2017**, *72*, 673–722. [CrossRef]

Article

Weather-Driven Scenario Analysis for Decommissioning Coal Power Plants in High PV Penetration Grids

Samuel Matthew G. Dumlao * and Keiichi N. Ishihara

Department of Socio-Environmental Energy Science, Graduate School of Energy Science, Kyoto University, Yoshidahonmachi, Sakyo-ku, Kyoto 606-8501, Japan; ishihara@energy.kyoto-u.ac.jp

* Correspondence: dumlao.samuelmatthew.k34@kyoto-u.jp

Abstract: Despite coal being one of the major contributors of CO_2, it remains a cheap and stable source of electricity. However, several countries have turned to solar energy in their goal to "green" their energy generation. Solar energy has the potential to displace coal with support from natural gas. In this study, an hourly power flow analysis was conducted to understand the potential, limitations, and implications of using solar energy as a driver for decommissioning coal power plants. To ensure the results' robustness, the study presents a straightforward weather-driven scenario analysis that utilizes historical weather and electricity demand to generate representative scenarios. This approach was tested in Japan's southernmost region, since it represents a regional grid with high PV penetration and a fleet of coal plants older than 40 years. The results revealed that solar power could decommission 3.5 GW of the 7 GW coal capacity in Kyushu. It was discovered that beyond 12 GW, solar power could not reduce the minimum coal capacity, but it could still reduce coal generation. By increasing the solar capacity from 10 GW to 20 GW and the LNG quota from 10 TWh to 28 TWh, solar and LNG electricty generation could reduce the emissions by 37%, but the cost will increase by 5.6%. Results also show various ways to reduce emissions, making the balance between cost and CO_2 a policy decision. The results emphasized that investing in solar power alone will not be enough, and another source of energy is necessary, especially for summer and winter. The weather-driven approach highlighted the importance of weather in the analysis, as it affected the results to varying degrees. The approach, with minor changes, could easily be replicated in other nations or regions provided that historical hourly temperature, irradiance, and demand data are available.

Keywords: scenario analysis; scenario generation; weather influence; coal decommissioning; high PV penetration; energy balance; CO_2 reduction

Citation: Dumlao, S.M.G.; Ishihara, K.N. Weather-Driven Scenario Analysis for Decommissioning Coal Power Plants in High PV Penetration Grids. *Energies* **2021**, *14*, 2389. https://doi.org/10.3390/en14092389

Academic Editors: T M Indra Mahlia and Islam Md Rizwanul Fattah

Received: 29 March 2021
Accepted: 19 April 2021
Published: 22 April 2021

1. Introduction

In 2013, signatories to the Paris Agreement committed to submit a national climate plan to mitigate climate change by reducing greenhouse gas emissions. Subsequently, one of the United Nations' Sustainable Development Goals, established in 2015, is focused on affordable and clean energy. These two global initiatives have motivated several nations to promote renewable energy sources such as wind, solar, and biomass into their energy mix. As a result, several "green energy transition" initiatives are ongoing in countries such as Germany and Denmark, and subnational jurisdictions such as California, Scotland, and South Australia [1]. Besides these major players, more than 150 countries have national targets for renewable energy in the power sector [2].

The Japanese government recently reiterated its commitment to the projected energy mix for 2030, where fossil fuel-based generation will be reduced to 46%, and renewable energy will comprise 22–24%, of which solar energy will have a 7% contribution [3]. There was a recent influx of solar PV installation mainly driven by the FIT program. The Kyushu region, located on Japan's western tip, is one of the country's leading regions in solar PV generation. Relative to the rest of the country, the region has higher solar power potential

103

and cheaper land, which has driven solar power investments. As of early 2021, the region has a total installed capacity of 10 GW, and additional plants, which will increase this capacity further to roughly 16 GW [4] by around 2027, are already approved. The share of solar PV generation has been steadily increasing in the region. In 2017, 2018, and 2019, solar PV generation accounted for 8.5%, 9.2%, and 10.1% of the total yearly generation, respectively. The International Energy Agency (IEA) classifies the impact of variable renewable energy (VRE) on the energy system's operation into four phases. Japan, as a country, is already in phase 2 where there is a minor to moderate impact on the system operation, whereas Kyushu, as a region, was categorized as phase 3, where VRE determines the operation pattern of the system [5]. This further shows that Kyushu is leading the country in terms of solar PV penetration and is already facing issues ahead of the rest of the country. Kyushu's situation lends itself as a viable case study in exploring the potential impact of solar energy in reducing CO_2 emissions by replacing traditional energy sources with solar energy.

Coal remains to be the cheapest and most economically stable source of electricity for many countries. However, it is also one of the major contributors of CO_2, which leads to global warming. Among the G7 countries, Germany (by 2038 [6]), France (by 2023 [7]), the United Kingdom (by 2024 [7]), Italy (by 2025 [7]), and Canada (by 2030 [8]) have presented their coal phase-out plans. Other European Union member countries have also developed their phase-out plans within the next two decades, and Austria and Belgium have already phased-out their coal plants [7]. Nonetheless, removing coal is a significant roadblock to the green energy transition in many countries, and as countries install increasing amounts of renewable energy, it might be time to consider reducing coal in the energy mix. Solar photovoltaics (PV) can be a green alternative to coal. However, the generation profile of solar energy is different from that of coal, which complicates the process of replacing coal with solar energy. Simultaneously, the variability of solar power requires another flexible source. Liquefied petroleum gas (LNG), given its flexibility, is often used to balance the VRE. Given these intertwined variables, it is necessary to understand the potential, limitations, and implications of using solar energy to replace coal, which are currently unclear.

Many countries see LNG as a bridge to a clean energy future that will pave the way for less coal in the energy mix [9]. It is still a fossil fuel, but it produces less CO_2, which is acceptable for now until a superior technology is available. Due to many countries' tendency to rely on LNG to reduce their CO_2 emissions, the demand for LNG has steadily been increasing, which threatens its supply and price. Shell reported in their LNG Outlook 2020 that global demand for LNG increased by 12.5% to 359 million tons in 2019, which they attributed mainly to the role of LNG in the low-carbon energy transition [10]. It has been reported that the price of LNG increased in October 2020 in anticipation of a colder winter in East Asia [11]. This shows the volatility of LNG's supply and price on the global market, which presents another factor for consideration in the analysis, since solar energy production needs LNG to a certain extent.

Aside from the potential CO_2 reduction benefits, reducing coal capacity could also reduce solar curtailment experienced by grids with high PV penetration. Kyushu started to suffer from curtailment in October 2018, which was explored in a previous study [12]. Several studies have also explored this recent issue in Kyushu. Bunodiere and Lee [13] explored several scenarios to mitigate solar curtailment in Kyushu using a logic-based forecasting method and concluded that reducing the region's nuclear capacity will reduce curtailment. However, in their approach, they considered coal and LNG as thermal generators as a whole due to data limitations. A coal station behaves like a nuclear plant, since these two technologies are considered baseload generators. By treating coal as separate from LNG and as a baseload generator, it could also be said that coal could reduce curtailment. Although Japan initially used their pump hydro energy storage (PHES) to improve the flexibility of nuclear power plants [14], it is now used to store excess solar electricity generation. Li et al. [15] conducted a techno-economic assessment of large-scale

PV integration with PHES and concluded that the PHES could effectively absorb some of the surplus PV production and could maintain low generation cost by using the surplus production. Since the available data regarding power generation in the region aggregate coal and LNG together, the understanding of coal generation in the energy balance is limited.

In order to fully understand the optimal conditions for coal, solar, and LNG production, it is necessary to conduct a power flow analysis to evaluate the impact of investing in more solar PV for driving coal decommissioning. This analysis will provide additional information about the energy balance, including information about solar power generation and curtailment, which are difficult to estimate. By gathering the generators' capacity and generation profiles and the demand profiles, the optimization can calculate the hourly energy balance and minimize the necessary coal capacity and generation. This insight provides the necessary understanding of the potential and limitations of solar energy in regard to replacing coal. However, to ensure the robustness of the analysis and the recommendation, the demand and solar power generation's stochasticity must be considered. It will be challenging to recommission a decommissioned plant due to an unforeseen circumstance; thus, careful analysis is necessary to account for potential variations.

Replacing part of coal's electricity production with solar electricity production, coupled with LNG electricity production, is a subset of the generation expansion planning (GEP) problem. Koltsaklis and Dagoumas [16] wrote a review article exploring the state-of-the-art generation expansion planning where they listed seven challenges to the GEP problem. One of the mentioned challenges is rooted in the risks involved in GEP. They enumerated several potential sources of risks and categorized them according to economic, political, regulatory, environmental, technical, social, and climate categories. Ioannou et al. [17] reviewed the risk-based methods for sustainable energy system planning and categorized the risks in the same manner. They identified seven risk-based methods: mean-variance portfolio theory, real option analysis, Monte Carlo simulation, stochastic optimization technique, multi-criteria decision analysis, and scenario analysis.

Santos et al. [18] conducted a study to identify uncertainties in the electricity system and demonstrated the corresponding impacts on the energy mix through scenario analysis. Their results highlighted that climate uncertainty represents primary risk sources for VRE, since it dictates the system's power generation. A review on the energy sector vulnerability to climate change [19] summarizes the authors' contributions on climate and energy, and they noted that climate change could affect variables that influence electricity generation from photovoltaics and concentrated solar power. The review highlighted that global solar radiation has increased in southeastern Europe [20] and decreased in Canada [21]. They also highlighted that power output calculations should account for air temperature, since it impacts the solar cell's efficiency [19].

Ioannou et al. [17] noted that energy planning has extensively used stochastic optimization techniques, and the stakeholder's motivation mainly drives the constraints. They also mentioned that the Monte Carlo simulation has many advantages, but it requires considerable data inputs to create probability density functions. Alternatively, scenario analysis evaluates the risks by creating potential future developments that range from the worst-case to the best-case scenario, which could then cover all the possible risks in the analysis. As highlighted by several authors [18–21], climate, and by extension weather, must be considered in modeling solar energy generation. Factors such as the changing solar irradiance and ambient temperature could influence solar panels' variability and efficiency.

By carefully identifying the test cases, scenario analysis is sufficient for ensuring the robustness of the analysis. The initial problem is then rooted in creating the scenarios representing the worst case, the best case, and the cases in between. The weather data analysis can provide the representative years that fit the scenario targets, such as warm summers, colder winters, extreme summers, and extreme winters. Although such data are limited, datasets could be synthesized based on the historical relationship between temperature and demand. Solar generation could be calculated from the irradiance and ambient temperature data. The robustness of the analysis and recommendation can be addressed by combining scenario analysis and past weather data.

Therefore, in this study, an hourly power flow analysis was conducted to understand the potential, limitations, and implications of using solar energy as a driver for decommissioning coal power plants. Understanding these factors can provide the necessary recommendations and precautions for energy planners. Since LNG scarcity is anticipated, LNG quota is one of the primary constraints. In order to ensure the robustness of the results, this study presents a straightforward weather-driven scenario generation that utilizes historical weather and electricity demand data processed through machine learning algorithms to generate scenarios that account for weather variations. Through the weather-driven approach, the study aims to reveal the impact of yearly variations in the factors that must be considered for long-term planning that reduce CO_2 emissions while ensuring grid reliability. The Kyushu region in Japan was used as a case study since (a) it is continuously increasing its solar capacity, (b) it has a fleet of coal power plants older than 40 years old ready for decommissioning, and (c) it has enough LNG power plant capacity to support the initial transition.

The code for the weather-driven approach used in this study is available through a public GitHub repository [22], where most of the code and diagrams used in this paper are documented in jupyter notebooks. The approach, with minor changes, could easily be replicated in other nations or regions provided that historical hourly temperature, irradiance, and demand data are available.

Section 2 discusses the methodology for the weather-driven approach, including data and data processing, weather-based data generation, and hourly simulation. The results are then presented in Section 3, and the implications are discussed in Section 4. Finally, the conclusions are drawn in Section 5.

2. Methodology

The overview of the proposed weather-driven approach can be seen in Figure 1, where it is divided into four stages. First, data were collected from Kyushu Electric Power Company (KyEPCO) [4] and Japan Meteorological Agency (JMA) [23] and were pre-processed to fit the intended applications. The weather-based data generation has three components. A weather selection metric was designed based on comfort-levels to identify the years that could represent the scenarios in the region. The *pvlib* Python library [24] was used to calculate the photovoltaic systems' generation under various weather conditions. A demand fingerprint was developed to generate synthetic demand for the selected years. These synthetic data were then used as input to the hourly power flow optimization done in Python for Power System Analysis (PyPSA) [25]. Finally, the simulation results were analyzed.

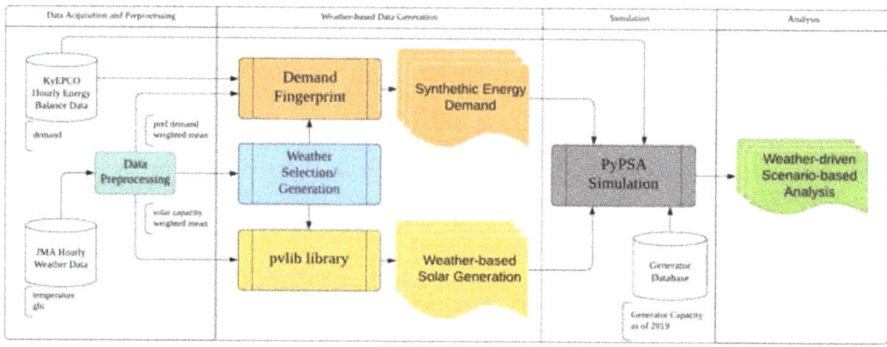

Figure 1. The proposed weather-driven scenario-based analysis approach capable of handling weather-related variations in electricity demand and solar energy production.

2.1. Data and Data Pre-Processing

2.1.1. Energy Demand

The energy data were collected from Kyushu Electric Company [4], where the hourly information about generation, transmission, and demand is published since April 2016. The data also include curtailment information for both solar and wind power. Transmission and pump hydro energy storage (PHES) could be positive or negative. For transmission, negative values represent energy export while positive values indicate electricity import. For PHES, negative and positive values represent the charging and generation phases, respectively. Although the data until December 2020 are already published, only data until March 2019 were used in the study since this represents four full fiscal years.

As seen in Figure 1, the energy data are used in both the demand fingerprint and simulation phase. Only the demand data were necessary for the demand fingerprint, while the other hourly data were used as parameters for the other generations and PHES.

2.1.2. Temperature and Irradiance

The temperature and irradiance data were collected from the Japan Meteorological Agency (JMA) [23], where the hourly weather data are published since 1946. For this study, 30 years of data were collected from 1990 to 2019 to serve as reference weather scenarios. A representative temperature was collected from each of the major cities' weather stations, as shown in Table 1.

Table 1. Weather stations in Kyushu.

Prefecture	Prefecture No.	Precinct Code	Block Code
Fukuoka	40	82	47,807
Saga	41	85	47,813
Nagasaki	42	84	47,817
Kumamoto	43	86	47,819
Oita	44	83	47,815
Miyazaki	45	87	47,830
Kagoshima	46	88	47,827

In order to represent the mean temperature and mean irradiance in the region, solar-capacity-weighted mean and monthly-demand-weighted mean were used for the solar generation calculation and demand generation, respectively. Using the consolidated data from [26], Table 2 shows the shares of the solar PV installation in Kyushu since 2012 and the shares in 2019 were used as the reference for the solar-capacity-weighted mean temperature and irradiance. The Ministry of Economy, Trade, and Industry (METI) publishes each prefecture's monthly energy demand since April 2016 [27]. Table 3 shows the mean of each prefecture's shares from 2016 until 2019. These values were used to calculate the monthly-demand-weighted temperature mean used in the demand fingerprint.

The solar capacity ratio was used for the solar power generation calculation because the power generation's distribution is influenced by the distribution of the capacity. It is necessary to use this weighted mean because the temperature where more solar panels are installed should have a more significant representation in the temperature used in the solar generation calculation. However, the temperature where there is greater demand should have more influence on the temperature used for demand calculations.

Table 2. Share of solar power installations (%) in the prefectures.

Prefecture	2012	2013	2014	2015	2016	2017	2018	2019
Fukuoka	34.83	24.98	25.16	25.14	24.62	23.86	23.06	22.26
Saga	8.48	8.04	7.22	6.96	6.98	6.78	6.85	6.50
Nagasaki	8.98	9.67	9.11	9.82	10.02	9.66	9.46	9.48
Kumamoto	15.87	13.31	14.52	14.26	14.50	14.46	14.75	14.63
Oita	11.40	15.07	13.26	13.08	12.54	12.00	12.51	12.34
Miyazaki	8.66	11.59	11.94	11.78	11.66	12.53	12.82	13.03
Kagoshima	11.77	17.34	18.79	18.96	19.68	20.72	20.55	21.76

Table 3. Average monthly demand share (%) in Kyushu from 2016 to 2019.

Prefecture	JAN	FEB	MAR	APR	MAY	JUN	JUL	AUG	SEP	OCT	NOV	DEC
Fukuoka	38.25	38.28	38.13	37.66	37.32	37.26	37.47	37.60	37.21	36.85	37.20	37.79
Saga	7.83	7.99	8.10	8.02	8.02	8.20	8.12	7.79	7.86	7.90	8.04	7.94
Nagasaki	9.55	9.51	9.35	9.42	9.40	9.25	9.28	9.60	9.48	9.23	9.22	9.39
Kumamoto	13.76	13.82	13.73	13.42	13.37	13.64	13.78	13.83	13.91	13.81	13.72	13.76
Oita	10.61	10.64	10.72	10.97	11.18	10.92	10.58	10.27	10.38	10.81	10.93	10.94
Miyazaki	8.49	8.27	8.45	8.75	8.77	8.72	8.63	8.54	8.54	8.81	8.73	8.55
Kagoshima	11.50	11.49	11.52	11.77	11.94	12.00	12.13	12.37	12.62	12.60	12.16	11.63

2.2. Weather-Based Data Generation

2.2.1. Representative Weather Selection

Since the scenarios are weather-driven, it is necessary to identify the years representing the various scenarios in the region. With this in mind, a metric system was created based on the notion that comfortable temperature levels are between 18 and 22 °C. Under 18 °C, people will start using their heaters, and above 22 °C, they will start using their air-conditioners.

Therefore, the metric system considers the mean and peak deviations per month from these values. Since the focus is to cover extreme cases, the years were ranked based on the summer-warmness and winter-coldness. Based on the rankings, six years were selected, as seen in Table 4. The frequency of occurrence was calculated based on the highest R^2 compared to the representative year from 1990 to 2019. The legends are used mainly in the figures in the results section.

Table 4. Representative years of different weather scenarios in Kyushu.

Summer	Winter	Representative	Legend	Frequency *	Comment
Mild	Mild	2014	14MM	2	Low variability
Mild	Severe	1991	91MS	6	Colder year
Severe	Mild	2016	16SM	8	Warmer year
Severe	Severe	2018	18SS	5	High variability
Extreme	-	2013	13E-	5	Extreme Summer
-	Extreme	2012	12-E	4	Extreme Winter

* Occurrence in the past 30 years from 1990 to 2019.

It can be seen in Figure 2 that around August, 2013 has the highest peak and mean positive temperature deviation, while around February, 2012 has the highest peak and mean negative temperature deviation as intended by the sampling. 2014 has the lowest deviation overall since it is the lowest in summer and in the winter. The other representative years fall between these extreme cases.

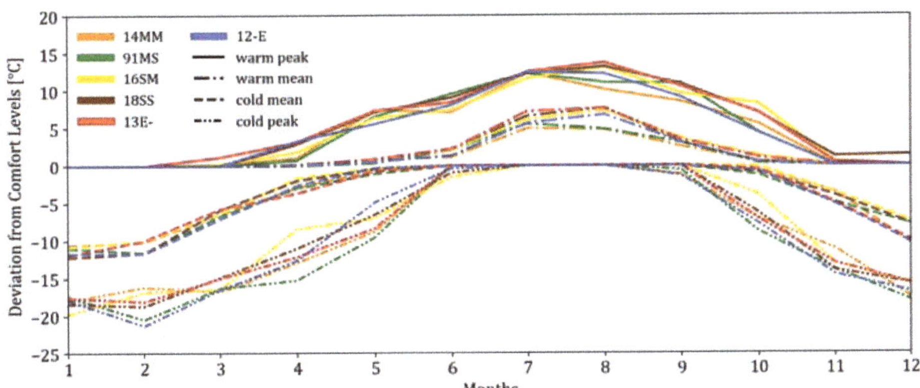

Figure 2. Temperature deviation of the representative year from the comfortable temperature of 18 °C to 22 °C.

2.2.2. Weather-Based Solar Generation

The weather-based solar power generation calculation was mainly based on the TMY to power tutorial written by the developers of *pvlib* [28]. Since the approach requires both the direct-normal irradiance (DNI) and diffuse horizontal irradiance (DHI), and JMA only provides the global horizontal irradiance (GHI), the built-in function *pvlib.irradiance.erbs* was used to estimate the DNI and DHI. The Erbs model [29] estimates the diffused fraction of GHI to calculate DHI and uses the solar zenith to calculate DNI. By providing the timezone, longitude, latitude, and altitude data along with the hourly GHI data from JMA, the DNI and DHI were calculated using *pvlib*'s built-in functions. Besides the irradiance data, the power generation calculation also requires temperature data to account for the impact of temperature on solar cells' efficiency. The solar-capacity-weighted mean was used for both the GHI and temperature since the solar power generation distribution is proportional to the generation capacity of each prefecture. Subsequently, the power generation values of a 208 W Kyocera Solar Panel and an ABB Micro 250 W micro-inverter were calculated using *pvlib.pvsystem.sapm* and *pvlib.inverter.sandia*. The resulting hourly annual generation was scaled by 208 W to represent the maximum power output for PyPSA.

2.2.3. Demand Fingerprint and Synthetic Demand Generation

Figure 3 shows the creation of the four parameters necessary to generate energy demand. The first two swim lanes in the flowchart show the identification of the cluster. Initially, the goal was to extract a demand shape or fingerprint from the data. Since human behavior through weekly routines greatly influences the demand, the energy demand data were split into weekly samples. The peaks and troughs of the demand patterns were identified to emphasize the demand's fingerprint. This process entailed moving the peaks and troughs to the following hourly locations: 0, 3, 6, 15, 18, 21, and 24, since it was discovered that the peaks and troughs occur at these times depending on the season. Actual values were selected for 9, 11, 12, and 13, since these values represent the midday demand dip that occurs due to the Japanese lunch hour, which is noticeable yearlong. After the alignment, since the goal was to extract the fingerprint, the demand's magnitude was scaled using z-transform. The resulting scaled value was then used as input to an FFT transform to extract the frequency components that comprise the fingerprint. Only the daily variations (multiples of 7 Hz) were selected as features for the clustering algorithm to reduce the noise. These feature values were scaled using z-transform to reduce the magnitude in the distance calculation.

These features were then clustered using the Kmeans algorithm through the *sklearn.cluster. Kmeans* method of the *sklearn* Python library. Several values of K were explored, and through experimentation K = 5 was identified as the number of clusters that could explain the data. The clusters' fingerprints are shown in the inset of Figure 4. Once these clusters were identified, the weekday datasets were combined for each cluster, and another FFT transform was done to extract the Fourier parameters necessary to represent the waveform. Combining the datasets, after clustering, emphasized the pattern for each cluster and removed the noise. As with the pre-clustering data, only the daily variations (different frequencies depending on the sample sizes) were selected.

A classifier must be developed to identify the appropriate fingerprint for each week during the energy demand generation. Through data exploration, the maximum weekly temperature, minimum weekly temperature, and the month of the week were identified as the features that could be used to classify each week. The *sklearn.neighbors.KneighborsClassifier* method of *sklearn* Python library was used with k = 5 to classify the weekly data. By running 80–20 training-test split 1000 times, the classification got an average accuracy score of 83%, with 66% and 95% as the minimum and maximum scores, respectively. This average accuracy score was deemed acceptable and this was used as the fingerprint classifier.

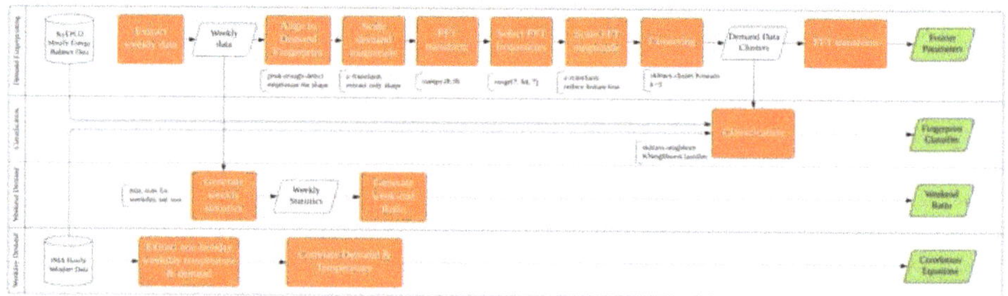

Figure 3. Generating the demand fingerprint based on the demand and temperature data.

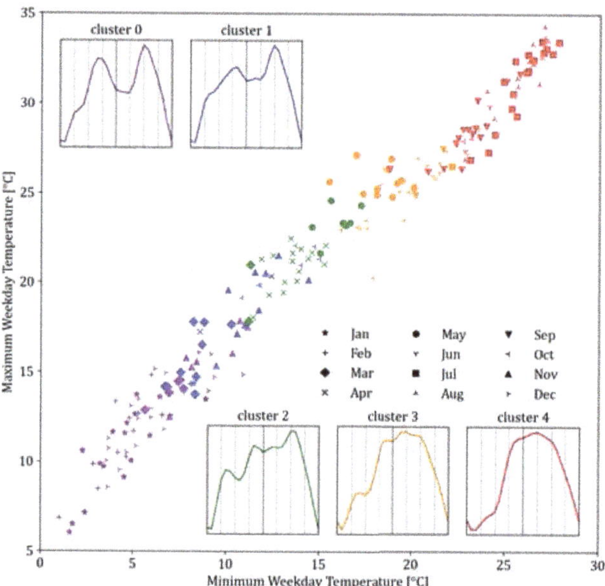

Figure 4. The weekly demand clusters of Kyushu from FY2016–2019.

While the first two swim lanes in Figure 3 provided the demand's fingerprint, the last two provide the minimum and maximum values that stretch or compress the fingerprint.

Through data exploration, it was observed that non-holiday weekday temperature and demand have a strong correlation; thus, it was extracted and fitted into known functions. Using *scipy.optimize.curve_fit* method of the *scipy* Python library, as seen in Figure 5, the minimum temperature and minimum demand were fitted to a quadratic curve with an R^2 of 0.80. The curve fitting for the maximum temperature and maximum demand required a piece-wise linear equation and was similarly fitted with an R^2 of 0.88. The weekend and holiday fitting were explored, but no meaningful functions were derived; thus, a simple weekday-to-weekend ratio was extracted by averaging all the known values. Seasonal variations in the ratio were initially explored, but no meaningful trend was seen; thus, the concept was dropped.

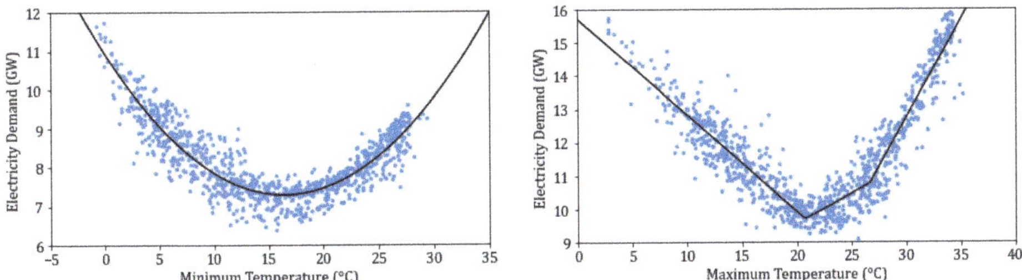

Figure 5. Correlation of Temperature and Demand in Kyushu.

Generating yearly demand based on temperature is shown in Figure 6 where the green input blocks represent the models, values, and functions generated from Figure 3, and the red input block represents the weekly temperature statistics from the selected year. Using these inputs, a fingerprint assignment and the minimum and maximum demand per day were identified. The fingerprint is then fitted to the daily min-max demand using *scipy.optimize.curve_fit* method and provides the A_0 and B_0 coefficient for the Fourier representation. This is done for all weeks of the year to generate the entire year. Testing this approach with the known values for 2017, 2018, and 2019, the synthetic demand approach could get R^2 of 0.8675, 0.8714, and 0.8177, respectively. A sample of the demand curve can be seen in Figure 7, where the demand were closely synthesized. The problem with holidays (e.g., new year) is noticeable and some weekends are not reproduced accurately. However, the general shape or *fingerprint* of the demand fits well with the actual values.

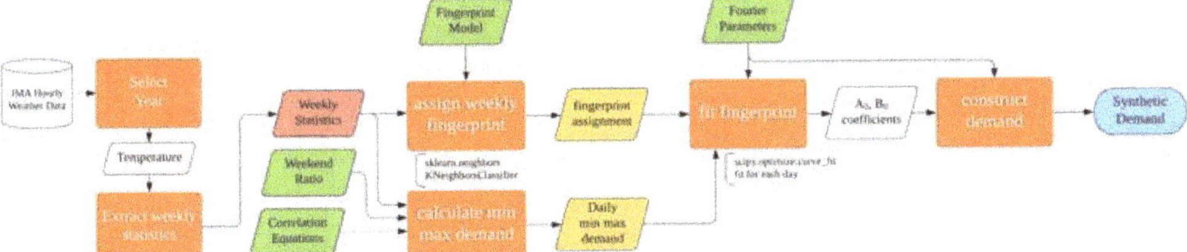

Figure 6. Temperature-dependent demand generation.

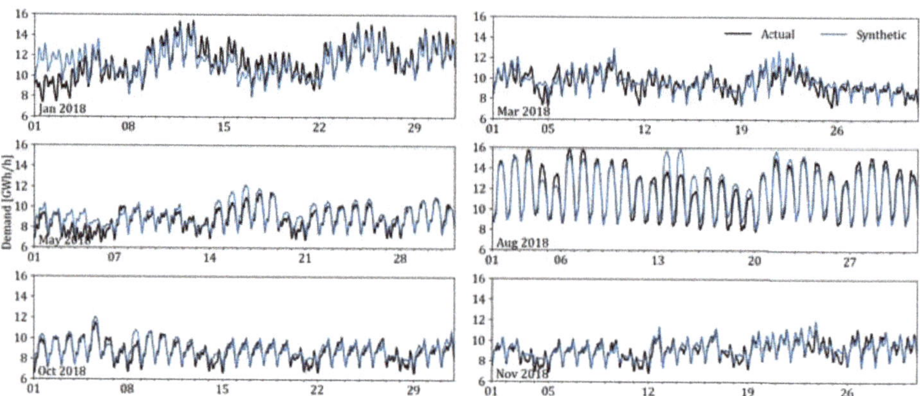

Figure 7. Sample demand synthesis for 2018.

2.3. Hourly Simulation and Scenario Analysis

The hourly simulation used Python for the Power System Analysis (PyPSA) Modeling Framework [25]. The PyPSA environment provides a framework for the buses, lines, loads, generators, and storage, units among many other parameters. In this simulation, since Kyushu was modeled as a single point, only one bus was used, and all the loads, generators, and storage units were connected directly to this bus. The single-bus network was constructed based on Figure 8, where the coal generator, the Kyushu demand, and the transmission demand were directly connected to a single bus. The rest of the generators was then connected to a sub-bus, which was then connected to the main bus. Creating the sub-bus ensured that the PHES could not charge from the coal generator, which prevented the optimizer from generating and storing more power from coal for later use. For this study, the synthetic load and solar power generation profile for the representative years (Table 4) were used iteratively during the optimization.

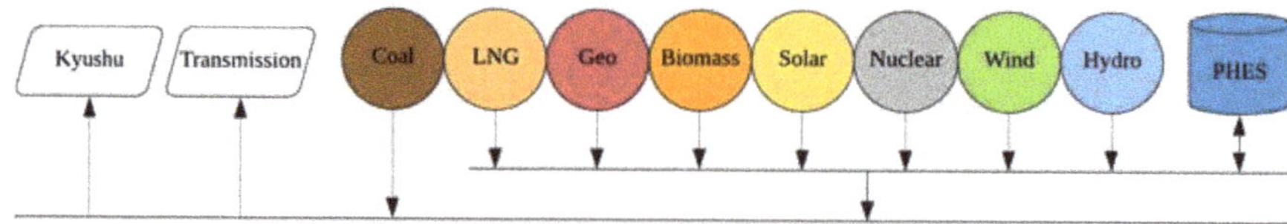

Figure 8. Configuration of the single-bus network used in the optimization.

The installed solar capacity was increased by 1 GW increments from 0 GW until 20 GW. The latest known capacities for the other generators as of FY2019 are shown in Table 5, which was consolidated based on various sources [27,30,31]. Although the nuclear, geothermal, and biomass could change within the year, as a baseload, they were fixed to their respective maximum capacities to provide consistency throughout the years under simulation. Hydropower generation was based on the daily dispatch capacity calculated using the total daily dispatch in the 2019 data. The simulator allocated the hourly dispatch based on the optimization. However, minimum and maximum dispatch were still considered based on the actual data. The PHES was treated as both a generator and a load with a maximum transfer capacity of 2.3 GW, a total capacity of 13.8 GWh, and round trip efficiency of 0.70% (0.84% one way).

Table 5. Generators in Kyushu as of FY2019.

Generator	Power (MW)	Carrier	Output$_{min}$ (%)	Ramp Limit (%)
Coal	7037	Coal	30	1
LNG	5250	Gas	15	40
Geothermal	160	Renewable	100	0
Biomass	450	Renewable	100	0
Solar	9000	Renewable	0	100
Nuclear	4140	Non-GHG	100	0
Wind	355	Renewable	15	40
Hydro	4000	Renewable	15	40

The optimization aims to minimize the coal capacity while ensuring energy balance. The hourly resolution was used due to the limitation of the available data. Solar energy is preferred as long as the minimum operating output or ramp limit seen in Table 5 for coal and LNG are satisfied. Since LNG might become a future bottleneck, LNG quota (in TWh) is used as a constraint in the simulation. LNG quota (LNG_{quota}^{TWh}) is defined as the maximum total annual electricity generation used in the optimization with the maximum generation capacity of 5.25 GW.

Using the recently published resource utilization data from KyEPCO [32], it was determined that the company generated 8 TWh from LNG in 2019 through their 4.625 GW LNG plants. This generation represents 20% LF for the company. An independent power producer owns the other 0.625 GW LNG power plants in the region, which are composed of mixed gas power generators. Assuming these IPP plants are running at a higher LF of 40%, it was determined that the LNG plants generated around 10 TWh in 2019. Using 10 TWh as the base case, the simulation gradually incremented the LNG quota by 20%, 60%, and 100%, yielding 12, 16, 20 TWh LNG quota. A report from Japan's Ministry of Economy, Trade, and Industry (METI) [33] showed that LNG is more economical than coal at LF < 60%; thus, the scenario analysis also explored 28 TWh (60% LF). Preliminary exploration also showed that increasing the quota further has a minor impact on the emission and cost unless more LNG capacity is installed; thus, this was not explored any further. Additional LNG quota could also increase energy security risk given the LNG market situation.

A summary of the LNG quota scenarios is shown in Table 6. The scenarios (LNG_{quota2}^{TWh}–LNG_{quota4}^{TWh}) reflects one way to reach each of the identified quota from the base case (LNG_{quota1}^{TWh}). For the 12, 16, and 20 TWh scenarios, KyEPCO could increase their efficient power plants' LF. Increasing it further would require KyEPCO to increase their steam LNG plants' LF plants and coordinate with the IPP to increase their production.

Table 6. LNG quota scenarios.

LNG Power Plant		LNG_{quota1}^{TWh}		LNG_{quota2}^{TWh}		LNG_{quota3}^{TWh}		LNG_{quota4}^{TWh}		LNG_{quota5}^{TWh}	
Category	Cap	LF	Gen	LF	Gen	LF	Gen	LF	Gen	LF	Gen
KyEPCO Steam	1800	20	3.15	20	3.15	20	3.15	20	3.15	60	9.46
KyEPCO CC	2825	20	4.95	27	6.68	44	10.89	60	14.85	62	15.34
IPP	625	40	2.19	40	2.19	40	2.19	40	2.19	60	3.29
Total	5250		10.29		12.03		16.23		20.19		28.09

Capacity (Cap) in MW; load factor (LF) in %; generation (Gen) in TWh.

2.4. Annual Generation Cost and CO$_2$ Emission Analysis

In 2015, Japan's Ministry of Economy, Trade, and Industry (METI) [34] reported and modeled the cost of electricity generation for 2014 and 2030. An Advisory Panel to the Foreign Minister on Climate Change (MOFA) [35] citing BloombergNEF presented their estimates on the cost of generation in 2018. Table 7 consolidates these reports along with the values used for the annual cost calculations. Generally, the cost in 2014 was used in the calculations except for wind and solar power, where it was averaged between the 2014 report and the 2030 model. Except for coal, the values are near the estimated values of BloombergNEF.

Table 7. Cost of electricity generation (JPY/kWh).

Technology	METI 2014	METI 2030	MOFA 2018 *	Applied **
Nuclear	10.1	10.1	-	10.1
Coal	12.3	12.9	6	12.3
LNG	13.7	13.4	10	13.7
Wind	21.9	13.9	10–22 (15)	17.9
Geothermal	19.2	19.2	-	19.2
Hydro	11.0	11.0	-	11.0
Biomass	12.6	13.3	-	12.6
Solar (Comm)	24.3	12.7	8–36 (17)	18.5
Solar (Home)	29.4	12.5	-	-

* Values in parentheses are the average values; ** used in the calculation; as of April 2021: 100 JPY = 0.92 USD = 0.77 EUR.

For the CO_2 emission analysis, the study mainly focuses on the CO_2 emission from fuel consumption, which does not cover the CO_2 emission during construction, maintenance, and disposal of the system. Therefore, the calculation assumes that, during generation, nuclear, geothermal, hydro, solar, and wind power do not generate CO_2 and biomass has net-zero CO_2 emissions. According to Japan's Ministry of Environment [36], depending on the technology, coal and LNG has a CO_2 emission of 0.95 kgCO2/kWh to 0.83 kgCO2/kWh and 0.51 kgCO2/kWh to 0.36 kgCO2/kWh, respectively. The average emission for coal (0.89 kgCO2/kWh) and LNG (0.44 kgCO2/kWh) were used in the analysis.

Since temperature leads to higher or lower demands, by calculating the levelized cost of generation and levelized CO_2 emissions, the relationship between cost and CO_2 becomes clearer. Although weather variations still have an impact, this impact is less when seen from a levelized perspective. The annual levelized cost of generation was calculated using Equation (1). The hourly simulation provides the annual generation per technology ($Generation_{tech}^{kWh}$). By multiplying the generation per technology to the corresponding cost of electricity generation ($Cost_{tech}^{JPY/kWh}$), the total cost per year could be calculated. The levelized cost of generation on that particular year can then be calculated by dividing the total annual cost by the total annual generation. Similarly, the levelized CO_2 emission was calculated Equation using (2) and the CO_2 emission per technology ($Emission_{tech}^{kG-CO2_2/kWh}$).

$$levelized\ cost\ of\ generation = \frac{\sum (Generation_{tech}^{kWh})(Cost_{tech}^{JPY/kWh})}{\sum Generation_{tech}^{kWh}} \qquad (1)$$

$$levelized\ CO_2\ emissions = \frac{\sum (Generation_{tech}^{kWh})(Emission_{tech}^{kG-CO2_2/kWh})}{\sum Generation_{tech}^{kWh}} \qquad (2)$$

3. Results

3.1. Demand and Solar Generation Profiles

3.1.1. Demand Duration Curve

The synthetic demands' duration curve can be seen in Figure 9, where the demands from 2013 and 2018 are noticeably higher than the rest of the representative years. It can also be seen that 2014 had a lower peak demand, as was intended by the selection of the representative years. Figure 10 focuses on the winter and summer months, and it is noticeable that winter still has a relatively lower demand compared to the peak summer demand in August. From this figure, it can be seen that the peak winter demand was represented well by 2012 in February. At the same time, 2013 represented the extreme case of summer demand in August. The demand barely reached 14 GW at peak for the other months, and most values were under 12 GW. This information is crucial in understanding the limitation of reducing the coal capacity since the generation capacity must be able to handle peak demands. For instance, although winter leads to higher demand, its impact is not as high as that of summer. Nonetheless, this does not translate to higher coal capacity since the generation profile of the other energy sources, in particular solar, also has seasonal variations.

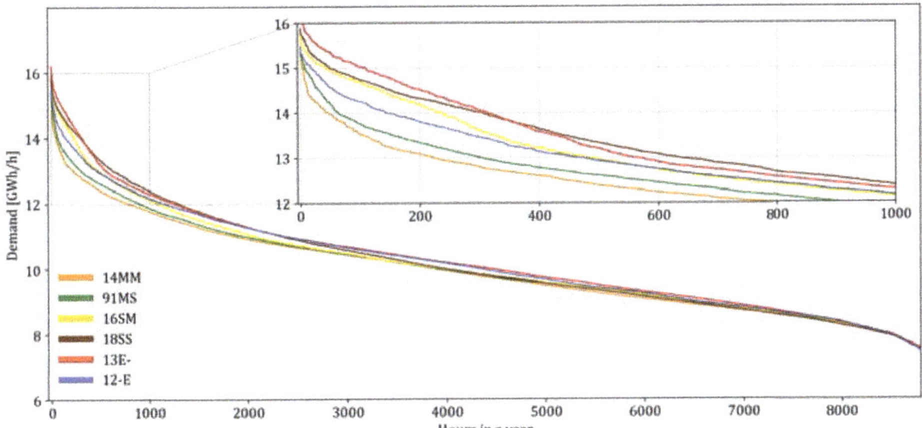

Figure 9. Duration curve of the synthetic demand.

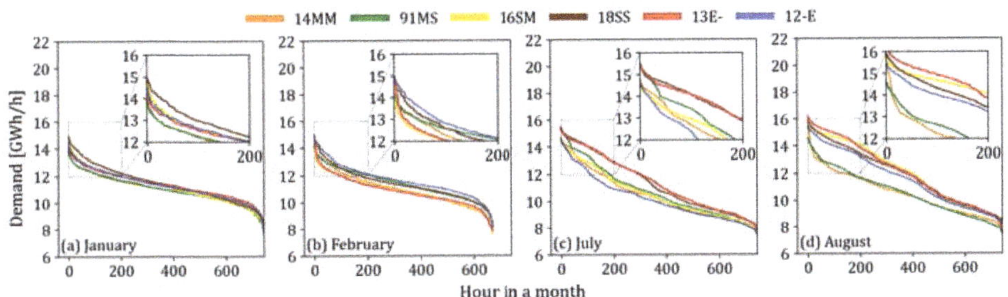

Figure 10. Monthly duration curve of the synthetic demand.

3.1.2. Solar Load Factor

The solar load factor for each month can be seen in Table 8. Since 2016 and 2018 have higher summer demand, August's higher load factor will be helpful, but the relatively lower load factor for 2013 will be an issue since this year has higher demand. The lower load factor in December could be a potential issue, but since the load factor increases by February, this could accommodate the increase in demand during this peak winter period.

Table 8. Monthly solar power generation load factor (%) using the irradiance of the representative years.

Year	JAN	FEB	MAR	APR	MAY	JUN	JUL	AUG	SEP	OCT	NOV	DEC
2014	13.28	11.47	14.89	14.59	17.03	9.94	11.45	9.05	11.38	13.80	11.61	10.15
1991	12.09	12.44	11.89	12.49	11.14	8.73	11.18	13.33	12.40	12.28	14.62	10.41
2016	7.94	12.40	15.34	11.84	13.59	10.67	13.75	15.90	10.52	9.18	12.64	11.93
2018	11.55	13.95	15.69	16.99	13.45	12.26	14.00	15.33	10.07	14.19	14.60	9.19
2013	11.03	13.77	14.84	16.34	16.09	9.53	14.70	14.70	14.73	13.29	12.01	10.67
2012	11.02	9.56	13.31	17.01	14.53	8.84	10.30	12.15	12.10	15.20	11.11	7.20
Mean	11.15	12.27	14.33	14.88	14.31	9.99	12.56	13.41	11.87	12.99	12.77	9.93

3.2. Coal Decommissioning Potential

Figure 11 shows the minimum coal capacity that could satisfy the demand for each of the 30 scenarios. The impact of yearly variations can be observed through the range of minimum coal capacity for each LNG quota scenario. As the LNG quota increases, the coal capacity could gradually be decommissioned without adding additional LNG capacity.

About 3.5 GW of the 7 GW coal capacity is older than 40 years old and should be decommissioned in the near future. However, based on the simulation results, this will be challenging if the LNG quota is not met. In the near term, where 10 GW of solar energy is already installed, the LNG quota must be at least 16 TWh. In the long term, where 16 GW of solar energy is already installed, the LNG quota must be at least 12 TWh. In both cases, as highlighted in Figure 12, around 400 to 600 MW of standby coal capacity is necessary to account for the yearly variations. As noted in the analysis of the demand duration curve, this standby capacity will be needed during the winter and summer periods, particularly in January, February, August, and September (Figure 10). Nonetheless, Figure 12 clearly shows the limitation for solar power in regard to decommissioning coal power plants beyond 12 GW installed capacity, since the minimum coal capacity no longer decreases despite additional solar generation.

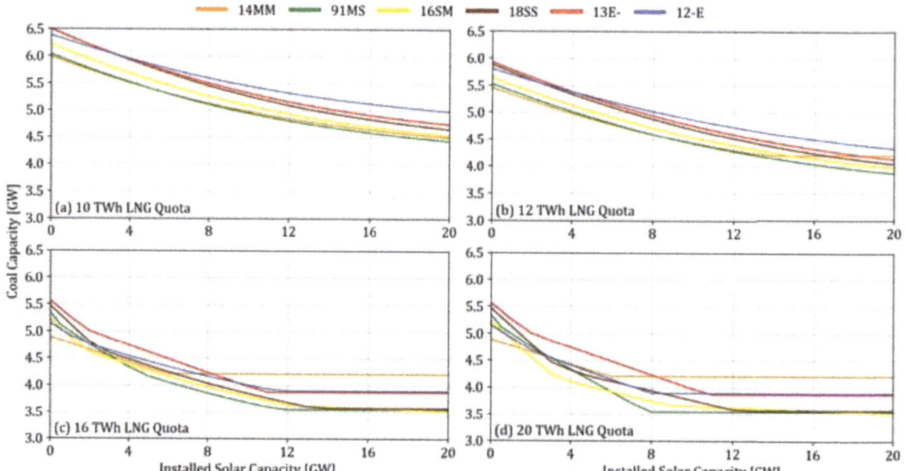

Figure 11. Minimum required coal capacity as installed solar capacity increases, and various LNG quotas.

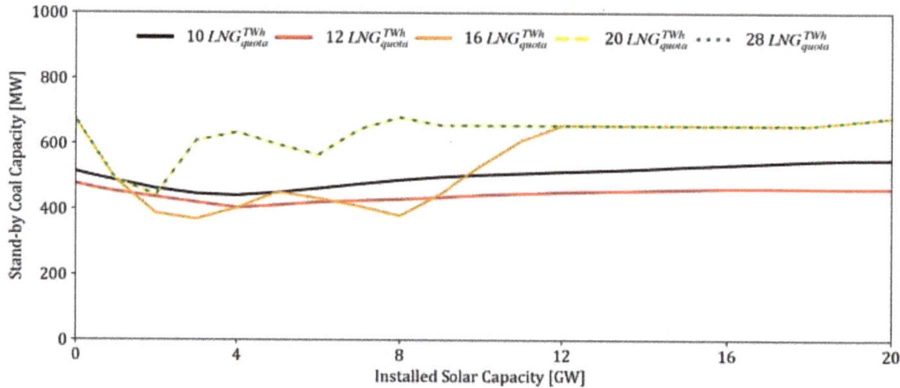

Figure 12. Standby coal capacity needed to ensure that the electricity grid can still handle weather-driven demand variations.

3.3. Coal Generation and Load Factor

Despite the floored impact of solar power on coal decommissioning, it can still reduce the coal generation and load factor. As seen in Figure 13, even beyond 12 GW, the coal generation is constantly decreasing in all LNG quota scenarios. Figure 14 further reveals that the load factor decreases as more solar capacity is installed into the grid. It should be noted that the load factors were computed using the minimum coal capacity as presented in Figure 11, which results in both a decrease in capacity and utilization rate.

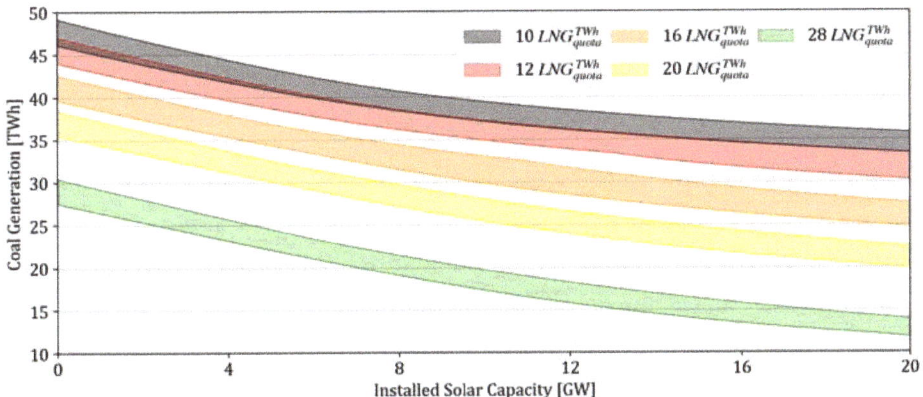

Figure 13. The coal generation in different LNG quota scenarios.

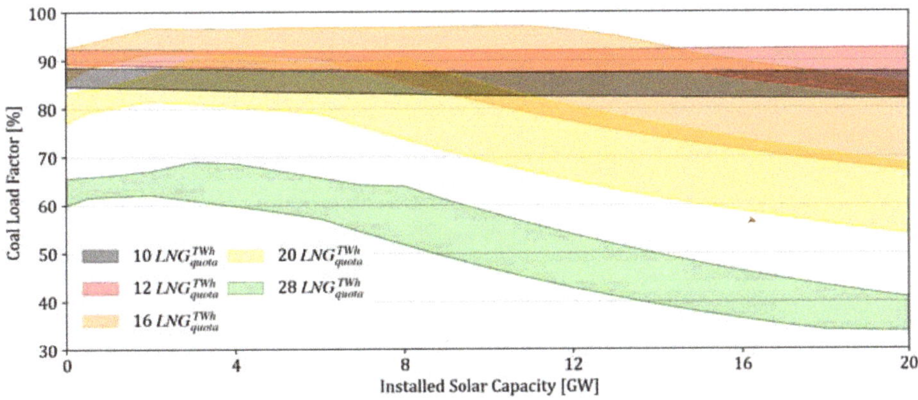

Figure 14. The coal load factor in different LNG quota scenarios.

3.4. Impact on Solar Curtailment Rate

As a consequence of the optimization that reduced the coal capacity and complementing solar power with a more flexible generator in the form of LNG, the curtailment was reduced to varying degrees. Figure 15 shows the range of curtailment rates based on the yearly variations and annual LNG quotas. Increasing the LNG quota from 12 TWh to 24 TWh, which in turn decreases the coal capacity, could reduce curtailment from 14% down to 3% in the worst-case scenario for the 10 GW installed solar capacity. Beyond the 28 TWh LNG quota, there are minor changes in the curtailment reduction. However, it could also be noted that the 20 TWh LNG quota can reduce the curtailment from 14% down to 3%. The curtailment reduction becomes more evident as the solar capacity increases.

However, beyond 16 GW installed capacity, even with sufficient complementary LNG, solar curtailment will always be greater than 10% at best and 30% at worst.

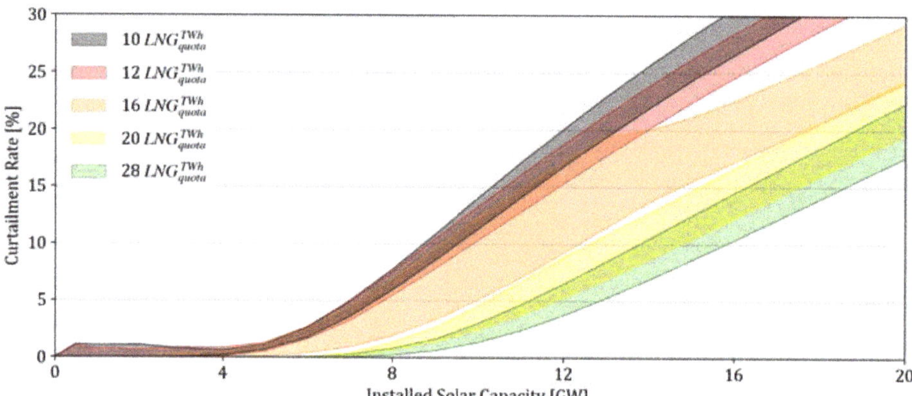

Figure 15. Projected curtailment rates assuming the optimal coal capacity was followed, with the corresponding annual LNG quotas.

3.5. Impact on Annual Cost and CO_2 Generation

Figure 16 shows the CO_2 emissions generated due to the fuel consumed by coal and LNG generators. The combination of solar and LNG electricity generation could cut the CO_2 emissions by half when comparing the "No solar + 10 TWh LNG" and "20 GW Solar + 28 TWh LNG" scenarios. The impact of solar power is also readily seen by comparing the values in each LNG quota scenario, which should be attributed to its capability to reduce coal generation even beyond 12 GW.

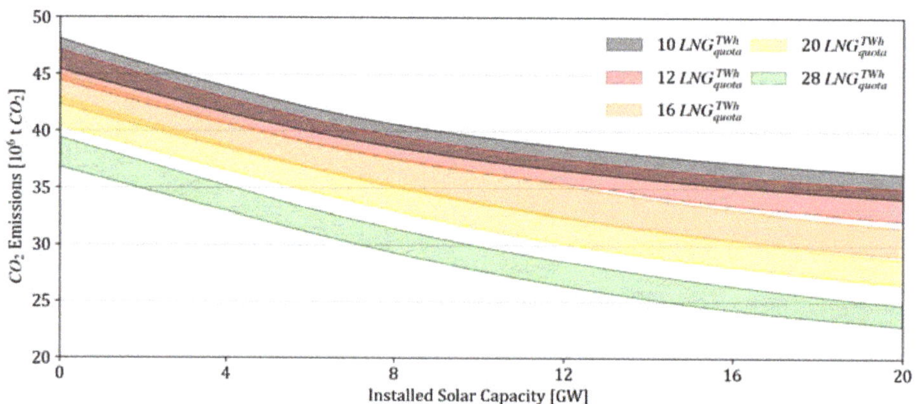

Figure 16. Range of CO_2 emissions for various scenarios in Kyushu.

As seen in Figure 17, since solar power has a higher generation cost than using LNG, its impact on the annual generation cost is more significant. The weather conditions greatly influence the annual generation cost: 2014 and 2013 represented the lowest and highest costs, respectively. The variations in the cost attributed to the LNG scenarios were more evident in 2013 followed by 2018 and 2016 caused by higher coal production during the extreme and severe summers. In the least costly year, the cost ranged from 1.22 to 1.36 trillion JPY (11.48% increase), and in the most costly year, the cost ranged from 1.26 to 1.41 (12% increase) trillion JPY (April 2021: 100 JPY = 0.92 USD = 0.77 EUR).

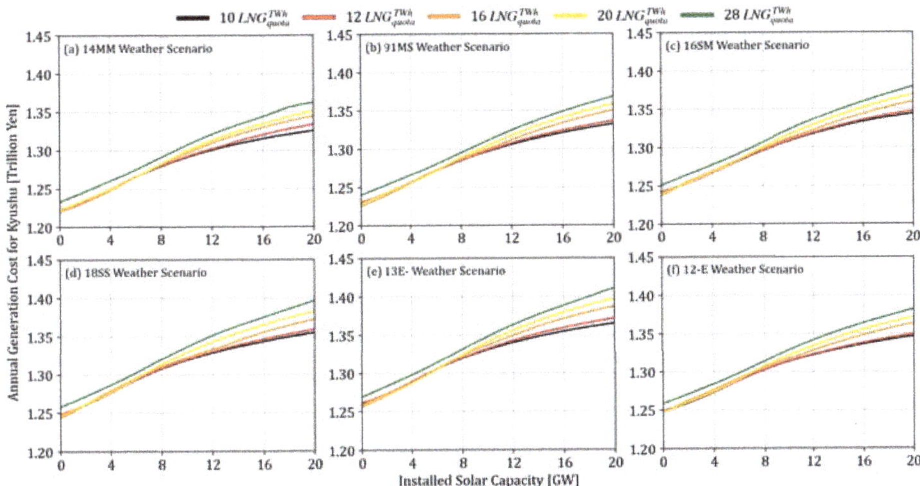

Figure 17. Annual generation costs for various scenarios in Kyushu.

Using 2016 as the representative, Figure 18 shows the impact of the installed solar capacity and the LNG quota on the levelized cost of generation and levelized CO_2 emissions. Currently, the Kyushu region already has 10 GW of installed solar capacity and generates around 10 TWh from LNG. This reference scenario is annotated as scenario 0 (S_0) in Figure 18 and values for the levelized cost and CO_2 emissions for the various weather conditions are shown in Table 9.

From this reference scenario, the company could further decrease their CO_2 emissions by having more LNG generation, adding more solar capacity, or both, but at the expense of increasing their generation cost. Five potential scenarios are annotated as S_1 to S_5 in Figure 18 and the impacts are tabulated in Tables 10 and 11. Initially, the LNG generation could be ramped up to 20 TWh to complement the solar capacity increase, as seen in S_1. This increased the generation cost by an average of 0.63% and decreased the CO_2 emissions by an average of 12.80% to 0.3226 kgCO2/kWh. From S_1, solar capacity could continuously increase, as seen in S_2 and S_3, or LNG could increase further as seen in S_3. The impact of S_2 and S_4 in reducing CO_2 emissions was the same, but the increase in cost was lower for S_4. S_5 represents the greenest yet feasible scenario that reduces the CO_2 emissions by an average of 37.31% but increases the cost by 5.60%.

Table 9. Cost and CO_2 emissions of the reference scenario.

	14MM	91MS	16SM	18SS	13E-	12-E	Mean
Levelized Cost (JPY/kWh)	12.5464	12.5266	12.5331	12.5819	12.5820	12.5160	12.5477
Levelized CO_2 (kgCO$_2$/kWh)	0.3640	0.3683	0.3714	0.3679	0.3723	0.3764	0.3700

As of April 2021: 100 JPY = 0.92 USD = 0.77 EUR.

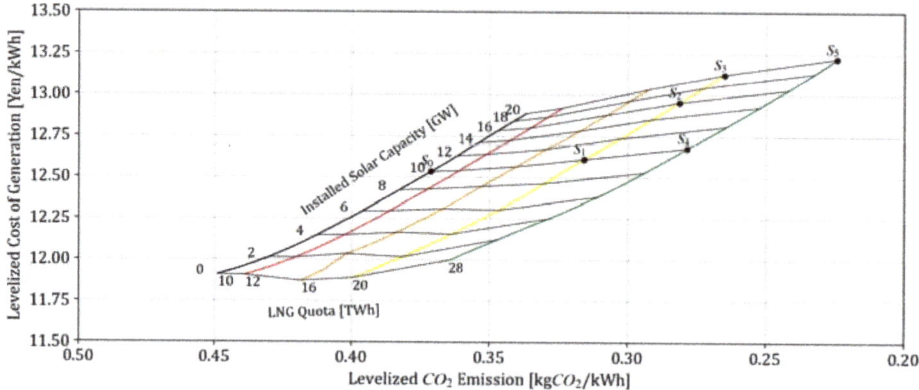

Figure 18. Levelized generation cost and CO_2 emissions for the warmer year (16SM) scenario. The CO_2 emissions (abscissa) are formatted in decreasing order to emphasize the trend. S_0 reflects the current situation, and S_1–S_5 are the potential future scenarios.

Table 10. Cost increase from the reference scenario.

	Scenarios				Increase in Cost per kWh (%)					
No	Name	$Solar_{cap}^{GW}$	LNG_{quota}^{TWh}	14MM	91MS	16SM	18SS	13E-	12-E	Mean
S_1	Increase LNG_{quota}^{TWh} 1	10	20	0.56	0.53	0.60	0.74	0.75	0.59	0.63
S_2	Increase $Solar_{cap}^{GW}$ 1	16	20	3.34	3.38	3.35	3.46	3.61	3.17	3.39
S_3	Increase $Solar_{cap}^{GW}$ 2	20	20	4.61	4.75	4.71	4.75	4.93	4.40	4.69
S_4	Increase LNG_{quota}^{TWh} 2	10	28	1.20	1.07	1.13	1.34	1.28	1.14	1.19
S_5	Increase both	20	28	5.51	5.52	5.49	5.86	6.00	5.22	5.60

Table 11. CO_2 emission decrease from the reference scenario.

	Scenarios				Decrease in CO_2 Emission per kWh (%)					
No	Name	$Solar_{cap}^{GW}$	LNG_{quota}^{TWh}	14MM	91MS	16SM	18SS	13E-	12-E	Mean
S_1	Increase LNG_{quota}^{TWh} 1	10	20	−13.50	−12.74	−12.55	−12.91	−12.94	−12.18	−12.80
S_2	Increase $Solar_{cap}^{GW}$ 1	16	20	−22.78	−22.53	−22.10	−22.66	−22.78	−21.02	−22.31
S_3	Increase $Solar_{cap}^{GW}$ 2	20	20	−27.05	−27.18	−26.62	−27.24	−27.29	−25.15	−26.75
S_4	Increase LNG_{quota}^{TWh} 2	10	28	−23.77	−23.24	−22.93	−23.37	−23.09	−22.27	−23.11
S_5	Increase both	20	28	−36.45	−38.32	−37.84	−37.97	−37.58	−35.70	−37.31

4. Discussion

4.1. Potential and Limitations of Solar PV in Coal Decommissioning

Although it cannot phase-out coal, the results show that solar energy has enough potential to be the driver for coal decommissioning with LNG's help. It has also been shown that the decommissioning potential is robust against yearly weather-driven demand, and standby-plants could be used for the colder and warmer periods of the year. Although solar power has limitations in reducing coal capacity, it continually decreases the necessary coal generation, thereby reducing the load factor of coal plants and the corresponding CO_2 emissions.

In Kyushu's case, given the 10 GW solar capacity along with a 16 TWh complementary LNG quota, 3.5 GW of the 7 GW coal power plants could be decommissioned. This configuration is already achievable by increasing the LF of the combined-cycle plants of KyEPCO from 20% to 44%. Beyond 12 GW installed solar capacity, solar power alone has no impact on reducing the coal capacity, but it could still reduce coal generation. Compared to the reference scenario, it was shown that CO_2 emissions could be reduced by 27% through 20 GW of solar power and a 20 TWh annual LNG quota. The reduction could reach 37% if all the LNG plants in the region are utilized at 60% LF. As a related consequence, reducing coal and introducing more LNG reduced solar curtailment. This potential and limitations show that energy planners should take the necessary precautions in adding solar energy to the grid since there is an appropriate balance. Solar can reduce coal capacity, but it alone cannot phase-out coal. As was shown in Kyushu's case, a thorough analysis of the situation that includes complementary energy sources should be considered in evaluating the potential of solar power in coal decommissioning.

4.2. Implications of Solar PV in Coal Decommissioning

Solar has its drawbacks in the form of cost and dependence on complementary flexible generators. The results show that in countries like Japan, where solar power remains to be more expensive than conventional generators—solar power presents an additional cost. Its dependence on flexible generators, which LNG currently fills, poses a threat to its ability to stand-alone. As the demand for LNG steadily increases, this will threaten its supply and price. The cost of LNG could exacerbate the cost problems of solar.

In Kyushu's case, increasing the solar capacity from 10 GW to 16 GW and 20 GW increases the levelized cost of generation by 3.39% and 4.69%, respectively. Increasing the LNG quota has a minor impact at the moment since the current LNG price is only about 12% higher than coal. In contrast, solar is still almost twice as expensive as coal. Solar prices around the world have been decreasing, and it might decrease in Japan in the future. The impact on CO_2 and cost now becomes a policy decision, and the ratio between these two factors presents several potential combinations between LNG quota and installed solar capacity that could yield identical cost or the same CO_2 targets as seen in S_2 and S_4. More LNG is necessary when cost is prioritized, but it will lead to more dependence on LNG. Alternatively, by investing more in solar capacity, it could lead the CO_2 reduction efforts and local power generation. This scenario entails lower dependence on both coal and LNG, which are both imported fuels. As with the previous results, the impact of weather on these values is evident, as seen in the variations in the levelized cost and levelized CO_2 emissions.

4.3. Potential Solutions beyond Solar PV

The supply and demand mismatch in winter and summer is one of the major roadblocks in the total phase-out of coal power plants through solar energy. Diurnal storage will be enough to solve the mismatch during summer, but seasonal storage or seasonal generation will be necessary for winter. Since there is still enough excess energy during peak solar production in summer, storage is the straightforward solution once these options become economically feasible. However, since there is less solar energy in winter, there is not enough excess solar energy for diurnal storage to work, which opens an opportunity for seasonal technology. Seasonal storage in the form of power-to-gas (P2G) could store the excess solar in autumn for winter. Combined heat and power (CHP) plants could be operated at a higher capacity in winter if local water heating is established.

4.4. Impact of Weather on Energy Transition Plans

The stochastic nature of demand and renewable energy sources was the primary motivation for developing the weather-driven approach since energy transition recommendations should consider scenarios that will test the limits of the planned energy mix. The variations are significant at 400 MW to 600 MW coal capacity, as seen from the results. In Japan's case, this translates to 1–3 coal power plants, but for smaller nations with smaller plants, this could be composed of more than five plants that should be on standby in the event of an extreme weather condition. Coordinating smaller plants will require more dialogue and agreements between the government and plant operators. Consequently, the government could also run standby plants to ensure the reliability of the system. It has also been shown that weather influences the potential for CO_2 reduction and the system's overall annual generation cost. Beyond coal decommissioning, weather will remain a necessary variable in energy planning since it influences the demand, which is the primary source of stochasticity in the analysis. As more VREs are added to the green energy transition, weather becomes a crucial variable for both wind and solar. Rainfall data could also influence hydropower generation, which was not explored in this study. It could also influence the viability of PHES since this requires sufficient water reservoirs affected by rain and water evaporation. Little is known about wave energy's potential, but the weather will also influence it since it depends on nature.

4.5. Importance and Limitation of the Proposed Approach

The proposed weather-driven scenario-based analysis revealed the importance of the LNG quota, demand variations, and solar generation through the annual hourly simulation. System reliability could be analyzed using the duration curve, but this does not show the hourly balance, which is greatly influenced by demand and solar generation's stochasticity. Through careful selection of representative years, the range of potential scenarios was identified and analyzed to ensure robust results. However, the approach is dependent on the yearly assignment and is limited by the probabilistic matching of weekends and holidays to high irradiance days. The former is influenced by human behavior, while the latter is non-deterministic. Thus, although the simulation considered the yearly variations, the probability of a low irradiance day being matched to a high-demand weekday was not covered by the approach. Nonetheless, the approach can be used to provide robust recommendations for green energy transition since it covers the stochastic nature of demand and variable renewable energy. In this study, the approach was used to determine the minimum coal capacity that can ensure the system's reliability, but it could also be used for energy storage assessments and capacity planning. This study only used a single-bus network, but it could be expanded to a national grid level by representing each region as a bus. The approach can then be used for grid expansion planning.

5. Conclusions

Driven by the idea of transitioning to a green electricity grid, an hourly power flow analysis was conducted to understand the potential, limitations, and implications of using solar energy as a driver for decommissioning coal power plants. The weather-driven scenario analysis ensured the robustness of the results and recommendations. The analysis revealed that solar power could reduce about half of Kyushu's coal capacity with the aid of LNG. Beyond 12 GW, solar power could not reduce the minimum coal capacity necessary to ensure the system's reliability, but it could still reduce the coal generation and the overall CO_2 emissions. The reduction in coal capacity comes at a cost, since solar power is still relatively more expensive in Japan. By installing 20 GW of solar PV systems and having 28 TWh of available LNG, the levelized CO_2 emissions could be reduced by 37%, but this would increase the levelized cost of generation by 5.6%. Most of the price increase is owed to the price of solar electricity generation, which remains high in Japan. In Kyushu's case, this change could be achieved without constructing additional power plants, since the LNG plants are operated at a low LF. However, additional planning is necessary to acquire more LNG. Countries that use LNG plants as peak-load generators share the same potential, and the results show that a minor change in the system could have a significant impact on emission goals.

The results emphasized that solar power with the aid of LNG could partially replace coal capacity, but it alone could not phase-out coal. For energy planners who are only starting to increase their solar capacity, insights from this work could help with understanding the interactions between coal, solar, and LNG electricity generation. For planners in countries with a considerable amount of solar power (>8%), the results from this study could serve as a precaution by highlighting the risks of further increasing the solar power penetration. Although solar power helped solve midday peak power, the problem remains because it simply shifted to periods where there is no solar energy. Summer and winter are challenging periods due to the increase in peak demand. Although it is counterintuitive, solar energy is not enough during summer, or, to be more precise, misaligned since the problem occurs in the late afternoon. Diurnal storage can address the misalignment in summer, but winter presents a more intricate problem, since the solar energy is insufficient. Thus, exploring other technologies that could further complement solar energy is necessary.

The weather-driven approach revealed the importance of weather in the analysis, as it affected the results to varying degrees. In addition, 400–600 MW of standby coal capacity is necessary due to the yearly fluctuations. Coal generation, coal load factor, curtailment rate, and CO_2 emissions vary by 7–18%, 8–27%, 0–5%, and 6–8%, respectively. Identifying the representative year is crucial since it should cover the worst case, best case, and the cases in between. Energy planners and policymakers should consider the weather when analyzing energy plans, as it could provide a range of values that can guide them in making the correct decisions. Since the approach can generate scenarios based on weather data, it could also be used for storage assessment and capacity planning. The approach could also be used for grid expansion planning by increasing the number of buses and modeling multiple demands. These energy planning topics could also benefit from the range of insights generated through the weather-driven approach.

Author Contributions: Conceptualization, S.M.G.D. and K.N.I.; methodology, S.M.G.D.; software, S.M.G.D.; validation, S.M.G.D. and K.N.I.; formal analysis, S.M.G.D.; investigation, S.M.G.D.; resources, K.N.I.; data curation, S.M.G.D.; writing—original draft preparation, S.M.G.D.; writing—review and editing, K.N.I.; visualization, S.M.G.D.; supervision, K.N.I. All authors have read and agreed to the published version of the manuscript.

Funding: This research received no external funding.

Data Availability Statement: The data and code presented in this study are openly available on GitHub at https://github.com/smdumlao/demandfingerprint/tree/main/papers/coaldecommissioning (accessed on 16 April 2021).

Acknowledgments: The authors are grateful for the financial support from the Ministry of Education, Culture, Sports, Science, and Technology of Japan for the doctoral studies of Samuel Matthew Dumlao at Kyoto University. The authors are also grateful to the Ambitious Intelligence Dynamic Acceleration (AIDA) Program of Kyoto University for the financial support.

Abbreviations

The following abbreviations are used in this manuscript:

CC	Combined Cycle
CHP	Combined Heat and Power
DHI	Diffuse Horizontal Irradiance
DNI	Direct-Normal Irradiance
GEP	Generation Expansion Planning
GHI	Global Horizontal Irradiance
IEA	International Energy Agency
IPP	Independent Power Producer
JMA	Japan Meteorological Agency
KyEPCO	Kyushu Electric Power Company
LNG	Liquefied Petroleum Gas
METI	Ministry of Economy, Trade, and Industry
MOFA	Ministry of Foreign Affairs
P2G	Power-to-Gas
PHES	Pump Hydro Energy Storage
PV	Solar Photovoltaics
pvlib	pvlib Python Library

PyPSA	Python for Power System Analysis
VRE	Variable Renewable Energy
14MM	2014 Mild Summer Mild Winter (Weather Scenario)
91MS	1991 Mild Summer Severe Winter (Weather Scenario)
16SM	2016 Severe Summer Mild Winter (Weather Scenario)
18SS	2018 Severe Summer Severe Winter (Weather Scenario)
13E-	2013 Extreme Summer (Weather Scenario)
12-E	2012 Extreme Winter (Weather Scenario)
LNG_{quota}^{TWh}	LNG Availability (TWh)
$Solar_{cap}^{GW}$	Solar Capacity (GW)
$Generation_{tech}^{kWh}$	Annual Generation (kWh)
$Cost_{tech}^{JPY/kWh}$	Cost of Electricity Generation (JPY/kWh)
$Emission_{tech}^{kG-CO_2/kWh}$	CO_2 Emission (kgCO2/kWh)

References

1. Martinot, E. Grid Integration of Renewable Energy. *Annu. Rev. Environ. Resour.* **2016**, 223–254. [CrossRef]
2. REN21. Renewables 2020 Global Status Report. 2020. Available online: https://www.ren21.net/gsr-2020/ (accessed on 8 October 2020).
3. Japan Ministry of Economy, Trade, and Industry. Japan's Fifth Strategic Energy Plan (Provisional Translation). 2018. Available online: https://www.enecho.meti.go.jp/en/category/others/basic_plan/5th/pdf/strategic_energy_plan.pdf (accessed on 8 October 2020).
4. Kyushu Electric Company. Supply and Demand Information (Japanese). 2019. Available online: http://www.kyuden.co.jp/wheeling_disclosure.html (accessed on 20 February 2021).
5. International Energy Agency. Status of Power System Transformation. 2019. Available online: https://www.iea.org/reports/status-of-power-system-transformation-2019 (accessed on 6 March 2021).
6. Deutscher Bundestag. Bundestag Passes the Coal Phase-Out Law. 2020. Available online: https://www.bundestag.de/dokumente/textarchiv/2020/kw27-de-kohleausstieg-701804 (accessed on 14 March 2021).
7. Europe Beyond Coal. Overview: National Coal Phase-Out Announcements in Europe. 2021. Available online: https://beyond-coal.eu/wp-content/uploads/2021/01/Overview-of-national-coal-phase-out-announcements-Europe-Beyond-Coal-January-2021.pdf (accessed on 14 March 2021).
8. Government of Canada. Coal Phase-Out: the Powering Past Coal Alliance. 2018. Available online: https://www.canada.ca/en/services/environment/weather/climatechange/canada-international-action/coal-phase-out.html (accessed on 14 March 2021).
9. Kumar, S.; Kwon, H.T.; Choi, K.H.; Cho, J.H.; Lim, W.; Moon, I. Current status and future projections of LNG demand and supplies: A global prospective. *Energy Policy* **2011**, *39*, 4097–4104. [CrossRef]
10. Shell Corporation. LNG Outlook 2020. 2020. Available online: https://www.shell.com/energy-and-innovation/natural-gas/liquefied-natural-gas-lng/lng-outlook-2020.html (accessed on 16 March 2021).
11. Nikkei Asia. Severe Winter Forecast Sparks Recovery in Asia's LNG Prices. 2020. Available online: https://asia.nikkei.com/Business/Markets/Commodities/Severe-winter-forecast-sparks-recovery-in-Asia-s-LNG-prices (accessed on 16 March 2021).
12. Dumlao, S.M.G.; Ishihara, K.N. Reproducing solar curtailment with Fourier analysis using Japan dataset. *Energy Rep.* **2020**, *6*, 199–205. [CrossRef]
13. Bunodiere, A.; Lee, H.S. Renewable Energy Curtailment: Prediction Using a Logic-Based Forecasting Method and Mitigation Measures in Kyushu, Japan. *Energies* **2020**, *13*, 4703. [CrossRef]
14. Guittet, M.; Capezzali, M.; Gaudard, L.; Romerio, F.; Vuille, F.; Avellan, F. Study of the drivers and asset management of pumped-storage power plants historical and geographical perspective. *Energy* **2016**, *111*, 560–579. [CrossRef]
15. Li, Y.; Gao, W.; Ruan, Y.; Ushifusa, Y. The performance investigation of increasing share of photovoltaic generation in the public grid with pump hydro storage dispatch system, a case study in Japan. *Energy* **2018**, *164*, 811–821. [CrossRef]
16. Koltsaklis, N.E.; Dagoumas, A.S. State-of-the-art generation expansion planning: A review. *Appl. Energy* **2018**, *230*, 563–589. [CrossRef]
17. Ioannou, A.; Angus, A.; Brennan, F. Risk-based methods for sustainable energy system planning: A review. *Renew. Sustain. Energy Rev.* **2017**, *74*, 602–615. [CrossRef]
18. Santos, M.J.; Ferreira, P.; Araújo, M. A methodology to incorporate risk and uncertainty in electricity power planning. *Energy* **2016**, *115*, 1400–1411. [CrossRef]
19. Schaeffer, R.; Szklo, A.S.; de Lucena, A.F.P.; Borba, B.S.M.C.; Nogueira, L.P.P.; Fleming, F.P.; Troccoli, A.; Harrison, M.; Boulahya, M.S. Energy sector vulnerability to climate change: A review. *Energy* **2012**, *38*, 1–12. [CrossRef]
20. Bartók, B. Changes in solar energy availability for south-eastern Europe with respect to global warming. *Phys. Chem. Earth Parts A/B/C* **2010**, *35*, 63–69. [CrossRef]
21. Cutforth, H.; Judiesch, D. Long-term changes to incoming solar energy on the Canadian Prairie. *Agric. For. Meteorol.* **2007**, *145*, 167–175. [CrossRef]

22. Dumlao, S.M.G.; Ishihara, K.N. Supplementary Materials: Weather-Driven Scenario Analysis for Decommissioning Coal Power Plants in High PV Penetration Grid. 2021. Available online: https://github.com/smdumlao/demandfingerprint/tree/main/papers/coaldecommissioning (accessed on 16 April 2021).
23. Japan Meteorological Agency. Past Regional Weather Data. Available online: http://www.data.jma.go.jp/risk/obsdl/index.php (accessed on 8 January 2021).
24. Holmgren, W.F.; Hansen, C.W.; Mikofski, M.A. pvlib python: a python package for modeling solar energy systems. *J. Open Source Softw.* **2018**, *3*, 884. [CrossRef]
25. Brown, T.; Hörsch, J.; Schlachtberger, D. PyPSA: Python for Power System Analysis. *J. Open Res. Softw.* **2018**, *6*. [CrossRef]
26. Knüpfer, K.; Dumlao, S.M.G.; Esteban, M.; Shibayama, T.; Ishihara, K.N. Analysis of PV Subsidy Schemes, Installed Capacity and Their Electricity Generation in Japan. *Energies* **2021**, *14*, 2128. [CrossRef]
27. Agency for Natural Resources and Energy. Electric Power Survey Statistical Data. Available online: https://www.enecho.meti.go.jp/statistics/electric_power/ep002/results_archive.html (accessed on 8 January 2021).
28. Holmgren, W.; Andrews, R. pvlib: TMY to Power Tutorial. Available online: https://nbviewer.jupyter.org/github/pvlib/pvlib-python/blob/master/docs/tutorials/tmy_to_power.ipynb (accessed on 8 January 2021).
29. Erbs, D.; Klein, S.; Duffie, J. Estimation of the diffuse radiation fraction for hourly, daily and monthly-average global radiation. *Sol. Energy* **1982**, *28*, 293–302. [CrossRef]
30. Ministry of Land, Infrastructure, Transport and Tourism. Japan Power Plant GIS Data. 2007. Available online: https://nlftp.mlit.go.jp/ksj/jpgis/datalist/KsjTmplt-P03.html (accessed on 8 June 2020).
31. Japan Beyond Coal. Japan Coal Plant Database. 2020. Available online: https://beyond-coal.jp/map-and-data/ (accessed on 29 September 2020).
32. Kyushu Electric Power Company. KYUDEN Group Environmental Data Book 2020. 2020. Available online: http://www.kyuden.co.jp/var/rev0/0270/6338/env_data_2020.pdf (accessed on 18 March 2021).
33. Ministry of Economy, Trade and Industry of Japan. Recent Situation of Thermal Power Generation. 2017. Available online: https://www.meti.go.jp/shingikai/enecho/shoene_shinene/sho_energy/karyoku_hatsuden/pdf/h29_01_04_00.pdf (accessed on 18 March 2021).
34. Ministry of Economy, Trade and Industry of Japan. Report on Power Generation Costs. 2015. Available online: https://www.enecho.meti.go.jp/committee/council/basic_policy_subcommittee/mitoshi/cost_wg/006/pdf/006_05.pdf (accessed on 10 March 2021).
35. Advisory Panel to the Foreign Minister on Climate Change. Promote New Diplomacy on Energy through Leading Global Efforts against Climate Change. 2018. Available online: https://www.mofa.go.jp/files/000337158.pdf (accessed on 10 March 2021).
36. Ministry of Environment of Japan. Evaluation of CO_2 Emissions of Thermal Power Generation. 2016. Available online: https://www.env.go.jp/council/06earth/y0613-16/ref06-16.pdf (accessed on 10 March 2020).

Article

The Energy Lock-In Effect of Solar Home Systems: A Case Study in Rural Nigeria

Olumide Hassan *, Stephen Morse and Matthew Leach

Centre for Environment and Sustainability, University of Surrey, Guildford GU2 7XH, UK;
s.morse@surrey.ac.uk (S.M.); m.leach@surrey.ac.uk (M.L.)
* Correspondence: o.hassan@surrey.ac.uk

Received: 23 November 2020; Accepted: 14 December 2020; Published: 17 December 2020

Abstract: Ongoing reductions in the costs of solar PV and battery technologies have contributed to an increased use of home energy systems in Sub-Saharan African regions without grid access. However, such systems can normally support only low-power end uses, and there has been little research regarding the impact on households unable to transition to higher-wattage energy services in the continued absence of the grid. This paper examines the challenges facing rural energy transitions and whether households feel they are energy 'locked in'. A mixed-methods approach using questionnaire-based household energy surveys of rural solar home system (SHS) users was used to collect qualitative and quantitative data. Thematic analysis and a mixture of descriptive and inferential statistical analyses were applied. The results showed that a significant number of households possessed appliances that could not be powered by their SHS and were willing to spend large sums to connect were a higher-capacity option available. This implied that a significant number of the households were locked into a low-energy future. Swarm electrification technology and energy efficient, DC-powered plug-and-play appliances were suggested as means to move the households to higher tiers of electricity access.

Keywords: solar home systems; rural households; energy transitions; energy access

---

1. Introduction

Energy plays a pivotal role in the social, economic and cultural development of any population [1]. Energy from carriers (e.g., electricity) is converted via end-use technology (e.g., televisions, mobile phones, bulbs and mechanical fans) into useful household energy services (e.g., entertainment, communication, lighting and space cooling). Energy also supports income-generating activities including construction, agriculture and manufacturing. This is especially relevant for developing nations, who have been shown to exhibit a much larger increase in the Human Development Index (HDI) for a corresponding increase in energy consumption per capita, compared to developed nations [2]. However, it is estimated that almost 800 million people lack any access to electricity, with approximately 550 million of them living in Sub-Saharan Africa [3]. Transitioning these people from traditional sources of energy (biomass and waste) to accessing modern energy such as that provided by electricity poses a major challenge. The importance of this challenge is encapsulated by the United Nation's (UN) Sustainable Development Goal (SDG) 7, which calls for universal access to affordable, sustainable, reliable and modern energy by 2030. There has been significant progress towards the goals of SDG7 at the country level in developing regions, with countries such as Kenya, Uganda and Sudan making the most progress, even though the current pace of global efforts has so far been insufficient [3].

While there is no single definition of energy poverty, it has been referred to as a lack of access to modern energy services [4] and a deficiency in the consumption of energy required to meet basic human needs [5–7]. Reddy [8] (p. 44) captures the multi-dimensional nature of energy poverty by defining it as "the absence of sufficient choice in accessing adequate, affordable, reliable, high-quality, safe and environmentally benign energy services to support economic and human development". This recognition of the multi-dimensional nature of energy poverty (and therefore access) helped shape development of the Multi-Tier Framework (MTF) formulated by the World Bank's Energy Sector Management Program (ESMAP) designed to provide a measurement framework for measuring energy access [9], although it does focus mainly on electricity. The MTF has been referred to as providing the most sophisticated energy measurement metric [10], and measures seven attributes of capacity, availability, reliability, quality, affordability, legality, and health and safety, an approach that takes into account the various factors that influence a household's ability to access energy. It also recognizes that energy access is not a binary 'have or have-not' condition, but rather that households can exist at varying levels of access depending on socioeconomic factors and other circumstances dictated by the seven attributes listed earlier. The levels of energy (electricity) access described in the MTF are referred to as 'tiers' and vary from Tier 0 (no access) to the highest level of access, which is Tier 5. Moving up these tiers can be seen as climbing up the so-called 'energy ladder'. A visual representation of the MTF, showing the various tiers, is shown in Figure 1, a matrix for measuring access to household electricity supply, and Figure 2, a matrix for measuring access to household electricity services.

For decades, rural electrification in developing regions has primarily been achieved via grid extension, and this is still integral to rural electrification policies in many parts of the developing world [11–13]. However, grid extension has been much slower than anticipated in many regions [12,14], and a key barrier to it is that it is not profitable for utilities to invest the capital required to deliver electricity to rural communities with relatively low patterns of consumption [15,16]. In the meantime, decentralized options such as solar home systems (SHSs) have proven to be popular for electrifying households and other small users due to their falling prices and renewable power source; in Kenya, more than 30% of off-grid households have a solar PV product at home [17].

A typical SHS consists of a PV module typically placed on a rooftop to capture sunlight, a battery for storing energy, and end-use appliances which provide energy services. Capacities of SHS can range from 10 to 200 W [18]. However, the typical SHSs used in rural Sub-Saharan Africa are in the 10–100 range [19,20]. While the upfront costs of SHSs, at up to $400 for an 80 W system [21], can act as a barrier to entry for many of the poor rural households, the introduction of Pay-As-You-Go (PAYG) service has helped improve its uptake. This is achieved by providing the SHS as a service, with the households making small, regular payments to access the service over an agreed time period, after which they own the SHS [22,23]. Companies such as M-KOPA and Bboxx in Kenya have successfully integrated mobile payment technologies with their SHS services, providing their customers with the convenience of making payments using their mobile phones, and increasing the flexibility of the amounts they pay and the payment intervals [24,25].

Multiple studies have shown that households moving from Tier 0 (i.e., unelectrified) to using SHS (typically Tiers 1 and 2) have seen an improvement in their quality of life [26–28]. SHS use has been linked to health benefits, as it displaces household use of kerosene and candles which expose users to a range of health risks including burns, child poisoning due to inadvertent consumption of fuels, and a variety of conditions linked to indoor air pollution from fine particulate matter, sulphur and nitric oxides. The most common use for SHSs is the improved lighting service it provides compared to kerosene or candles [29] with educational improvements for children who use the light to read and do homework as a key outcome [26,30]. SHS use has also reduced the need for household members to travel long distances for charging their mobile phones [31].

ATTRIBUTES		TIER 0	TIER 1	TIER 2	TIER 3	TIER 4	TIER 5
1. Peak Capacity	Power capacity ratings (in W or daily Wh)		Min 3 W	Min 50 W	Min 200 W	Min 800 W	Min 2 kW
			Min 12 Wh	Min 200 Wh	Min 1.0 kWh	Min 3.4 kWh	Min 8.2 kWh
	OR Services		Lighting of 1,000 lmhr/day	Electrical lighting, air circulation, television and phone charging are possible			
2. Availability (Duration)	Hours per day		Min 4 h	Min 4 h	Min 8 h	Min 16 h	Min 23 h
	Hours per evening		Min 1 h	Min 2 h	Min 3 h	Min 4 h	Min 4 h
3. Reliability						Max 14 disruptions per week	Max 3 disruptions per week of total duration <2 h
4. Quality							Voltage problems do not affect the use of desired appliances
5. Affordability						Cost of a standard consumption package of 365 kWh/year < 5% of household income	
6. Legality							Bill is paid to the utility, pre-paid card seller, or authorized representative
7. Health and Safety							Absence of past accidents and perception of high risk in the future

Figure 1. Multi-tier matrix for measuring access to household electricity supply (source: [9]).

	TIER 0	TIER 1	TIER 2	TIER 3	TIER 4	TIER 5
Tier criteria		Task lighting AND Phone charging	General lighting AND Phone Charging AND Television AND Fan (if needed)	Tier 2 AND Any medium-power appliances	Tier 3 AND Any high-power appliances	Tier 2 AND Any very high-power appliances

Figure 2. Multi-tier matrix for measuring access to household electricity services (source: [9]).

Access to, and initial consumption of electricity has been shown to lead to increases in consumption over time, as users purchase more appliances (if affordable) [32]. This is in line with the study by Opiyo [33], which demonstrated that rural households in Kenya, upon realising the socioeconomic benefits of their basic electricity access from SHS, began to desire more high-powered appliances such as televisions and cooling fans. However, some researchers, such as Lee et al. [34] and Stojanovski et al. [20], have observed that while rural households purchased these appliances to benefit from the services they provide, they could not be accommodated by their SHSs due to their relatively small energy capacity, and hence these appliances were unusable. Hence, despite their broad appeal, there are questions surrounding home solar products and the energy services they are able to provide to rural households. There is a paucity of research designed to explore the nature of the energy services home solar products provide to rural communities in the developing world and especially the views and aspirations of households regarding access to energy services. For example, do households feel that they may become 'trapped' into what they see as an 'inferior' source of energy? In this paper, the results of a household energy survey conducted across two locations in Nigeria are used to provide an empirically based analysis of SHS use in rural households and how those households view such systems vis-à-vis alternatives such as an extension of the grid supply. In particular, the research aims were to:

1. Assess whether the SHS technologies taken up by these rural communities met their non-cooking household energy service demands,
2. Whether households may feel 'energy trapped' by the inherent limited capacity of the SHS technology.

2. Methodology

2.1. Location of Research

Data collection for this research was carried out in rural communities across two states in Nigeria, a lower-middle-income country in West Africa [35]: Lagos State, located in the South West region, and the Federal Capital Territory (FCT), which is located in the geographical centre of the country. See Figure 3 for a map of Nigeria, including the locations used for data collection.

Nigeria is a multi-ethnic and culturally diverse country, with 36 states and the Federal Capital Territory [36]. Rural locations across the country differ in terms of wealth, culture and weather, all of which can affect energy use. Hence, the locations were chosen in part to reflect a range of socioeconomic conditions. The selected communities were in what can be termed the 'urban fringe' of two major cities in Nigeria–Lagos (Lagos State) and Abuja (FCT). Lagos used to be the Federal Capital of Nigeria, but that role has now been adopted by Abuja, and the communities selected for the research have been involved in projects designed to increase energy provision via SHS. Indeed, the choice of communities was based on advice from Nigerian experts in rural electrification, including providers of SHS. A total of 150 households (50 in Lagos State and 100 in FCT), the majority without any type of access to grid electricity were surveyed face to face in six communities across both locations, and this was undertaken between May and June 2019. The households were selected using the following criteria:

- Community designation: All the households considered had to be rural. As described earlier, the communities were located on the 'urban fringe' of the cities of Lagos and Abuja.
- Location: All communities visited were to be in locations accessible with a car, and also could not be located in regions designated as "advise against all but essential travel" or "advise against all travel" on the UK Foreign and Commonwealth Office travel advice website, at the time of the research.
- Age/status: Head of households were preferred as respondents. In Nigeria, households are generally headed by men [37], who make most of the household decisions, including energy purchases. When they were not available, the next senior family member (usually a spouse or sibling) could be interviewed. Respondents under the age of 18 years were excluded, as age can be a powerful symbol of inequality in many Nigerian cultures [38]. They were also excluded for ethical concerns.
- Wattage: The SHSs used by the households were to be in the range of 10 to 100 Wp.
- Duration of ownership of SHS: Households should have been using their SHS for no less than 6 months prior to the interviews. This is so they would have had time to have gotten familiar with using the SHS.

Note that a few of the households surveyed had varying classifications of access to the grid. Some, particularly in Lagos State, lived 'under the grid', i.e., people who live in an area where electricity is available, but they are not able to access it for any number of reasons. Others had a connection, but the duration of electricity per day would not qualify them as having access under the MTF tiered system (see Figure 1).

The communities selected for the research are shown in Table 1. The sample for each location was drawn from three communities and while the intention was also to have 100 households for Lagos State this was not possible due to logistical constraints.

2.2. Participant Selection, Data Collection and Analysis

The selected households had SHS products with ratings of 12 Watts-peak (Wp) in FCT, and 50 Wp in Lagos State, both of which possess 4 LED lighting points and providing lighting service for at least 6 h a day. They also powered a small radio, charged mobile phones and LED torchlights, while the higher capacity systems in Lagos State were connected to a DC-powered television with the ability to pick up selected satellite broadcasts. At the time the interviews were held, the households had owned their SHSs between 6 months and 4 years, with a median ownership period of 29 months. The households made monthly PAYG payments for accessing the SHS via an authorised agent.

Data collection was via face-to-face interviews and respondents (one per household, usually the household head unless unavailable) were first asked basic demographic-type questions, and due to the seasonal nature of their occupation (mostly farmers and fishermen), they were not asked about their income; instead the researchers used the numbers of household members, size of the house, i.e., number of rooms, and the amount spent on non-cooking energy purchases, as proxy indicators for wealth. Respondents were then asked about their non-cooking energy mix, including their current expenditure patterns and willingness to pay for grid electricity. They were also asked about the electrical appliances already owned by the household, and the appliances they desired to own. The survey questionnaire is included in Appendix A.

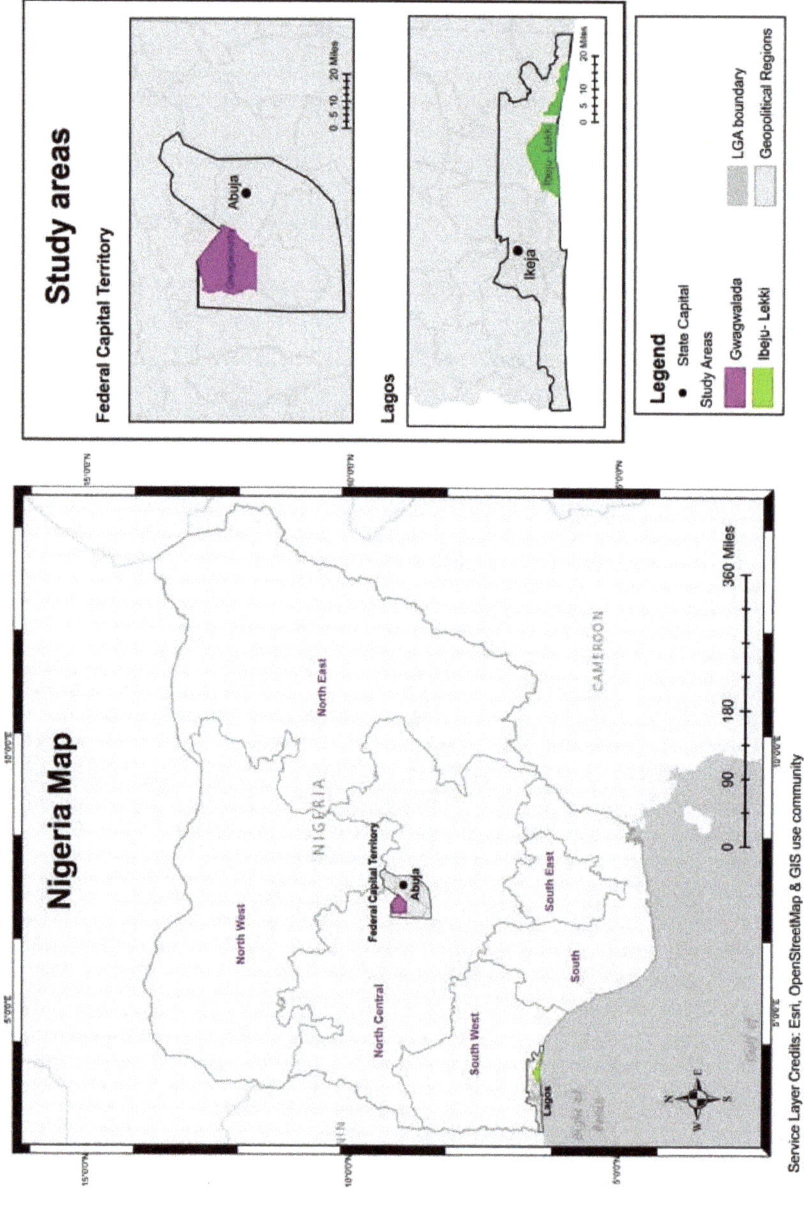

Figure 3. Map of Nigeria showing the states and local government areas (LGA) where the energy surveys were performed. Note that the FCT does not have LGAs, but instead is made up of Area Councils, of which Gwagwalada is one.

Table 1. Communities selected for the survey along with the number of households (HH) in each community.

Lagos State			Federal Capital Territory (FCT)		
Community	**Number of Households**	**%**	**Community**	**Number of Households**	**%**
Magbon	23	15.3	Wuna	55	36.7
Eleko	17	11.3	Rafin Zurfi	30	20.0
Akodo	10	6.7	Dukpa	15	10.0
Total	**50**	**33.3**		**100**	**66.7**

The questions were mostly 'closed' in nature where respondents were provided with a series of options and asked to tick the ones which applied to them. The options provided to the respondents were based in part on published research designed to explore energy use by rural households but also on a series of interviews held with key informants in both Lagos and FCT. There were a few open-ended questions to allow the respondents to provide answers in their own words rather than select from a range of options. In both locations, research assistants familiar with the local languages were recruited to help with the interviewing, as many of the respondents did not understand, or speak English. The questionnaire was written in English and the research assistants verbally translated the questions into the local languages in situations where the respondents did not understand English. While translation between languages risked introducing errors in the results, the structured nature of the questionnaire (which, other than 3 open-ended questions, consisted of closed questions and predefined answers), mitigated this risk.

Data analysis was performed using both quantitative and qualitative methods. The approach to the analysis of the qualitative data was a deductive one and thematic analysis method was used in analysing the data, with the transcription and coding of the interviews performed in Microsoft Word and Nvivo 12, respectively. The quantitative data were analysed using a mixture of descriptive and inferential statistics. As the quantitative data comprised counts of responses, analysis was undertaken using non-parametric statistical tests to explore associations between categories. The Chi-square test was used to determine whether potential associations between different variables were in fact significant. In cases where at least 80% of the expected frequencies of the data are less than 5, the Fisher's exact test was used instead. For post hoc testing, the odds ratio (for 2×2 tables) method was used, while the adjusted standardized residual method proposed by Agresti [39] was used for all other tables. The software Statistical Package for the Social Sciences (SPSS) was used to perform the statistical analyses. The statistical tests for FCT and Lagos State households were performed separately and are presented in the subsequent tables.

All interviewed participants gave their informed consent for inclusion before they participated in the study. The data collection protocol involving human subjects was vetted and approved by the University Ethics Committee (UEC) of the University of Surrey (UEC ref: UEC 2018 039 FEPS).

3. Results

3.1. Sample Description and Household Fuel Sources

The breakdown of the sample in terms of age group, gender of respondent, number of permanent household members and education of respondent is shown as Table 2. Most of the sample respondents were between 18 and 45 years of age, and most were male (especially in FCT). As mentioned earlier, most household heads in Nigeria tend to be male and they make many of the key decisions for the household, so the preponderance of males in the sample is not surprising. However, there were two female household heads in the sample, and both were in the Lagos State communities. Household sizes tended to be in the 1 to 7 range for Lagos State communities (82% of households were in this category) but the figure was higher for the FCT communities with over a third of the households having 14 or more members. This was reflected in the number of rooms across both locations, with the majority of

Lagos State households having 1–4 rooms (84% of the households) while the households in the FCT communities had a higher number of rooms on average, with 65% having greater than 4 rooms per households. Most of the respondents in both areas were educated to at least secondary level.

Table 2. Sample description in terms of age, gender, number of rooms in the house, size of household and education.

	Frequency		Percentage of Total Sample	
(a) Age group of respondents				
	Lagos State	FCT	Lagos State	FCT
18–30	12	39	24	39
31–45	24	43	48	43
46+	14	18	28	18
Total	**50**	**100**	**100**	**100**
(b) Gender of respondents				
Male	37	91	74	91
Female	13	9	26	9
Total	**50**	**100**	**100**	**100**
(c) Number of permanent household members				
1–7	41	29	82	29
8–13	6	36	12	36
14+	3	35	6	35
Total	**50**	**100**	**100**	**100**
(d) Number of rooms in the household				
1–4	42	35	84	35
5–8	8	33	16	33
9+	0	32	0	32
Total	**50**	**100**	**100**	**100**
(e) Maximum education level of respondent				
Male respondents				
Primary	5	23	14	25
Secondary	20	37	54	41
Tertiary	10	21	27	23
No formal education	2	10	5	11
Total	**37**	**91**	**100**	**100**
Female respondents				
Primary	1	2	7.7	22
Secondary	10	5	77	56
Tertiary	1	0	7.7	0
None	1	2	7.7	22
Total	**13**	**9**	**100**	**100**

Table 3a shows the households used a variety of energy sources as their main non-cooking fuel. The patterns of use were observed to vary across both locations. The majority (86%) of households surveyed in Lagos State used a solar home system (SHS) as their main energy source for providing lighting, followed by grid electricity, and then petrol (for powering a generator). Solar home systems also provided the main source of home lighting in FCT communities surveyed; however, the percentage of users was lower, at 66%, compared to the Lagos State households. The next largest lighting energy source in the surveyed FCT households were dry cell batteries at 30%, with kerosene at 4% completing the list.

Table 3. Main energy source for non-cooking (mostly lighting) purposes and satisfaction with the capacity of the main household non-cooking energy source.

| | Main Household Lighting Energy Source | | | |
| | Lagos State | | FCT | |
	Frequency	Percent	Frequency	Percent
(a) Main household non-cooking energy source				
Electricity-SHS	43	86	66	66
Petrol	3	6	0	0
Dry cell batteries	0	0	30	30
Electricity grid	4	8	0	0
Kerosene	0	0	4	4
Total	**50**	**100**	**100**	**100**
(b) Household satisfaction with the capacity of the main household non-cooking energy source				
	Satisfied	Not Satisfied	Satisfied	Not Satisfied
Electricity-SHS	0	43	0	66
Petrol	1	2	0	0
Dry cell batteries	0	0	20	10
Electricity grid	4	0	0	0
Kerosene	0	0	1	3
Total	**5**	**45**	**21**	**79**

Table 3b is a frequency table showing the satisfaction of the surveyed households with the energy capacity of their main non-cooking household energy sources. Across both locations, most respondents (90% in Lagos State and 79% in FCT) indicated not being satisfied with the capacity of their main non-cooking energy source. In the FCT, the only energy source mentioned by households as satisfactory, capacity wise, were dry cell batteries. In Lagos State, 2% of households were satisfied with the energy capacity of their petrol-powered generators, while 8% were satisfied with the capacity provided by the grid. None of the households across both locations indicated being satisfied with the energy capacity of their SHS.

Table 4 is a frequency table showing the spending on non-cooking energy sources by the households across both Lagos State and FCT. These data show that the surveyed Lagos State households spend more money on non-cooking energy than the surveyed FCT households; 52% of Lagos State households mentioned spending at least $20 per month, while the figure for FCT households was 7%. The median spending of both locations was calculated using SPSS, and the values were rounded to $20 for the Lagos State households and $6 for the FCT households. The monthly spend on non-cooking energy variable is used as a proxy for household wealth in the sections covering the inferential statistical analyses, and the calculated median spending for both locations are used as thresholds to indicate the wealthier and less wealthy households.

Table 4. Household monthly spend on non-cooking energy across both locations.

| Monthly Spend on Non-Cooking Energy, USD ($) | Lagos State | | FCT | |
	Frequency	Percent	Frequency	Percent
0–4.99	0	0	42	42.0
5.00–9.99	11	22.0	29	29.0
10.00–14.99	11	22.0	15	15.0
15.00–19.99	2	4.0	7	7.0
20.00–24.99	4	8.0	4	4.0
25.00–29.99	8	16.0	0	0
30+	14	28.0	3	3.0
Total	**50**	**100**	**100**	**100**

The following sections will present the associations between the outcome variables and the monthly spend on non-cooking energy across both locations.

3.2. Ownership of Household Appliances

Table 5 (FCT) and Table 6 (Lagos State) are contingency tables with the association between the households' ownership of various household appliances and their monthly spend on non-cooking energy. The results from Table 5 show a statistically significant association between the FCT households' ownership of various household appliances and their monthly spend on non-cooking energy. Households that spend at least $6 monthly on non-cooking energy owned more mechanical fans than would be expected, and the post hoc analysis using the odds ratio (OR) suggested that these households were 3.62 times more likely to own a mechanical fan compared to those spending less than $6 monthly on non-cooking energy. A similar result was obtained with the ownership of televisions, i.e., households spending at least $6 dollars monthly on their non-cooking energy owned more televisions than could be explained by chance, and the OR implied that these households were 7.65 times more likely to own a television compared to those spending less than $6 per month. Ownership of a CD/DVD player was also significantly associated with the monthly spend on non-cooking energy. The OR suggested that households spending at least $6 monthly on non-cooking energy were 8.7 times more likely to own a CD/DVD player compared to those spending less than $6 monthly.

Table 5. Association between the ownership of household appliances and the monthly spend on non-cooking household energy, FCT.

Household Appliance	Ownership	Monthly Spend on Non-Cooking Energy, USD ($)		Total
		Less than 6	6 and Greater	
Mechanical fan (for indoor cooling)	No	40 (33.5)	27 (33.5)	67
	Yes	9 (15.5)	22 (15.5)	31
	Chi square	7.97 ** (df = 1, N = 98)		
	Odds ratio	3.62 (95% CI: 1.45, 9.05)		
Television	No	40 (29)	18 (29)	58
	Yes	9 (20)	31 (20)	40
	Chi square	20.45 *** (df = 1, N = 98)		
	Odds ratio	7.65 (95% CI: 3.03, 19.35)		
CD/DVD player	No	42 (31)	20 (31)	62
	Yes	7 (18)	29 (18)	36
	Chi square	21.25 *** (df = 1, N = 98)		
	Odds ratio	8.70 (95% CI: 3.26, 23.23)		

Figures provided in the table are counts of respondents in the categories while figures in parentheses are expected counts. Results of the Chi-square statistical tests are provided, while the post hoc test results are presented using the odds ratio. ** $p < 0.01$; *** $p < 0.001$.

The results in Table 6 also show that there is a significant association between the ownership of household appliances in Lagos State households, and their monthly spend on non-cooking energy. The households were asked about their ownership of selected household appliances, and just as with the FCT results, the predictor variables here were used as proxies for household wealth. The results show that there was a significant positive association between the ownership of the household appliances and household monthly spend on non-cooking energy, suggesting the wealthier households owned more appliances in their households, even when these appliances could not be powered by their SHS. The OR values indicates wealthier Lagos State households, i.e., those spending at least $20 monthly on non-cooking energy, are almost 124 times more likely to own a mechanical fan and almost 6 times more likely to own a CD/DVD player, when compared to the households spending less than $20 monthly.

Table 6. Association between the ownership of household appliances and the monthly spend on non-cooking energy, Lagos State.

Household Appliance	Ownership	Monthly Spend on Non-Cooking Energy, USD ($)		Total
		Less than 20	**20 and Greater**	
Mechanical fan (for indoor cooling)	No	17 (8.16)	0 (8.84)	**17**
	Yes	7 (15.84)	26 (17.16)	**33**
	Chi square	27.90 *** (df = 1, N = 50)		
	Odds ratio	123.67 (95% CI: 6.63, 2306.11)		
CD/DVD player	No	15 (10.08)	6 (10.92)	**21**
	Yes	9 (13.92)	20 (15.08)	**29**
	Chi square	7.22 ** (df = 1, N = 50)		
	Odds ratio	5.56 (95% CI: 1.62, 19.03)		

Figures provided in the table are counts of respondents in the categories while figures in parentheses are expected counts. Results of the Chi-square statistical tests are provided, while the post hoc test results are presented using the odds ratio. ** $p < 0.01$; *** $p < 0.001$.

3.3. Preference for Grid Connection and Satisfaction with SHS Capacity

Table 7 sets out the association between amount households are willing to pay for grid connection and the monthly spend on non-cooking energy, FCT. The results indicate a statistical association between the amount the FCT households were willing to pay for a one-off connection fee, and their monthly spend on non-cooking energy. For the households willing to pay a connection fee for the grid, their willingness to pay, in monetary terms, was significantly associated with their monthly spend on non-cooking energy. This association was driven by households willing to pay amounts greater than $28 for this connection—households that spent at least $6 monthly on non-cooking energy were more likely to be willing to pay a minimum grid connection fee of $28, compared to those spending less than $6 monthly. Note that other outcome variables were tested for an association with 'monthly spend on cooking energy' to gauge the households' desire for grid electricity (e.g., preference for stable grid connection for 4–6 h/day to current SHS connection, willingness to pay a one-off grid connection fee, maximum monthly tariff households are willing to pay for this connection, and their satisfaction with daily duration of SHS). However, none provided a significant result.

Table 7. Association between amount households are willing to pay for one-off grid connection fee and the monthly spend on non-cooking energy, FCT.

One-Off Grid Connection Fee Household is Willing to Pay, USD ($)	Monthly Spend on Non-Cooking Energy, USD ($)		Total
	Less than $6	**$6 and Greater**	
Less than $11	17 (15.7) [0.59]	13 (14.3) [−0.59]	**30**
Between $11 and $28	22 (18.31) [1.62]	13 (16.69) [−1.62]	**35**
Greater than $28	6 (10.99) [−2.51]	15 (10.01) [2.51]	**21**
Chi square	6.53 * (df = 2, N = 86)		
Total	**45**	**41**	**86**

Figures provided in the table are counts of respondents in the categories while figures in round parentheses are expected counts. Results of the Chi-square test is presented, * $p < 0.05$. The figures in square parentheses are the adjusted standardized residual values.

The dissatisfaction with SHS capacity, and desire for grid electricity (or any technology capable of powering larger appliances) was also mentioned in the respondents' answers to questions about how they felt about their SHS, for example ('FCT-R' stands for 'FCT respondent'):

'The only problem I have with it is that it cannot (power) my television or my fridge and my fan, that is the only problem I have with the solar' (FCT-R1).

'I would prefer a bigger one that can take a fridge and television' (FCT-R4).

'I really want NEPA, if it were available, I would pay for NEPA. But since there is no NEPA, that is why we went for solar' (FCT-R7).

NEPA is an acronym for the National Electric Power Authority, a now defunct institution that once managed the electricity supply system in Nigeria and is commonly used as a metaphor for grid electricity in Nigeria.

Table 8 shows that the preference of the Lagos State households to connect to the grid, and to pay a one-off connection fee, was significantly associated with their monthly spend on non-cooking energy. The results suggested that wealthier households (spending more than $20 monthly on non-cooking energy) were 13 times more likely to prefer a stable grid connection with 4–6 h of electricity access per day, and were 9 times more willing to pay a connection fee to access this stable grid, compared to the less wealthy households spending less than $20 monthly on non-cooking energy. Table 8 also shows that the maximum amount these households were willing to spend as monthly tariffs for this connection was significantly associated with their monthly spend on non-cooking energy. The results implied that the wealthier households were more likely to pay higher sums (in excess of $40), as monthly tariffs for this stable connection to the grid compared to the less wealthy households, while the less wealthy households were more likely to pay smaller sums (less than $30) as monthly tariffs for the stable grid electricity.

Table 8. Association between the preference for a stable grid connection to SHS, willingness to pay for one and the satisfaction with SHS duration, and monthly spend on non-cooking energy in Lagos State households.

Preference for Grid over SHS, Willingness to Pay for Access to a Grid Connection & Satisfaction with Duration of SHS		Monthly Spend on Non-Cooking Energy, USD ($)		Total
		20 and Greater	**Less than 20**	
Preference for stable grid connection for 4–6 h/day to current SHS connection	Yes	22 (15.08)	7 (13.92)	**29**
	No	4 (10.92)	17 (10.08)	**21**
	Chi square	15.57 *** (df = 1, N = 50)		
	Odds ratio	13.36 (95% CI: 3.35, 53.20)		
Willing to pay a connection fee to access stable grid?	Yes	23 (17.68)	11 (16.32)	**34**
	No	3 (8.3)	13 (7.7)	**16**
	Chi square	10.42 ** (df = 1, N = 50)		
	Odds ratio	9.06 (95% CI: 2.13, 38.49)		
Maximum regular monthly payment households are willing to make for access to stable grid electricity, USD ($)	$10–$29	3 (8.12) [−3.93]	9 (3.88) [3.93]	**12**
	$30–$40	7 (6.09) [−0.76]	2 (2.91) [0.76]	**9**
	$41–$83	13 (8.79) [−3.17]	0 (4.21) [3.17]	**13**
	Fisher's exact test	16.63 *** (N = 34)		
Are you satisfied with the daily duration of your SHS?	Yes	16 (10.08)	5 (10.92)	**21**
	No	8 (13.92)	21 (15.08)	**29**
	Chi square	11.53 *** (df = 1, N = 50)		
	Odds ratio	8.40 (95% CI: 2.31, 30.60)		

Figures provided are counts of respondents in the categories while figures in parentheses are expected counts. Results of the statistical tests are provided using the Chi-square and Fisher's exact test. Post hoc tests are performed using the odds ratio and adjusted standardized residuals, which are presented in square parentheses. ** $p < 0.01$; *** $p < 0.001$.

The preference for grid-based electricity was also suggested by the respondents' answers to questions about how they felt about their SHS, with responses such as the following ('LG-R' stands for 'Lagos State respondent'):

'I prefer NEPA because, you know the solar cannot (power) a fan, but NEPA can (power) the fan, fridge, television' (LG-R3).

'If NEPA (is available) I would prefer it, so I can have fridge and fan' (LG-R12).

'I just think it (SHS) is a waste of money (…) it does not have fan, it cannot (power) a freezer, meanwhile in one week you spend N2,500 ($7) on it (…) If there is NEPA, nobody in this environment will use it' (LG-R11).

The households were asked if they were satisfied with the daily duration of their SHS, and the results for the Lagos State communities are also shown in Table 8. There was a significant association between their response to this question and the household monthly spend on non-cooking energy. The results suggest that the less wealthy households (spending less than $20 monthly on non-cooking energy) were more likely to be satisfied with its duration while the wealthier households (spending more than $20 monthly on non-cooking energy) were less likely to be satisfied, than could be explained by chance. The OR value suggests that less wealthy households were 8.4 times more likely to be satisfied by the SHS duration, compared to the wealthier households.

In summary, the results show that while most of the households across both FCT and Lagos State use SHS as their main household lighting energy source, none of them were satisfied with its capacity. It also suggested that the wealthier households in both locations were likely to own appliances that could not be accommodated by their SHSs, when compared to the less wealthy households. Furthermore, the results imply that among the FCT households, the wealthier ones were significantly more willing to pay for a connection to the grid, compared to their less wealthy neighbours. The Lagos State results were similar, with wealthier households more likely to prefer a grid connection and be more willing to pay for it, compared to the less-wealthy households.

4. Discussion

Four themes emerged from the data obtained across both locations. First, differences were observed in the main lighting energy source used by the households across both locations. SHS was the dominant source used for lighting across both locations, 66% using it as their main source in FCT while 86% used it as a main source in Lagos State. Further, 30% of households in FCT used dry cell batteries for their main lighting energy source, while no household in Lagos State reporting this. The dominant nature of SHS for lighting in the households is to be expected, considering research has shown that once solar PV technologies are introduced into households, they displace kerosene used for lighting [40–42]. The difference in SHS use observed across both locations could be explained by the fact the FCT households, on average, have a greater number of rooms compared to the Lagos State households (see Table 2). Both types of SHS used in the households surveyed provide only 4 lighting points, hence many of the FCT households (65% with greater than 4 rooms) would require separate energy sources to provide lighting for the other rooms. This likely explains their relatively high use (30%) of dry cell batteries, compared to Lagos State households (84% with less than 5 rooms).

Second, as set out in Tables 5 and 6, multiple households owned various appliances that could not be accommodated by their SHS and hence were, for all intents and purposes, unusable. While this behaviour seemed baffling, it has been documented in previous research. In their survey in rural Kenya, Lee et al. [34] (p. 91) observed a similar situation and referred to it as "aspirational" purchases, while Stojanovski et al. [20], in their survey of SHS-using rural Kenyan and Ugandan households, implied that these purchases might provide their owners with a sense of having higher social status. Stojanovski et al. [20] also indicated the households believed, without any proof, that the national grid would soon be extended to their communities, which would allow them to use these aspirational appliances.

Third, studies on communities using SHS as their primary lighting providers have shown that many view it as more of a pre-electrification process, and are willing to make payments to a value that would be economically sustainable for a grid-type utility [13,43]. This was observed in the results of this study as seen in Tables 7 and 8, with 86% of FCT households and 68% of Lagos State households stating a willingness to pay a one-off connection fee for access to grid electricity. This follows on from the earlier observation, i.e., a significant number of households purchase electrical appliances that cannot be accommodated by their SHS, in the hope of getting a grid connection in the future and make use of the energy services (space cooling, entertainment, etc.) these appliances provide. Finally, while the energy services afforded to the households by their SHSs differed across both locations (FCT households had lighting and mobile phone charging, while the Lagos State households' SHS included a television) none of the households in either location indicated being satisfied with the energy capacity of their SHS.

These themes show that while both locations had similar issues with their lack of electrification, there were some differences in the way households used energy sources, and their desire to connect to the grid. This suggests it might be pertinent for policy makers and private institutions involved with rural electrification to tailor their interventions as opposed to applying a blanket approach to these rural communities. For example, the levels of subsidies or grants that might be offered for rural electrification projects could be made to be dependent on the relative wealth of the communities to be electrified, based on metrics that indicate their wealth and willingness to pay for electricity access. However, further research is required to determine whether these observed differences are statistically significant.

SHSs are often seen as a means to get households onto the modern energy ladder [33,44,45]. However, the results show that for many it was a less desirable substitute for a grid-based service. Their inability to connect as many appliances to the SHS as they would have liked seemingly led to the statistically significant associations between the wealth of a household and (i) their ownership of various household appliances with wattages too high for their SHS, and (ii) their desire to connect to, and pay for, a grid-type electricity supply. This confirms the intuition that wealthier households are more likely to purchase these high wattage appliances and therefore will be more inclined to prefer a grid connection. The minimum wage of approximately $50 per month in Nigeria [46] at the time of conducting the survey provides some context regarding the FCT households' willingness to spend in excess of $28 for a one-off grid connection fee, and the Lagos State households' willingness to make monthly $40-plus electricity tariff payments for ongoing access to the grid. However, it is important to note that a household stating they are willing to pay for a grid connection does not necessarily equate to their ability to pay for said connection or indeed a decision to actually do it if the opportunity arises. Nonetheless, the findings demonstrate that a significant number of the households are unable to move up the energy ladder (or in the case of the MTF, up the tier levels) regardless of their willingness or ability to pay, as they do not have access to an electricity connection that can meet their demands. In this situation, the household can be referred to as being 'energy locked in'.

There are other ways to get rural households up the energy ladder ahead of expansion of the national grid, and ahead of alternative major infrastructure investments in mini-grids. One approach is so-called "swarm electrification", to connect a number of SHS units across a community, which will act to smooth the total electricity demand through diversity effects and share the generation capacity from all individual systems [47], as a sort of 'micro-grid'. This bottom-up approach to electrification eliminates the typical oversizing in the design of mini-grids and grid extension projects while enabling households to transition to a higher energy access tier using their legacy SHS [44,47]. Another approach is the development of a diverse and competitive market for energy efficient, DC-powered plug-and-play appliances. As described by Narayan et al. [48], these consume a fraction of the energy required by their mainstream counterparts running on AC. Some SHS providers bundle these efficient appliances with their service, and in the households surveyed for this research the SHS came with four energy efficient, DC-powered LED bulbs at 1 W each, a torchlight at 1.1 W, and (in Lagos State) a television at 10.8 W. Creating more energy efficient appliances, such as the Youmma refrigerator which has a power rating of 17.8 W [49] and having them come with a standardized, compatible plug-and-play facility, can enable SHS-using households to move up the MTF tiers. However, in order for this energy efficient appliance market to grow, quality standards need to be created and adopted by countries involved with SHS electrification programs. Further, traditional household electronics manufacturers, with the know-how of making these appliances, could be encouraged to get involved in the production of energy-efficient DC-powered appliances. This would help provide a greater selection of appliances, and more competition in the market.

A natural progression of this research would be to expand the numbers of surveys to include communities using a wider variety of off-grid technologies, for example from simple solar lanterns to solar mini-grids, and observe the changes, if any, to the results of this research. Further, expanding the number of locations to include more remote rural communities that are a further distance from major cities/urban centres than in this research, might result in energy-use patterns that provide different results. It should also be noted that the choice of interviewing the heads of households led to the majority of the respondents in the study being male. The traditional role women have with the collection of cooking fuel, food preparation and cooking due to cultural gender norms [50–52] would likely reduce the time they can spend using certain energy services provided by the SHS, for example watching a television. It is therefore possible women might have different opinions to men with regards to their views and aspirations about their non-cooking energy service demands and the ability of a SHS to meet them, and the significance of these gender-based differences, if any, could be the subject of future research.

5. Conclusions

Energy is a crucial input for socioeconomic development, and it follows that the ability to not only access modern energy, but to transition to a wide range of energy services, is instrumental to the development of any society. This research set out to examine energy transitions in rural Nigeria, with the following objectives:

- Examine if the SHS technologies distributed to these rural communities met their non-cooking household energy services demand.
- Examine if energy lock-in effects occur in these rural households using SHS technologies.

The results showed that none of the households were satisfied with the energy capacity of their SHS, and that the wealthier households were willing to pay relatively large sums to connect to the grid so they could use energy services unavailable to them with their SHS. However, they were unable to do so due to the unavailability of any electricity supply technology that would power their appliances. While the goal of SDG7 may have been achieved across these communities with the provision of SHS, many of the households clearly exhibited some frustration and dissatisfaction with the actual level of energy access, as they do not see a clear transition pathway up the energy ladder.

Two methods were suggested to help rural households in this situation move up the MTF energy access tiers: swarm electrification using the existing SHS technologies, and the provision of a greater selection of energy efficient, plug-and-play DC-powered appliances. Public and private entities in Nigeria involved in rural electrification with SHS could use the findings shown here to better understand the energy transition needs of rural households, and provide the interventions required to move them up the ladder. For example, they could provide funding into researching swarm electrification technology in rural Nigeria.

Author Contributions: Conceptualisation, O.H., S.M., and M.L.; methodology, O.H., S.M., and M.L.; software, O.H.; validation, O.H., S.M., and M.L.; format analysis, O.H., investigation, O.H.; resources, O.H., S.M., and M.L.; data curation, O.H.; writing—original draft preparation, O.H.; writing—review and editing, O.H., S.M., and M.L.; visualisation, O.H.; supervision, S.M. and M.L. All authors have read and agreed to the published version of the manuscript.

Funding: This research received no external funding.

Acknowledgments: This work was supported by the Centre for Environment and Sustainability, University of Surrey.

Conflicts of Interest: The authors declare no conflict of interest.

Appendix A. Survey Questionnaire

SECTION 1: RESPONDENT DETAILS

1.1   **Age of respondent** □ 18–21 (1) □ 22–25 (2) □ 26–30 (3) □ 31–35 (4) □ 36–40 (5)

□ 41–45 (6) □ 46–50 (7) □ 51–55 (8) □ 56–60 (9) □ 61+ (10)

1.2   **Gender of respondent** □ Female (2) □ Male (1)

SECTION 2: HOUSEHOLD INFORMATION

2.1   **What is the highest level of formal education by respondent?** □Nursery (1) □Primary (2) □ Secondary (3)

□Undergraduate (4) □Postgraduate (5)

□Other (please explain) (6) ___________________

□None of the above (7)

2.2   **Is the respondent the head of the household ?**   □Yes (1) □No (2)

2.3   **If (no) to 2.2, what is the relationship of respondent to head of household?**   □ Brother (1) □ Sister (2) □ Friend (3)

□ Husband (4) □ Wife (5) □ Father (6)

□ Mother (7) □ Child (8)

□ Other (please explain) (9)

2.4   **What is the main occupation of the respondent?** □ Store owner (1) □ Farmer (2) □ Street Hawker (3)

□ Civil servant (4) □Factory worker (5) □ Teacher (6)

□ Trader (7) □ Mechanic (8) □ Hairdresser (9)

□ Others (specify) (10) _______________________

2.5   **What type of dwelling is the household?** □ separate house (1) □ semi-detached house (2) □ Hut (3)

□ Tent (4) □ Compound house (5) □ Flat (6)

□ Other (specify) (7) _______________________

2.6   **What material is the roof mostly made of?** _______________________

2.7   **What material is the floor mostly made of?** _______________________

2.8.   **What material are the walls mostly made of?** _______________________

2.9   **How many 'family' or living spaces does the home have?** _______________________

2.10   **How many rooms are in the household?** _______________________

2.11   **Does the respondent keep any livestock?**   □ Yes (1) □ No (2)

If yes, please list all and the number of each

Type ___________ Number ___________

Type ___________ Number ___________

Type ___________ Number ___________

Type ___________ Number ___________

Type ___________ Number ___________

2.12   **Do you grow your own crops?**   □ Yes (1) □ No (2)

If yes, please list all and approximate area of **total** farmland (plot)

Type ___________ Area of farm ___________

Type ___________

Type ___________

Type ___________

Type ___________

2.13   **What proportion of your income would you say you spend**

     on household energy (non-cooking)?   ______________________________

2.14   **How many people live in this household?**   ______________________________

2.15   **Of the people in (2.14), how many live in the house at least**

     18 days or more per month?   ______________________________

SECTION 3: HOUSEHOLD ENERGY MIX/CONSUMPTION

3.1   **Which of these is your main household fuel** *for lighting only*? **Please select** *no more than one*

     □ Candles (1) □ Kerosene (2) □ LPG (3) □ firewood/biomass (4) □ Car batteries (5) □ Dry cell batteries (6)

     □ Electricity (grid) (7) □ electricity (mini-grid) (8) □ Electricity (SHS) (9) □ Petrol (10) □ Diesel (11)

     □ Other (please specify) (12)   ______________________________

3.2   **Reasons for using the above fuel?**

     □ Familiar fuel (1) □ No other fuel available (2) □ Easy to use (3) □ Easily available (4) □ Cheap (5)

     □ Other fuels are not affordable (6) □ Other (please explain) (7) ______________________________

3.3   (a) **Are you satisfied using this (main) fuel?**

     □Yes (1)   □ No (2)

     (b) **If No to (3.3), please give reason why. Select up to 3 reasons**

     □ Not selected (0) □ Not safe (1) □ Expensive (2) □ Health concerns (3) □ Other (please explain) (4)_________

3.4   **What other fuel sources do you use at home for lighting?** *Please select at least one*

     □ Candles (1) □ Kerosene (2) □ LPG (3) □ firewood/biomass (4) □ Car batteries (5) □ Dry cell batteries (6)

     □ Electricity (grid) (7) □ electricity (mini–grid) (8) □ Electricity (SHS) (9) □ Petrol (10) □ Diesel (11)

     □ Mobile phone (12) □ Other (please specify) (13) ______ □ None (14) **(if none, skip to question 3.6)**

3.5   **When do you use this other fuel source for lighting?**

     □ Main fuel too expensive (1) □ Main fuel not available (2) □ Main fuel capacity inadequate (3)

     □ Other (please explain) (4)______________________________ □ not applicable (5)

3.6   **What other fuel source do you use at home for other non-lighting & non-cooking activities? (radio, television, fan, fridge, etc.)**

     □ Car batteries (1) □ Dry cell batteries (2) □ Electricity (grid) (3) □ electricity (mini-grid) (4) □ Electricity (SHS) (5)

     □ Petrol (6) □ Diesel (7) □ Other (please specify) (8) ______________ □ None (9)

3.7   **Other sources of fuel which are available for purchase (for any non-cooking purpose) that are not used, and why (please tick if relevant):**

Available Fuel	Not Safe (1)	Expensive (2)	Smoky (3)	Unreliable (4)	Health Concerns (5)	Redundant (6)	Not Available (7)
Candles							
Kerosene							
LPG							
Electricity (grid)							
Electricity (mini grid)							
Electricity (SHS)							
Petrol							
Diesel							
Other (specify)							

3.8   **Who determines the household choice of fuels (for non-** □ Head of household (1) □ Other (please specify) (2)

     cooking activities only, e.g., lighting, fan, fridge, etc.)? ______________________________

3.9   **Energy Access Attributes: Make notes of the units used (per week, per month, kg, bag, etc. to allow for consistency). If individual respondents prefer to use weekly or monthly basis, make a note of which one they used.**

Household Consumption Questions	Candles (i)	Kerosene (ii)	LPG (iii)	Electricity (Grid) (iv)	Electricity (Mini Grid) (v)	Electricity (SHS) (Hours) (vi)	Petrol (Litres) Q (vii)	Diesel (viii)	Car Batteries (ix)	Dry Cell Batteries (x)	Other (xi)
(a). How much do you use, on average, on a weekly basis for household fuel?											
(b). How long does it take to purchase, or how far do you have to travel to purchase?											
(c). * How often do you charge it, on a weekly basis?											

Notes: For (a), measuring the consumption of grid electricity might be difficult to ascertain. Calculate based on how much they are charged in their tariffs. Further, the questions with "*" is for car batteries only.

3.10

Reliability Questions	Candles (i)	Kerosene (ii)	LPG (iii)	Electricity (Grid) (iv)	Electricity (Mini Grid) (v)	Electricity (SHS) (Hours) (vi)	Petrol (Litres) Q (vii)	Diesel (viii)	Car Batteries (ix)	Dry Cell Batteries (x)	Other (xi)
(a). In the past 1 month, would you say you have experience shortages/interruptions with this fuel?											
(b). In the past 1 month, how many times have you experienced shortages/interruptions with this fuel?											

Notes: For the electricity fuels, number of shortages would be measured by how many hours per day there was no electricity. For other fuels, they would involve number of times these fuels have not been available to purchase in that month period.

3.11

Affordability Questions	Candles (i)	Kerosene (ii)	LPG (iii)	Electricity (Grid) (iv)	Electricity (Mini Grid) (v)	Electricity (SHS) (Hours) (vi)	Petrol (Litres) Q (vii)	Diesel (viii)	Car Batteries (ix)	Dry Cell Batteries (x)	Other (xi)
(a). Over the past 1 month, how much on average, per unit, do you spend on purchasing the fuel?											
(b). Would you say that you can afford to purchase the fuel:											
1. All the time											
2. Only some of the time											
3. Can barely or cannot afford it											

Notes: Per unit here means per kg, per litre, per bag, etc. Ensure the "per unit" is defined by the respondent. It does not have to be an exact figure, but an approximation of size/quantity should be obtained. 'Per unit' is different for SHS users as they would most likely be paying a flat rental rate per week/month, hence their 'unit' would be a period of time.

SECTION 4: HOUSEHOLD ELECTRIFICATION

4.1   (a) **Is this household connected to the national grid supply for electricity? (only national provider should be considered, i.e., PHCN)**

☐ Yes (1)   ☐ No (2)

(b) **If No, please say why you are not connected**

☐ High cost of connection (1) ☐ High cost of tariff/monthly payments (2) ☐ not reliable (3) ☐ not available (4)

☐ other (please explain) (5)

4.2   (a) **Does this household receive electricity from the grid (defined as defined as 12+ hours of grid electricity per week)**

☐ Yes (1)   ☐ No (2)

(b) **If No, then how is electricity supplied to the household?**

☐ PV solar panel (1) ☐ Mini-grid (2) ☐ Car battery (3) ☐ Dry cell battery (4) ☐ Personal generator (5)

☐ Communal generator (6) ☐ Other (please specify) (7) ☐ not applicable (8)

4.3   **Do you own a solar home system (SHS)?**

☐ Yes (1)   ☐ No (2) **(if "no", skip to question 4.19)**

(if "yes", skip questions 4.19 & 4.20)

4.4   (a) **Did you purchase the SHS, or was it provided for free?**

☐ Purchase (1)   ☐ Free (2)   ☐ Not applicable (3)

(b) **If "free", would you still have gotten the SHS if it was not free?**

☐ Yes (1)   ☐ No (2)   ☐ Not applicable (3)

4.5   **How long ago (in months) did the household first get the SHS?**

_____________ months

4.6   **What is the capacity of your SHS (in Watts?)** (*If they do not know the answer to this, offer to check the unit*)

_____________ watts

4.7   **Which of the following appliances do you want to own, already own, do not want/care to own, or cannot use? (Tick as appropriate)**

Appliances	Want to Own (1)	Already Own (2)	Do Not Need/Want (3)	Own, But Cannot Use (4)	Not Applicable (5)
LED bulb					
Mechanical fan					
Radio					
Television					
Fridge/freezer					
Water Pump					
Mobile Charger					
Rechargeable Torch					
Electric Iron					
CD/DVD Player					
Air conditioning					
Electric Cooker					
Other (please specify)					

4.8 **If you ticked any appliance under "want to own", why haven't you purchased the equipment?**

☐ Too expensive to purchase (1)   ☐ Not available to purchase (2)   ☐ capacity of power source not adequate (3)

☐ other (please explain) (4)

4.9 **In the past 3 months, how many hours a day would you say you use the SHS system?**

☐ <1 (1) ☐ 1 (2) ☐ 2 (3)   ☐ 3 (4) ☐ 4 (5)   ☐ 5 (6) ☐ 6 (7)   ☐6+ (8) ☐ n/a (12)

4.10 **What is the duration, in hours/day, you would prefer to have the supply of electricity, as opposed to what you have now?**

a ☐ From ________ to __________ (1)   **b** ☐ Happy with the current duration (2) **c** ☐ n/a (3)

4.11 **What has been the most important use of your SHS for your household? Please tick in descending order of importance, with 4 being the most important and 1 the least important.**

	Entertainment (television, radio) (a)	Security (as a result of better light outside) (b)	Using it for work (mobile phone charging business, etc.) (c)	For general household lighting (d)
4				
3				
2				
1				
n/a				

4.12 **In the past 1 month, has your SHS failed to provide electricity for more than half the time it did when you got it?**

☐ Yes (1)   ☐ No (2)   ☐ n/a (3)

4.13. **Do you feel the capacity of your SHS limits your ability to fully enjoy the benefits of electricity, i.e., would you enjoy it more if it were able to power more electrical appliances?**

☐ Yes (1)   ☐ No (2)   ☐ n/a (3)

4.14 **(a) Do you prefer a mini-grid/national grid connection guaranteed for, say 4 to 6 h per day, (use NEPA or PHCN acronym if it helps them understand) than the SHS for the whole day?**

☐ Yes (1)   ☐ No (2)   ☐ n/a (3)

(b) If "Yes, what are your reasons?

____

____

____

(c) If "No", what are your reasons?

____

____

____

4.15 **(a) Would you be willing to pay a part of the connection/set up cost if it meant you would get connected to a national/mini grid that promises to provide uninterrupted supply of electricity?**

☐ Yes (1)   ☐ No (2)   ☐ n/a (3)

(b) If "yes", how much would you be willing to pay? ____________

(c) If "yes", what is the minimum number of hours per day of electricity you would expect from the grid to be worth your payment?

Amount ____________

4.16 **How much more compared to your SHS payments would you be willing to pay regularly on a monthly basis for electricity, if you can upgrade to the national grid/mini grid?** (*If SHS use is free, then just ask how much they would be willing to pay monthly to use the grid, as opposed to comparing to SHS payments*)

Amount: ___________

4.17 **Has your SHS lived up to the expectations you had before you purchased it. Please explain**

____

4.18 **If the answer to 4.3 is "no", why do you not own a SHS?**

☐ High cost of connection (1)   ☐ High cost of tariff/monthly payments (2)   ☐ not reliable (3)   ☐ not available (4)

☐ Not needed (5)   ☐ other (please explain)   (6)   ☐ n/a (7)   ☐ capacity too small (8)

4.19 **(a) If the answer to 4.3 is "no", have you previously owned a SHS?**

☐ Yes (1)   ☐ No (2)

(b) If "yes", why do you no longer have one? Please explain

____

(c) If "yes", what has been your experience since you have switched to using other fuel sources for your household? Please explain _______________________________

____

References

1. Oyedepo, S.O. Energy and sustainable development in Nigeria: The way forward. *Energy Sustain. Soc.* **2012**, *2*, 1–17. [CrossRef]

2. Martínez, D.M.; Ebenhack, B.W. Understanding the role of energy consumption in human development through the use of saturation phenomena. *Energy Policy* **2008**, *36*, 1430–1435. [CrossRef]

3. IEA; United Nations Statistics Division; International Renewable Energy Agency; World Bank; World Health Organization. *Tracking SDG 7: The Energy Progress Report 2020*; Washington, DC, USA, 2020; Available online: https://www.iea.org/reports/tracking-sdg7-the-energy-progress-report-2020 (accessed on 4 November 2020).

4. Zubi, G.; Fracastoro, G.V.; Lujano-Rojas, J.; El Bakari, K.; Andrews, D. The unlocked potential of solar home systems; an effective way to overcome domestic energy poverty in developing regions. *Renew. Energy* **2019**, *132*, 1425–1435. [CrossRef]

5. Bednar, D.J.; Reames, T.G. Recognition of and response to energy poverty in the United States. *Nat. Energy* **2020**, *5*, 432–439. [CrossRef]

6. González-Eguino, M. Energy poverty: An overview. *Renew. Sustain. Energy Rev.* **2015**, *47*, 377–385. [CrossRef]

7. Jessel, S.; Sawyer, S.; Hernández, D. Energy, Poverty, and Health in Climate Change: A Comprehensive Review of an Emerging Literature. *Front. Public Health* **2019**, *7*, 357. [CrossRef]

8. Reddy, A.K.N. Energy and Social Issues. In *World Energy Assessment: Energy and the Challenge of Sustainability*, 1st ed.; Goldenberg, J., Ed.; UNDP: New York, NY, USA, 2000; pp. 39–61.

9. Bhatia, M.; Angelou, N. *Beyond Connections: Energy Access Redefined*; The World Bank: Washington, DC, USA, 2015.

10. Culver, L.C. Energy Poverty: What You Measure Matters. In Proceedings of the Reducing Energy Poverty with Natural Gas: Changing Political, Business and Technology Paradigms Symposium, Stanford, CA, USA, 9–10 May 2017; pp. 1–25.

11. Bhattacharyya, S.C.; Palit, D. Introduction. In *Mini-Grids for Rural Electrification of Developing Countries*, 1st ed.; Bhattacharyya, S.C., Palit, D., Eds.; Springer International Publishing: Cham, Switzerland, 2014; pp. 1–10.

12. Palit, D.; Bandyopadhyay, K.R. Rural electricity access in South Asia: Is grid extension the remedy? A critical review. *Renew. Sustain. Energy Rev.* **2016**, *60*, 1505–1515. [CrossRef]

13. Palit, D.; Chaurey, A. Off-grid rural electrification experiences from South Asia: Status and best practices. *Energy Sustain. Dev.* **2011**, *15*, 266–276. [CrossRef]

14. Bos, K.; Chaplin, D.; Mamun, A. Benefits and challenges of expanding grid electricity in Africa: A review of rigorous evidence on household impacts in developing countries. *Energy Sustain. Dev.* **2018**, *44*, 64–77. [CrossRef]

15. Agenbroad, J.; Carlin, K.; Ernst, K.; Doig, S. *Minigrids in the Money: Six Ways to Reduce Minigrid Costs by 60% for Rural Electrification*; Rocky Mountain Institute: Boulder, CO, USA, 2018.
16. Deichmann, U.; Meisner, C.; Murray, S.; Wheeler, D. The economics of renewable energy expansion in rural Sub-Saharan Africa. *Energy Policy* **2011**, *39*, 215–227. [CrossRef]
17. Carr-Wilson, S.; Pai, S. Pay-As-You-Go: How a Business Model Is Helping Light Millions of Rural Kenyan Homes with Solar. *Case Stud. Environ.* **2018**, *2*, 1–10. [CrossRef]
18. Sawin, J.L.; Sverrisson, F.; Rickerson, W. *Renewables 2015 Gobal Status Report*; REN21: Paris, France, 2015.
19. Baurzhan, S.; Jenkins, G.P. Off-grid solar PV: Is it an affordable or appropriate solution for rural electrification in Sub-Saharan African countries? *Renew. Sustain. Energy Rev.* **2016**, *60*, 1405–1418. [CrossRef]
20. Stojanovski, O.; Thurber, M.; Wolak, F. Rural energy access through solar home systems: Use patterns and opportunities for improvement. *Energy Sustain. Dev.* **2017**, *37*, 33–50. [CrossRef]
21. Ashden. Mobisol/Electrifying East Africa. Available online: https://www.ashden.org/winners/mobisol#continue (accessed on 10 January 2018).
22. Barrie, J.; Cruickshank, H.J. Shedding light on the last mile: A study on the diffusion of Pay as You Go Solar Home Systems in Central East Africa. *Energy Policy* **2017**, *107*, 425–436. [CrossRef]
23. Adwek, G.; Boxiong, S.; Ndolo, P.O.; Siagi, Z.O.; Chepsaigutt, C.; Kemunto, C.M.; Arowo, M.; Shimmon, J.; Simiyu, P.; Yabo, A.C. The solar energy access in Kenya: A review focusing on Pay-As-You-Go solar home system. *Environ. Dev. Sustain.* **2020**, *22*, 3897–3938. [CrossRef]
24. Bisaga, I.; Puźniak-Holford, N.; Grealish, A.; Baker-Brian, C.; Parikh, P. Scalable off-grid energy services enabled by IoT: A case study of BBOXX SMART Solar. *Energy Policy* **2017**, *109*, 199–207. [CrossRef]
25. IRENA. *Pay-as-You-Go Models: Innovation Landscape Brief*; International Renewable Energy Agency: Abu Dhabi, UAE, 2020.
26. Azimoh, C.L.; Klintenberg, P.; Wallin, F.; Karlsson, B. Illuminated but not electrified: An assessment of the impact of Solar Home System on rural households in South Africa. *Appl. Energy* **2015**, *155*, 354–364. [CrossRef]
27. Mondal, A.H.; Klein, D. Impacts of solar home systems on social development in rural Bangladesh. *Energy Sustain. Dev.* **2011**, *15*, 17–20. [CrossRef]
28. Rahman, S.M.; Ahmad, M.M. Solar Home System (SHS) in rural Bangladesh: Ornamentation or fact of development? *Energy Policy* **2013**, *63*, 348–354. [CrossRef]
29. Lemaire, X. Solar home systems and solar lanterns in rural areas of the Global South: What impact? *Wiley Interdiscip. Rev. Energy Environ.* **2018**, *7*, e301. [CrossRef]
30. Diallo, A.; Moussa, R.K. The effects of solar home system on welfare in off-grid areas: Evidence from Côte d'Ivoire. *Energy* **2020**, *194*, 1–12. [CrossRef]
31. Opiyo, N.N. Impacts of neighbourhood influence on social acceptance of small solar home systems in rural western Kenya. *Energy Res. Soc. Sci.* **2019**, *52*, 91–98. [CrossRef]
32. Gustavsson, M. Educational benefits from solar technology—Access to solar electric services and changes in children's study routines, experiences from eastern province Zambia. *Energy Policy* **2007**, *35*, 1292–1299. [CrossRef]
33. Opiyo, N.N. How solar home systems temporally stimulate increasing power demands in rural households of Sub-Saharan Africa. *Energy Transit.* **2020**, 1–13. [CrossRef]
34. Lee, K.; Miguel, E.; Wolfram, C. Appliance Ownership and Aspirations among Electric Grid and Home Solar Households in Rural Kenya. *Am. Econ. Rev.* **2016**, *106*, 89–94. [CrossRef]
35. World Bank. *Nigeria on the Move: A Journey to Inclusive Growth. Systematic Country Diagnostic*; World Bank: Washington, DC, USA, 2020.
36. Adegbami, A.; Uche, C.I.N. Ethnicity and Ethnic Politics: An Impediment to Political Development in Nigeria. *Public Adm. Res.* **2015**, *4*, 59–67. [CrossRef]
37. Oginni, A.; Ahonsi, B.; Ukwuije, F. Are female-headed households typically poorer than male-headed households in Nigeria? *J. Socio Econ.* **2013**, *45*, 132–137. [CrossRef]
38. Van Den Berghe, P.L. *Power and Privilege at an African University*, 1st ed.; Schenkman Publishing Company, Inc.: Cambridge, MA, USA, 1973.
39. Agresti, A. *An Introduction to Categorical Data Analysis*, 2nd ed.; John Wiley & Sons, Inc.: Hoboken, NJ, USA, 2007.

40. Chaurey, A.; Kandpal, T.C. Carbon abatement potential of solar home systems in India and their cost reduction due to carbon finance. *Energy Policy* **2009**, *37*, 115–125. [CrossRef]
41. Rehman, I.H.; Malhotra, P.; Pal, R.C.; Singh, P.B. Availability of kerosene to rural households: A case study from India. *Energy Policy* **2005**, *33*, 2165–2174. [CrossRef]
42. Sarker, S.A.; Wang, S.; Adnan, M.K.M.; Anser, M.K.; Ayoub, Z.; Ho, T.H.; Tama, R.A.Z.; Trunina, A.; Hoque, M.M. Economic Viability and Socio-Environmental Impacts of Solar Home Systems for Off-Grid Rural Electrification in Bangladesh. *Energies* **2020**, *13*, 679. [CrossRef]
43. Alam, M.; Bhattacharyya, S. Are the off-grid customers ready to pay for electricity from the decentralized renewable hybrid mini-grids? A study of willingness to pay in rural Bangladesh. *Energy* **2017**, *139*, 433–446. [CrossRef]
44. Narayan, N.; Chamseddine, A.; Vega-Garita, V.; Qin, Z.; Popovic-Gerber, J.; Bauer, P.; Zeman, M. Quantifying the Benefits of a Solar Home System-Based DC Microgrid for Rural Electrification. *Energies* **2019**, *12*, 938. [CrossRef]
45. Narayan, N.; Chamseddine, A.; Vega-Garita, V.; Qin, Z.; Popovic-Gerber, J.; Bauer, P.; Zeman, M. Exploring the boundaries of Solar Home Systems (SHS) for off-grid electrification: Optimal SHS sizing for the multi-tier framework for household electricity access. *Appl. Energy* **2019**, *240*, 907–917. [CrossRef]
46. Nwanne, C. New Minimum Wage of N30,000 Takes Effect. Available online: https://search.proquest.com/docview/2330910722 (accessed on 28 December 2019).
47. Koepke, M.; Groh, S. Against the Odds: The Potential of Swarm Electrification for Small Island Development States. *Energy Procedia* **2016**, *103*, 363–368. [CrossRef]
48. Narayan, N.; Qin, Z.; Popovic-Gerber, J.; Diehl, J.; Bauer, P.; Zeman, M. Stochastic load profile construction for the multi-tier framework for household electricity access using off-grid DC appliances. *Energy Effic.* **2020**, *13*, 197–215. [CrossRef]
49. Youmma. Reinvent Access. Available online: https://www.yoummasolar.com/ (accessed on 4 November 2020).
50. Atagher, P.; Clifford, M.; Jewitt, S.; Ray, C. What's for dinner? Gendered decision-making and energy efficient cookstoves in benue state, north central nigeria. *WIT Trans. Ecol. Environ.* **2017**, *224*, 101–111.
51. Abdullahi, A.A. An Analysis of the Role of Women in Curbing Energy Poverty in Nigeria. *J. Sustain. Dev. Stud.* **2017**, *10*, 45–60.
52. Jewitt, S.; Atagher, P.; Clifford, M. "We cannot stop cooking": Stove stacking, seasonality and the risky practices of household cookstove transitions in Nigeria. *Energy Res. Soc. Sci.* **2020**, *61*, 101340. [CrossRef]

Publisher's Note: MDPI stays neutral with regard to jurisdictional claims in published maps and institutional affiliations.

energies

Article

On Approaching Relevant Cost-Effective Sustainable Maintenance of Mineral Oil-Filled Electrical Transformers

Ramsey Jadim [1],*, Mirka Kans [1], Jesko Schulte [2], Mohammed Alhattab [3], May Alhendi [3] and Ali Bushehry [3]

[1] Department of Mechanical Engineering, Faculty of Technology, Linnaeus University, 351 95 Växjö, Sweden; mirka.kans@lnu.se

[2] Department of Strategic Sustainable Development, Blekinge Institute of Technology, 371 79 Karlskrona, Sweden; jesko.schulte@bth.se

[3] Primary Substation Maintenance Department, Ministry of Electricity and Water, Safat 13001, Kuwait; mohammed.alhattab@mew.gov.kw (M.A.); may.alhendi@mew.gov.kw (M.A.); ali.hussain@mew.gov.kw (A.B.)

* Correspondence: ramsey.jadim@lnu.se

Abstract: Fire and explosion accidents of oil-filled electrical transformers are leading to negative impacts, not only on the delivery of energy, but also on workplace health and safety as well as the surrounding environment. Such accidents are still being reported, regardless of applying the regular maintenance strategy in the power plants. The purpose of this paper is to integrate a sustainability perspective into the maintenance strategy. The problem addressed is: how can we approach the relevant cost-effective sustainable maintenance for oil-filled electrical transformers? For this purpose, an empirical study in a power plant in Kuwait was introduced. The first stage was to carry out a sustainability assessment using the ABCD procedure. In this procedure, gaps to approach sustainability were identified and actions prioritized to close these gaps were demonstrated. Applying this procedure yielded an early fault diagnosis (EFD) model for achieving cost-effective sustainable maintenance using a fault trend chart based on a novel numerical method. Implementing this model resulted in an extension of the lifetime of transformers with suspected failure propagation, leading to a deferral of the replacement investment costs. The principal conclusion of this paper is the importance of viewing the maintenance strategy of transformers from a strategic sustainability perspective, in order to approach relevant cost-effective sustainable maintenance.

Keywords: maintenance strategy; sustainable maintenance; ABCD procedure; strategic sustainable development; transformer failures

check for **updates**

Citation: Jadim, R.; Kans, M.; Schulte, J.; Alhattab, M.; Alhendi, M.; Bushehry, A. On Approaching Relevant Cost-Effective Sustainable Maintenance of Mineral Oil-Filled Electrical Transformers. *Energies* **2021**, *14*, 3670. https://doi.org/10.3390/en14123670

Academic Editors: T M Indra Mahlia and Islam Md Rizwanul Fattah

Received: 20 May 2021
Accepted: 18 June 2021
Published: 20 June 2021

Publisher's Note: MDPI stays neutral with regard to jurisdictional claims in published maps and institutional affiliations.

1. Introduction

The basic components of an oil-filled electrical transformer are copper windings and an iron core. The windings contain copper coil turns bundled together and wrapped with insulating paper. These components are entirely immersed in a mineral insulating oil in the transformer tank [1]. Overheating of the insulating oil in-service can occur due to increased current load, oxidation, and corrosion deposits on windings, or unexpected electrical and mechanical faults resulting in the formation of partial discharge (corona) and arcing phenomena, which in turn can generate flammable gases. The energy in these phenomena can be very high and develop temperatures in hundreds of degrees, though for a short time, causing fire and explosion accidents [2–5]. The transformer fires that contain several tons of mineral insulating oil can not only impact on the delivery of energy, but also on workplace health and the surrounding environment [6,7]. In [8], statistical data analysis of the failure rate of European substation transformers between 2000 and 2010 showed that around 9.5% of the total failures caused fire accidents and 3.3% caused explosion accidents. Another source of failure rates from 2015 [9] showed a significant increase in fire and explosion accidents in USA. These accidents occurred regardless of using the regular maintenance strategy during the transformer's useful life [2,10].

151

The common regular maintenance strategy used for high voltage substation power transformers is condition-based maintenance (CBM) [3,11] which is defined according to EN 13306:2017 [12] as "preventive maintenance, which includes assessment of physical conditions, analysis and the possible ensuing maintenance actions". Oil analysis technology is commonly used in the condition monitoring (CM) parameters to detect faults, such as partial discharge, arcing, oxidation, and corrosion [2,13–15]. Fault detection is based on the concentration value level of the measurable variables, such as hydrogen gas, acetylene gas, acidity, hydrogen sulfide gas, toluene, and more, related to the standard caution limit (the maximum value that indicates fault incidence). The corrective action is carried out once the measured value of a measurable variable exceeds its caution limit [13–17]. Onsite electrical testing technology is also used in the maintenance strategy to identify the type and location of the fault. However, this technology is usually recommended at the transformer shutdown time for scheduled maintenance or when a fault is suspected [3,18]. Assessment of the transformer's condition, based on CM parameters, is a crucial key in the CBM. In this context, several mathematical models are utilized in modelling the deterioration status in order to assess the overall condition of transformers [11,19,20]. The drawbacks of these models are modelling the deterioration with a single path without considering the complexity of the deterioration in transformers [21], and the assessment of overall condition without providing information about the root cause of the fault [20].

Sustainable maintenance of transformers is considered a new challenge that requires an understanding of the process of sustainable development and the integration between social and environmental aspects [11,22,23]. The vital objectives of sustainable maintenance are extending the lifetime of the transformers and reducing the negative impacts in a way that contributes to the organization's own competitiveness [24]. Sustainable maintenance is defined according to [22] as "proactive maintenance operations striving for providing balance in social, environmental, and financial dimensions". Nowadays, investigations spotlight three technical methods for achieving sustainable maintenance. The first method [25], is selecting energy-efficient green transformers from manufacturers that perform best with sustainability aspects. The concept of "green" according to [26] is related to a consideration of the environmental aspect as well as the economic benefit. An example of this method is provided by qualified raw resources such as amorphous materials used in the core design, yielding a core with higher hardness compared to current common iron core (silicon steel laminations), resulting in reduced energy loss and greenhouse gas emissions [25]. The main shortcoming of this method is its availability only for the newly manufactured transformers. The second method [27–29] is replacing the mineral insulating oil with natural ester oil. The strength in using natural ester oil is the excellent specifications, such as higher cooling efficiency, lower inflammation risk due to high flash point value, and lower toxicity than mineral oil. The disadvantages of this oil are the low electrical resistivity, which reduces the insulation capability of the oil, and the high investment costs comparing with mineral oil [28]. The last method [30] is the waste oil recycle process to remove the hazard contamination. One of the vital techniques in this process is using the reduction procedure to remove the toxic chemical, such as polychlorinated biphenyl (PCB) from oil waste, which has negative impacts on human and environmental health [31]. In general, the most critical challenge for achieving sustainable maintenance is the investment costs of these available technical methods and the return on investment [25,27,30]. Therefore, it is crucial and necessary to establish a relevant sustainable maintenance with regard to minimized costs, i.e., relevant cost-effective sustainable maintenance. The purpose of this paper is to integrate sustainability perspective into the maintenance strategy.

Sustainability assessment in many organizations is still unclear for stakeholders [22,24]. The main reason for this challenge is the complications of carrying out the assessment due to the lack of a well-organized procedure and lack of competence in the integration of sustainability in the maintenance strategy. However, the sustainability assessment can be addressed by utilizing ABCD procedure of the Framework for Strategic Sustainable

Development (FSSD) [32–34]. The outcome of applying this procedure in a power plant was creating a model for early fault diagnosis to achieve the prioritized action for improving the maintenance strategy toward sustainability.

2. Methods and Materials

2.1. Methodological Approach

An empirical study was applied in this study to explore how a strategic sustainability perspective could be integrated into the maintenance strategy of transformers in order to approach relevant cost-effective sustainable maintenance. According to [35], an empirical study design should include five elements. The first element is the formulation of relevant questions related to the problem, which can be used in the first interview with the organization team. The second element is study propositions to be examined in the research. The third element is the identification of the problem to be studied. In this step, the problem should be defined and bounded. These three elements can help in identifying the data to be collected. The fourth element is linking the collected data to the study propositions. The last is establishing criteria for interpreting the findings. The last two elements elucidate the suggested process after the data have been collected. These elements are handled in this empirical study by using the ABCD procedure structure to assess sustainability. Therefore, the ABCD procedure acts both as a structure for achieving the study design and as a theoretical framework to be tested for its applicability in the context of maintenance strategy development.

2.2. The ABCD Procedure

A long-term strategic perspective is crucial when working with sustainability to ensure that actions lead in the right direction towards full sustainability and contribute to the own organization's competitiveness. In order to be strategic, it is necessary to know the desired vision of success. This comes with the challenge of defining sustainability in a way that is neither too broad, not providing any guidance in practice, nor too specific, making it difficult for people to agree and risking becoming obsolete in the face of, for example, new technology advances. Within the Framework for Strategic Sustainable Development (FSSD), the following principle-based definition of sustainability is used: In a sustainable society, nature is not subject to systematically increasing (1) concentrations extracted from the Earth's crust, (2) concentrations of substances produced by society, (3) degradation by physical means; and people are not subject to structural obstacles to (4) health, (5) influence, (6) competence, (7) impartiality, and (8) meaning-making [33,34,36]. These eight principles represent the root-causes of unsustainability, up-stream in cause-and-effect chains. Thereby, the myriad unsustainability-related impacts in social and ecological systems, such as climate change, biodiversity loss, ozone depletion, discrimination, etc., can be related to a few comprehensive mechanisms of destruction. These eight principles can be used as boundary conditions for the re-design of socio-technical systems. They can be used to create that vision, which is necessary to be able to be strategic. This provides the basis for backcasting, i.e., starting the planning process with the end in mind, asking what is necessary to happen in order to reach the vision over time. Thereby, the direction of change can be anticipated, enabling decision-making that makes sure to be on the right track and giving benefits to the organization, e.g., by being ahead of legislation or meeting new customer demands [37]. The funnel metaphor of the FSSD can be used to understand the sustainability challenge and the self-benefit of strategic sustainability thinking based on backcasting from a vision framed by the eight sustainability principles [34,36,38] (see Figure 1).

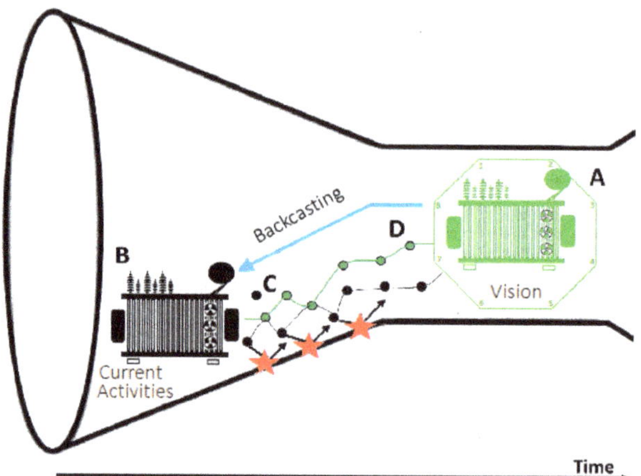

Figure 1. Funnel metaphor for transformers and the ABCD procedure for strategic sustainable development. The violations of the sustainability principles are represented here as red stars, which are hitting the funnel wall. These violations should be evaluated to identify the hotspots of sustainability impact.

The decreasing cross-section of the funnel represents the systematic degradation of the social and ecological systems. Over time, a situation must be achieved where this is no longer the case, and the funnel turns into a cylinder—this represents a sustainable state. However, society is currently moving towards the funnel walls, which represents a collapse of the social and ecological systems. This unsustainable development comes with important dynamics from a business perspective. The closer society comes to the funnel walls, the greater the need for sustainable solutions and to stop unsustainable practices. Therefore, there will be increasing business opportunities for companies that can contribute towards developing a sustainable path, while there will be increasing threats for companies that enforce unsustainable development. Such threats and opportunities, i.e., risks, can be related to, for example, legislative change, brand and reputation, shifting customer needs, raw material and waste costs, the ability to attract and retain talented employees, and more [37]. An organization must identify solutions that can work as flexible platforms on the way towards the sustainable vision and that help to exploit business opportunities and avoid threats [39].

To operationalize this strategic sustainability thinking, the FSSD includes the ABCD procedure. Step A includes setting up a shared understanding of the sustainability challenge and defining the overall vision within the restrictions of the sustainability principles. In this step, the organization must take a long-term perspective and think of an ideal success vision. Besides the sustainability principles, there are no other constraints such as the feasibility from today's perspective. The vision created in this step provides the basis for backcasting. In step B, an assessment of the organization's current reality and activities are carried out and analyzed through the lens of the proposed vision. Thereby, hots-spots of sustainability impact are identified, and a baseline is created. In step C, possible actions are identified and listed to close the gaps between the vision defined in step A and the current reality that was analyzed in the step B. In step D, strategic guidelines are applied to analyze the possible actions in the step C and select the prioritized ones to be utilized in the organization's plan of moving to sustainability. According to the basic strategy guidelines of the FSSD, the prioritized actions should be flexible platforms for future steps and achieve a balance between the rate of improvement toward the sustainable vision and return on investment [33,34,36]. The organizations need to ensure that the return on investment of applying the prioritized actions toward sustainability is satisfactory to continue the process of moving toward sustainability. However, when the cost of actions is very high, it is recommended to wait for more studies and investigations in order to find appropriate economic solutions. The actions that can be easily implemented with a good return on investment should be started first to set up a concrete structure for the future process for the other actions [33,34]. The ABCD procedure has been applied in various contexts to support strategic planning for sustainability, including in the context of sustainable product development [32].

2.3. Empirical Study Design

The ABCD procedure was applied in the Primary Substation Maintenance Department (PSMD) power plant, an organization of the Ministry of Electricity and Water in Kuwait State. The PSMD power plant is responsible for moving toward sustainability, primarily to prevent fire and explosion accidents of transformers, which recently increased. According to statistical data from PSMD power plant, around 1.5% of 3950 transformers were reported as out of service due to fire and explosion accidents during the last two decades, regardless of applying the regular maintenance strategy. In addition, there is also a growing problem of oil waste due to the replacement of many old and scraped transformers with new ones. The PSMD power plant was selected among others in this study due to the availability of good historical data of a massive quantity of transformers, and the appreciated cooperation of the maintenance team. Two online workshop meetings were organized by the first author with the PSMD team from maintenance, safety, and economic departments. The first workshop took around three hours, where relevant questions were addressed to the PSMD team. The main questions discussed the vision, current condition monitoring practices, actions need for sustainable maintenance, and current sustainability challenges. The problem of the increasing rate of fire and explosion accidents was also discussed. The second workshop took around two hours, where the problem was defined and bounded, and the propositions were studied. Accordingly, the capability to implement the ABCD procedure to integrate a strategic sustainability perspective in the current maintenance strategy was proposed. Besides, the need for the development of an effective diagnosis model for early fault detection was also proposed. The PSMD team approved to provide access to oil analysis data of 972 nominated oil-filled electrical transformers, with medium and large power rates, from the laboratory's application software in order to evaluate the transformers' condition and study the current activities from a sustainability perspective. The outcome of the gathered data of the oil analysis was analyzed and categorized according to the fault type (see Figure 2). The importance of the categorized data is identifying the fault types and defining the category that can be utilized in the ABCD procedure.

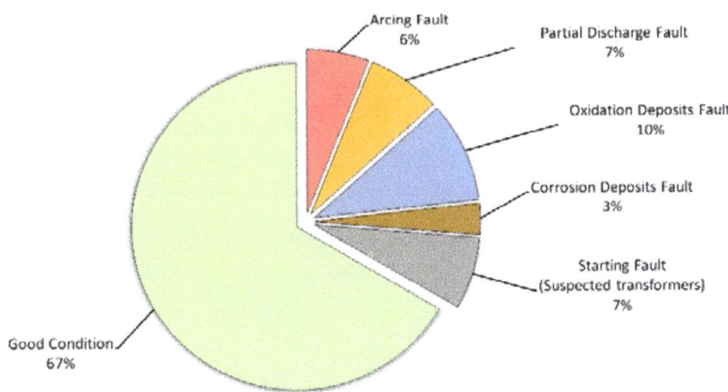

Figure 2. Fault type mode analysis of 972 transformers of the PSMD power plant.

The category of "Arcing Fault" is at 6% of the fault mode. The arcing fault can be indicated by detecting a high concentration value above the caution limit of the two gases, hydrogen and acetylene. This arcing can generate extremely high temperatures over 700 °C, though for a short time, in the oil, leading to fire accidents [4,13,16]. The category "Partial Discharge Fault" is at 7% of the fault mode. Partial discharge can be indicated by detecting a high concentration value above the caution limit of the hydrogen gas in the oil. This fault can occur at any temperature over 200 °C [4,13,16]. The category of "Oxidation Deposits Fault" is at 10%, the most dominant fault mode. Oxidation products such as acidity may convert to ester and thereafter form sludge deposits on the transformers' windings. This fault can be indicated by detecting a high concentration value above the caution limit of the acidity in the oil [5,15]. The category of "Corrosion Deposits Fault" is at 3% of the fault mode. The corrosion deposits are in the form of copper sulfide on the windings of the transformers. This fault can be indicated by detecting a high concentration value above the caution limit of the hydrogen sulfide gas and toluene compound in the oil [2]. The category "Starting Fault" is at 7% of the fault mode. This category represents the transformers with suspected failure propagation, that have a fault in progress and need corrective action. The investigation in this study focused on the "Starting Fault" category to apply a model of early fault diagnosis in order to follow up the fault before progression to a risky level. The last category "Good Condition" represents the transformers which have not any fault's indication.

3. Results

In this section, the results of applying the ABCD procedure on the transformers of the PSMD power plant are presented, as well as a model for early fault diagnosis, which was identified as a prioritized action.

3.1. ABCD Procedure for PSMD Power Plant

As seen from the fault type mode in Figure 2, the percentage of the total fault incidences was around 26% of the addressed transformers. These faults have already occurred and need proper corrective actions to avoid further progress. However, the recommended corrective actions for these faults are demonstrated in the discussion section. The category that needs to be further investigated from a sustainability perspective is "Starting Fault", to detect faults in the initial stage and find solutions to avoid transformer failures.

3.1.1. Step A: Building a Shared Understanding and Vision

The main vision of the PSMD power plant is to provide sustainable power for customers in the future with regard to minimized costs and implied integration between maintenance and sustainability aspects; social, environmental, and economic. The other visions are preventing transformers failures and catastrophic accidents, extending the lifetime of the transformers, and finding a solution to the hazardous oil waste.

3.1.2. Step B: Assessing the Current Reality

In this step, the current activities in the maintenance of transformers of the PSMD power plant were investigated. Furthermore, the violations of the sustainability principles were evaluated to identify hotspots of sustainability impact. In this study, oil analysis technology is exclusively used in the condition monitoring (CM) parameters. The electrical testing was excluded in this study due to the shortcoming in this technology to detect a fault in the initial stage [3,18]. Measurable variables of oil analysis are demonstrated in Table 1.

Table 1. Measurable variables of oil analysis related to fault type used in the PSMD power plant.

Measurable Variable (Oil Analysis)	Fault Type	Standard Method
Hydrogen & Acetylene Gases	Arcing	IEC 60567 [40]
Hydrogen Gas	Partial Discharge	IEC 60567 [40]
Acidity	Oxidation Deposits	IEC 62021-1 [41]
Hydrogen Sulfide Gas & Toluene	Copper Corrosion Deposits	ASTM D5504 [42] & D5580 [43]

Three significant violations of the sustainability principles were recorded in the PSMD power plant. The first hotspot is the usage of several tons of insulating oil in the transformers. That presents a violation of sustainability principles one and two, since it contributes to systematically increasing concentrations of substances from the Earth's crust (fossil oil) and substances produced by society (chemical products in the oil, which can be persistent and bio-accumulating). The second hotspot is the existence of a huge quantity of oil waste, which can include toxic chemicals used as additives or generated as by-products of chemical reactions in the oil during the transformer's useful life. That can be a violation of sustainability principle two. The third hotspot is the possibility of exposing the workforce or people near the fire and explosion accidents to negative impacts. That can be a violation of sustainability principle four, since it can present a structural obstacle to health.

3.1.3. Step C: Brainstorming Actions to Close the Gaps

The main gap in the current maintenance strategy is the shortcoming in detecting faults in the initial stage during the transformers' useful life to prevent failures. In this context, the interpretation and evaluation of the analysis results are currently based on the standard guidelines that recommends corrective action when the measured value of a measurable variable exceeds its caution limit. For example, standard guideline IEC 60422 [15] recommends starting corrective action such as oil regeneration or oil replacement as soon as the acidity value, which is considered the indication key of oxidation deposits on the windings, exceeds its caution limit. However, this corrective action can be no more useful to solve the issue early because the deposits may have already, partially or entirely, formed. Another gap is the lack of the maintenance team's competence in the implementation of sustainable maintenance. Accordingly, a list of possible actions can be indicated to close most of the mentioned gaps. The suggested possible actions in order to comply with the sustainability principles and to approach the vision are:

1. Improving the maintenance strategy for more effectiveness being able to detect early faults and carry out corrective actions at the correct time. Hence, it is necessary to develop an effective fault diagnosis model to provide information that can trigger an alarm for carrying out corrective action before fault incidence, on the other hand, before the measured value of a measurable variable exceeds its caution limit.
2. Providing sustainable maintenance training to enrich the maintenance team of the PSMD power plant with the knowledge of sustainability and improve the company's effectiveness.
3. Recycling waste oil to reduce the impacts of toxic compounds on the environment.

3.1.4. Step D: Prioritization

One or more prioritized actions can be selected among the possible actions indicated in step C. Analyzing all these actions shows that the third action received a lower priority due to the requirements of an economic plan to evaluate the return on investment of recycling several tons of waste oils. The first and second actions are the highest priority to the PSMD power plant in order to accomplish the sustainable vision, due to the following strategic guidelines:

- Possibility to apply in today's scenario.
- Flexible platforms for future development if required.
- It can reduce the PSMD power plant misalignments with the sustainability principles.

- Cost-effective and good return on investment.

For achieving the first prioritized action, a numerical method for early fault detection was developed to define the correct time of carrying out corrective action before the measured value of a measurable variable exceeds its caution limit (see Section 3.2 for details). Furthermore, a model for early fault diagnosis was created based on this numerical method to follow up the fault's progression during the useful life of the transformers, see sub-Section 3.3.

Regarding the second prioritized action, the lack of competence in sustainable maintenance can be a source for improper maintenance. According to a case study in [35], 40% of the breakdowns of eight Swedish manufactures were due to improper maintenance. It is important for the PSMD power plant to provide training on sustainability for the maintenance team to increase their competence in the sustainable maintenance.

3.2. Development of Numerical Method

There are few studies related to the use of mathematical theories to improve the effectiveness of maintenance strategy. An investigation in the study [11] applied Markov prediction model (MPM) to assess transformer condition based on available CM parameters under various pre-determined maintenance repair rates, and to ascertain the corrective action for maintenance, repair, and replacement. The MPM provides a simulation of the asset's deterioration using health index % (HI) to assess the condition as good, fair, poor, etc. HI is defined in [11,19] as a method that uses historical condition data in a single measurable index in order to provide a comprehensive assessment of the current state of transformers. In another study [19], a statistical distribution model (SDM) was utilized to predict the HI of transformers using the data of CM-parameters. The main outcome of applying SDM was establishing a modelling framework for future health index of transformers with even limited historical CM-parameters data. MPM is also applied in an investigation [20] based on CM-parameters to model the future deterioration in the transformers. However, all the mentioned models provided relevant information about the overall current condition status of transformers, and predicted the condition in the future, but none of them integrated a method to detect early faults.

In this study, a novel numerical method was created in order to create a model for early fault diagnosis. The numerical method aims to calculate relative alarm threshold (RAT) and relative fault detection value (RFDV) of a measurable variable before a measured value exceeds its caution limit (CL) in order to detect faults in the initial stage. In addition, the daily trend (DT%) of a measured value was calculated to investigate the increased amount percentage of this value per day after a period.

1. Relative alarm threshold (RAT): RAT is estimated as a critical threshold for any measurable variable that can be used in the indication of the probability of having a fault. According to the international standard IEC 60599 [13], the probability of having a failure may increase significantly at values much above typical concentration levels. The situation is then considered critical, for even though a fault may never occur at these high levels, the risk of having one is high. To calculate the RAT, three limits are required. The first limit is the caution limit (CL), that can be recorded from standard guidelines such as IEC 60599 [13], IEC 60422 [15], IEEE C57.106 [44], ASTM D3487 [45], CIGRE TB 771 [46], as well as limits reported by experimental investigations [2]. Selecting one of these standard guidelines is based on the organization's maintenance plan. Table 2 displays the CLs and standard guidelines used in the PSMD power plant. The second limit is the warning limit (WL), which was estimated here by supposing the starting of a fault's progression can occur when a measured value exceeds a typical value such as 50% of the caution limit (CL) value.

$$\text{WL} = \text{CL} \times 0.50 \tag{1}$$

The third limit is alarm limit (AL) based on 80% of the CL. Estimation of 80% is based on an experimental investigation carried out by [47], which revealed that the relative error in the oil analytical method could be until 20% of the measured value. The sources of the error can be from oil sampling procedure, instrument inaccuracy, result deviation, human error, etc.

$$AL = CL \times 0.80 \tag{2}$$

So, the relative alarm threshold (RAT) can be calculated based on the difference between AL and WL relative to WL (see Equation (3)).

$$RAT = (AL - WL)/WL \tag{3}$$

As seen in Table 2, the RAT value for any measurable variable was found to be 0.60 at different CLs.

Table 2. Relative Alarm Threshold (RAT) of measurable variables.

Measurable Variable (Oil Analysis)	CL	WL	AL	RAT
Hydrogen Gas	150 ppm [1]	75	120	0.6
Acetylene Gas	20 ppm [2]	10	16	0.6
Acidity	0.150 mgKOH/g [3]	0.075	0.12	0.6
Hydrogen Sulfide gas	1 ppm [4]	0.5	0.8	0.6
Toluene	2 ppm [5]	1	1.6	0.6

[1,2] according to standard guideline IEC 60599 [13], [3] according to standard guideline IEC 60422 for power transformer HV $\geq$ 170 kV [15], [4,5] according to experimental investigation [2].

2. Relative fault detection value (RFDV): RFDV is calculated based on the difference between the first measured value ($w1$) of a measurable variable and WL relative to the WL, as demonstrated in Equation (4):

$$RFDV = (w1 - WL)/WL \tag{4}$$

where $w1$ is the first measured value of a measurable variable.

If the value of the RFDV < RAT (0.60), the transformer is still in a good condition. Whereas, if REDV $\geq$ 0.60, that is an indication of the probability of having a fault. Hence, a new oil sample from the same transformer should be analyzed after a period, i.e., one month as recommended in [13] to record the second measured value ($w2$).

3. Daily trend (DT%): DT is calculated according to the following equation:

$$\text{Daily Trend, DT\%} = ((w2 - w1)/w1) \times 100)/n_d \tag{5}$$

where $w1$ is the first measured value, $w2$ is the second measured value after a period, and n_d is the number of days between $w1$ and $w2$.

If the DT $\geq$ 0.33%, there is a strong indication of starting a fault in the transformer. Accordingly, it is recommended to immediately carry out a corrective action to prevent the progression of the fault to a risky level. The factor 0.33% is the critical percentage of increasing the rate of measured value per day according to the standard IEC 60599 [13], which considered the increase in the measured value more than 10% per one month (~0.33% per day) as an indication of an active fault. In contrast, if DT < 0.33% this is an indication of not having a significant trend in the measured values. Accordingly, the transformer can be considered still in a safe condition.

3.3. Early Fault Diagnosis (EFD) Model—Concept and Verification

The suggested early fault diagnosis (EFD) model is established to approach relevant cost-effective sustainable maintenance based on a fault trend chart that shows the faults' progression during the useful life of the transformers. This chart is created based on the outcome of the numerical method for early fault detection. In order to verify the outcome of the numerical method throughout the fault trend chart, four transformers with suspected failure propagation, of the PSMD power plant from the category "Start Fault" were selected based on covering all fault types to follow up their progression. The measurable variables that used in this chart were hydrogen gas, acetylene gas, acidity, hydrogen sulfide gas, and toluene to detect anticipated arcing, partial discharge, oxidation, and corrosion faults. The standard methods used for analysis of these measurable variables are demonstrated in Table 1. The uncertainty ($U_{exp.}$) of five repeated measurements of each measurable variable; hydrogen gas, acetylene gas, hydrogen sulfide gas, toluene, and acidity, was ± 3.6 of mean value 104.4 ppm, ± 0.37 of mean value 10.1 ppm, ± 0.035 of mean value 0.51 ppm, ± 0.063 of mean value 1.51 ppm, and ± 0.031 of mean value 0.10 mgKOH/g, respectively. Furthermore, the relative standard deviation (RSD%) for same measurements was 3.2%, 3.8%, 3.1%, 3.6% and 3,2%, respectively. The historical data of these four transformers has been gathered since December 2018. Measured values for each transformer were recorded with three months interval, in March, June, September, and middle of November and December 2019 (see Figure 3).

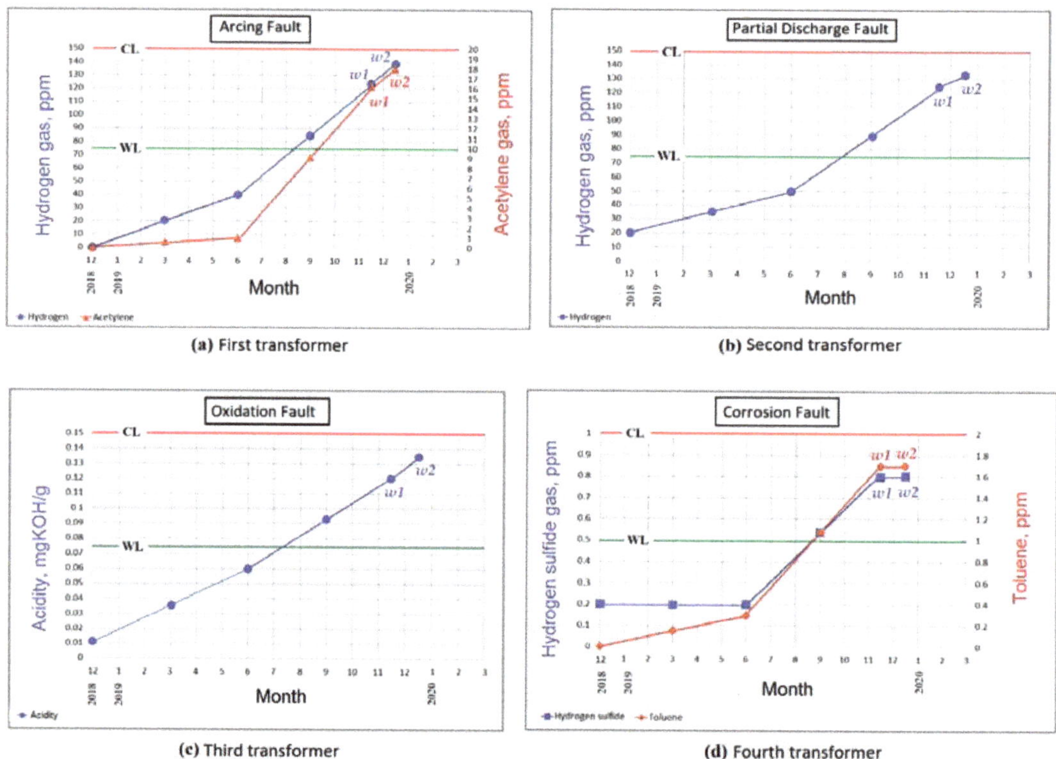

Figure 3. Fault trend chart of the four transformers with suspected failure propagation. CL= Caution Limit; WL= Warning Limit; $w1$ = First measured value; $w2$ = Second measured value. (**a**) Chart of "Arcing Fault" for the generation of the gases hydrogen and acetylene in the first transformer registered serial number 107772 (Power rate = 20 MVA, HV = 33 kV); (**b**) Chart of "Partial Discharge Fault" for the generation of the hydrogen gas in the second transformer registered serial number 20081807 (30 MVA, 132 kV); (**c**) Chart of "Oxidation Deposits Fault" for the generation of the acidity in the third transformer registered serial number 3247PG1548 (300 MVA, 275 kV); (**d**) Chart of "Corrosion Deposits Fault" for the generation of the hydrogen sulfide gas and toluene compound in the fourth transformer registered serial number 07MD970101 (30 MVA, 132 kV).

4. Discussion

The purpose of applying ABCD procedure in the PSMD power plant was to assess the sustainability level of the transformers and find the gaps to approach the vision of providing sustainable power with regard to minimized costs. The outcomes of this investigation can be summarized as follows:

1. The suggested early fault diagnosis (EFD) model using the fault trend chart, which shows the faults' progression based on the novel numerical method, was proved as a relevant model for approaching cost-effective sustainable maintenance. By utilizing this finding, the gaps to achieve the PSMD power plant vision were fulfilled, resulting in the extension of the lifetime of the four transformers with suspected failure propagation. A foremost challenge connected to this suggested model is the application on old transformers with insufficient historical data of oil analyses.

2. As seen in charts a–d of Figure 3, the measured values in March and June 2019 were less than warning limit (WL), which indicates that the transformers were in good condition. In September, the values exceeded the WL, but the calculated RFDV was <0.60, which indicates that the transformers were still in good condition. In November, the measured values ($w1$) for all transformers were significantly increased and accordingly the calculated RFDV was >0.60, which indicates the probability of having a fault in the initial stage. Hence, the analysis of the second measured values ($w2$) was carried out one month later, December 2019, to calculate the daily trend% as follows:

   - Figure 3a represents the generation of hydrogen and acetylene gases before exceeding the caution limits (CLs) of 150 and 20 ppm, respectively. The daily trend after one month was >0.33%, which indicates that the arcing fault is in progress. Accordingly, carrying out arcing measurement was recommended to find the source of overheating. In the arcing measurement, a high-frequency current transducer clamping on the transformer's grounding cable was used to detect the source and site of the arcing [48].
   - Figure 3b represents the generation of hydrogen gas before exceeding the CL of 150 ppm. The daily trend after one month was <0.33%, which indicates that the partial discharge fault is not in progress. However, if there is an indication of a partial discharge progression, the recommended corrective action could be to carry out a partial discharge test [49,50], utilizing infrared thermal imaging technology [51] to find the source of overheating and the site of the expected flashpoint in the transformer.
   - Figure 3c represents the formation of acidity before exceeding the CL of 0.15 mgKOH/g. The daily trend after one month was >0.33%, which indicates that the oxidation fault is in progress. Accordingly, adding an antioxidant additive in the insulating oil was recommended to suppress the oxidation reaction [52].
   - Figure 3d represents the generation of hydrogen sulfide gas and toluene compound before exceeding the CLs of 1 and 2 ppm, respectively. The daily trend after one month was <0.33%, which indicates that the corrosion fault is not in progress. However, if there is an indication of a corrosion progression, the recommended corrective action could be adding an anticorrosion additive in the insulating oil to suppress the corrosion reaction [53].

3. An example of applying the numerical method is demonstrated here to elucidate the calculation process. Suppose a maintenance team in a power plant observed a variation in the transformer performances when the measured value of hydrogen gas, a key indicator of partial discharge, was high. An oil sample from this transformer was analyzed for the first measured value of hydrogen gas ($w1$) and was found 125 ppm. The question: Is there any indication of partial discharge fault in progress?

As the measured value of the hydrogen was 125 ppm in the example, which is below caution limit (CL= 150 ppm) but above the alarm limit (AL= 120 ppm) (see Table 2), the relative fault detection value (RFDV) of hydrogen gas value should be calculated according to Equation (4),

$$\text{RFDV} = (w1 - \text{WL})/\text{WL} = (125 - 75)/75 = 0.67$$

The REDV was >RAT (0.60), so it was recommended to take another oil sample after one month (30 days) for analysis of the second measured value ($w2$). The value of $w2$ was found 138 ppm. Then the daily trend should be calculated according to Equation (5),

$$\text{DT} = ((w2 - w1)/w1) * 100)/n_d = ((138 - 125)/125) * 100)/30 = 0.35\%$$

As the DT of hydrogen gas value was >0.33%, there is a strong indication of starting a partial discharge fault. Accordingly, it is recommended to start an investigation to solve the problem before fault's progression to a risky level. This example shows that the corrective action could be carried out before a measured value exceeds its caution limit. The flow diagram in Figure 4 elucidates the process steps of using the numerical method in the maintenance strategy.

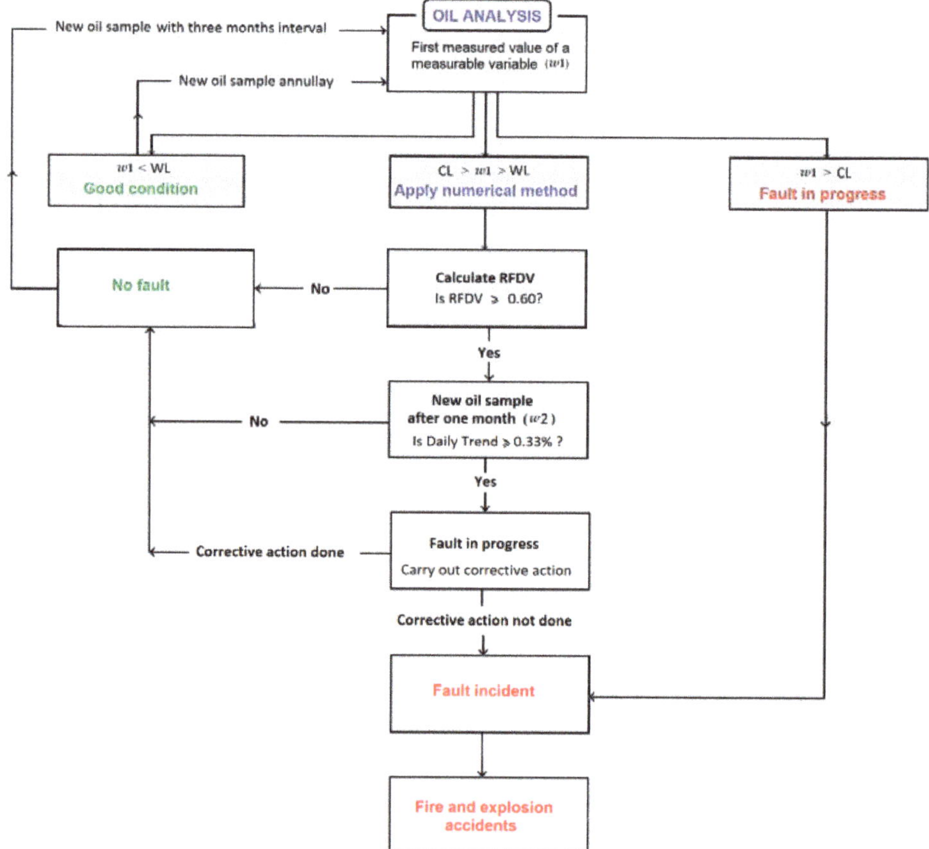

Figure 4. Flow diagram of the numerical method. $w1$ = First measured value, $w2$ = Second measured value CL = Caution Limit, WL = Warning Limit, and RFDV = Relative Fault Detection Value.

The flow diagram starts with oil analysis of the first measured value ($w1$) representing any related measurable variable. There are three alternative evaluation outcomes for the $w1$. The first is $w1$ less than WL, which reveals good condition of the transformer, hence a new oil sample could be recommended annually according to [15] or according to the organization maintenance plan. The second is $w1$ more than CL, which indicates a fault in progress, i.e., fault has already occurred and may lead to fire and explosion accidents. The last is $w1$ more than WL and below CL, which reveals the probability of having a fault in the initial stage. In this last alternative, the numerical method could be applied to assess the fault level and carry out appropriate corrective action to prevent fault incidence. If the fault is not in the risky level, the oil sample could be recommended for analysis with three months interval according to [2], in order to gather sufficient data for an accurate evaluation of the fault's progression. The generic procedure described in Figure 4 provides a structured process for CBM purposes using any measurable variable at any caution limit. The main advantage of applying the numerical method is the possibility of early faults detection before progression to a risky level. The disadvantage is the requirement of historical data to track the fault level.

4. By applying the fault trend chart using the numerical method, the faults were detected early before progression to a risky level, resulting in the extension of the lifetime of the transformers with suspected failure propagation. Hence, the gap in the implementation of sustainable maintenance is fulfilled.

5. The outline of capabilities and shortcomings in the early fault diagnosis (EFD) model comparing with other technical methods of sustainable maintenance are demonstrated in a matrix schedule (see Table 3).

Table 3. Matrix of capabilities and shortcomings in the technical methods of sustainable maintenance.

Methods of Sustainable Maintenance	Efficiency	Cost Investment	Ease of Apply	Limitation
Energy-efficient green transformers	Good	High	No	Only for new transformers
Natural ester oil	Good	High	No	Low electrical resistivity
Waste oil recycle	Good	High	No	High investment cost
Early Diagnosis Method (EFD) Model	Good	Low	Yes	Needs historical data

6. For further verification and validation, the model will be applied for tracking the fault type "copper corrosion" on 84 transformers at the PSMD power plant. For future work, the application of the EFD model can be extended to another field, such as any asset containing oil in production manufacturing. Future work could also involve investigations to find a solution to the insufficient historical oil analysis data of transformers to be used in the EFD model.

5. Conclusions

The principal conclusion of this paper is the importance of viewing the maintenance strategy of transformers from a strategic sustainability perspective, in order to approach relevant cost-effective sustainable maintenance. This was proven in the empirical study by detecting the faults in the initial stage using the early fault diagnosis (EFD) model. Several conclusions can also be highlighted as follows:

- The suitability of applying the ABCD procedure to assess the sustainability aspects for moving toward sustainable maintenance. Due to its intelligible guidance structure and its flexible process, the ABCD procedure is a useful and well-organized tool for integrating sustainability into transformer maintenance, which was demonstrated by the empirical study.
- The suggested EFD model for the empirical study has succeeded in filling the gap to approach relevant cost-effective sustainable maintenance.

- The importance of defining the correct time of carrying out corrective actions before the measured value of a measurable variable exceeds its caution limit. This was illustrated throughout tracking the faults in the fault trend chart.
- The benefits of integrating the EFD model in the CBM strategy compared with other models detecting faults in the initial stage, defining the correct time of corrective actions, easily modelling any measurable variable in the fault trend chart at any caution limit, and providing a transformer diagnosis at specific and overall fault parameters.

Author Contributions: Conceptualization, visualization, investigation, methodology, writing—original draft preparation, R.J.; supervision, reviewing and editing of the maintenance part, M.K.; reviewing and editing of the sustainability part, J.S.; providing data, participating in the workshops, and reviewing the outcome of the workshops, M.A. (Mohammed Alhattab), M.A. (May Alhendi), and A.B. All authors have read and agreed to the published version of the manuscript.

Funding: This research received no external funding.

Institutional Review Board Statement: Not applicable.

Informed Consent Statement: Not applicable.

Data Availability Statement: Not applicable.

Acknowledgments: Authors acknowledge the support of the PSMD-power plant, Kuwait Ministry of Electricity and Water, for providing the analysis data and resources. Authors also wish to acknowledge the support of SAECO laboratory of Saudi Arabian Engineering Company for analysis service.

Conflicts of Interest: The authors declare no conflict of interest. The funders had no role in the design of the study; in the collection, analyses, or interpretation of data; in the writing of the manuscript, or in the decision to publish the results.

Abbreviations

AL	Alarm Limit
BTA	Benzo Triazole
CBM	Condition Based Maintenance
CCD	Covered Conductor Deposition
CL	Caution Limit
CM	Condition Monitoring
DBDS	Dibenzyl disulfide
DT	Daily Trend %
EFD	Early Fault Diagnosis
FSSD	Framework for Strategic Sustainable Development
HI	Health Index %
H_2S	Hydrogen sulfide gas
Irgamet 39	Toluiltriazole-dialkylamine
MPM	Markov Prediction Model
PCB	Polychlorinated biphenyl
PPM	Part per million
PSMD	Primary Substation Maintenance Department
RAT	Relative Alarm Threshold
RFDV	Relative Fault Detection Value
RSD	Relative Standard Deviation %
SDM	Statistical Distribution Model
U_{exp}	Expanded uncertainty
WL	Warning Limit

References

1. Kulkarni, S.; Khaparde, S. *Transformer Engineering Design and Practice*, 1st ed.; Marcel Dekker: New York, NY, USA, 2004.
2. Jadim, R.; Kans, M.; Rehman, S.; Alhems, L.A. A relevant condition monitoring of corrosive sulphur deposition on the windings of oil-filled electrical transformers. *IEEE Trans. Dielectr. Electr. Insul.* **2020**, *27*, 1736–1742. [CrossRef]
3. Jadim, R.; Ingwald, A.; Al-Najjar, B. A Review Study of Condition Monitoring and Maintenance Approaches for Diagnosis Corrosive Sulphur Deposition in Oil-Filled Electrical Transformers. In Proceedings of the New Paradigm of Industry 4.0, Studies in Big Data 64, Bhubaneswar, India, 27 September 2018.
4. Akshatha, A.; Anjana, K.; Ravindra, D.; Vishwanath, G.; Rajan, J. Study of Copper Corrosion in Transformers due to Sulphur in Oil Using Chemical Methods. In Proceedings of the International Conference IEEE on Electrical Insulation and Dielectric Phenomena (CEIDP), Montreal, QC, Canada, 14 October 2012.
5. Liland, K.; Kes, M.; Glomm, M.; Lundgaard, L.; Christensen, B. Study of oxidation and hydrolysis of oil impregnated paper insulation for transformers using a microcalorimeter. *IEEE Trans. Dielectr. Electr. Insul.* **2011**, *18*, 2059–2068. [CrossRef]
6. El-Harbawi, M.; Al-Mubaddel, F. Risk of fire and explosion in electrical substations due to the formation of flammable mixtures. *Sci. Rep. Nat. Res.* **2020**, *10*, 6295. [CrossRef] [PubMed]
7. Allan, J. Fires and Explosions in Substations. In Proceedings of the International Conference and Exhibition IEEE/PES on Transmission and Distribution, Yokohama, Japan, 6 October 2002.
8. Vahidi, F.; Tenbohlen, S. Statistical Failure Analysis of European Substation Transformers. In Proceedings of the Conference ETG-Fachbericht—Diagnostik Elektrischer Betriebsmittel, Berline, Germany, 25 November 2014.
9. Heinz-Peter, B.; Nicole, F. Reliability and vulnerability of transformers for electricity transmission and distribution. *J. Pol. Saf. Reliab.* **2015**, *3*, 15–23.
10. Amaro, P.S.; Facciotti, M.; Holt, A.F.; Pilgrim, J.A.; Lewin, P.L.; Brown, R.C.D.; Wilson, G.; Jarman, P. Tracking Copper Sulfide Formation in Corrosive Transformer Oil. In Proceedings of the Annual Report Conference on Electrical Insulation and Dielectric Phenomena: Proceedings of the IEEE Conference, Chenzhen, China, 20 October 2013.
11. Yahaya, M.S.; Azis, N.; Selva, A.M.; Ab Kadir, M.Z.A.; Jasni, J.; Hairi, M.H.; Ghazali, Y.Z.Y.; Talib, M.A. Effect of pre-determined maintenance repair rates on the health index state distribution and performance condition curve based on the Markov prediction model for sustainable transformers asset management strategies. *Sustainability* **2018**, *10*, 3399. [CrossRef]
12. EN 13306. Standard for Maintenance-Maintenance Terminology. Available online: https://infostore.saiglobal.com/en-gb/Standards/BS-EN-13306-2017-232245_SAIG_BSI_BSI_543804/ (accessed on 1 January 2017).
13. IEC 60599, Mineral Oil-Filled Electrical Equipment in Service—Guidance on the Interpretation of Dissolved and Free Gases Analysis. In Proceedings of the International Electrotechnical Commission, Geneva, Switzerland, 16 September 2015.
14. Wada, J.; Ueta, G.; Okabe, S. Method to evaluate the degradation condition of transformer insulating oil—establishment of the evaluation method and application to field transformer oil. *IEEE Trans. Dielectr. Electr. Insul.* **2015**, *22*, 1266–1274. [CrossRef]
15. IEC 60422. Mineral Insulating Oils in Electrical Equipment—Supervision and Maintenance Guidance. In Proceedings of the International Electrotechnical Commission, Geneva, Switzerland, 10 January 2013.
16. Bustamante, S.; Manana, M.; Arroyo, A.; Castro, P.; Laso, A.; Martinez, R. Dissolved gas analysis equipment for online monitoring of transformer oil: A review. *Sensors* **2019**, *19*, 4057. [CrossRef]
17. Mharakurwa, E.T.; Nyakoe, G.N.; Akumu, A.O. Power transformer fault severity estimation based on dissolved gas analysis and energy of fault formation technique. *J. Electr. Comput. Eng.* **2019**, *2019*, 1–10. [CrossRef]
18. Mendler, R.; Osborne, G.D.; Thevenet, T.; Janak, M. Electrical Equipment Maintenance 101. In Proceedings of the IEEE Petroleum and Chemical Industry Conference (PCIC), Toronto, ON, Canada, 19 September 2011.
19. Mohd Selva, A.; Azis, N.; Shariffudin, N.S.; Ab Kadir, M.Z.A.; Jasni, J.; Yahaya, M.S.; Talib, M.A. Application of statistical distribution models to predict health index for condition-based management of transformers. *Appl. Sci.* **2021**, *11*, 2728. [CrossRef]
20. Mohd Selva, A.; Azis, N.; Yahaya, M.S.; Ab Kadir, M.Z.A.; Jasni, J.; Yang Ghazali, Y.Z.; Talib, M.A. Application of Markov model to estimate individual condition parameters for transformers. *Energies* **2018**, *11*, 2114. [CrossRef]
21. Liang, Z.; Parlikad, A. A Markovian model for power transformer maintenance. *Electr. Power Energy Syst.* **2018**, *99*, 175–182. [CrossRef]
22. Jasiulewicz-Kaczmarek, M. The Role and Contribution of Maintenance in Sustainable Manufacturing. In Proceedings of the International Conference 7th IFAC on Manufacturing Modelling, Management, and Control, International Federation of Automatic Control, Saint Petersburg, Russia, 19 June 2013.
23. Hjortha, P.; Bagheria, A. Navigating towards sustainable development: A system dynamics approach. *Futures* **2006**, *38*, 74–92. [CrossRef]
24. Vladimir, V.; Jasiulewicz-Kaczmarek, M. Maintenance in sustainable manufacturing. *Sci. J. Logist.* **2014**, *10*, 273–284.
25. Yurekten, S.; Kara, A.; Mardikyan, K. Energy Efficient Green Transformer Manufacturing with Amorphous Cores. In Proceedings of the International Conference on Renewable Energy Research and Applications, Madrid, Spain, 20 October 2013.
26. Saettaa, S.; Caldarellia, V. Lean production as a tool for green production: The green foundry case Study. *ScienceDirect* **2020**, *42*, 498–502. [CrossRef]

27. Pompili, M.; Calcara, L.; Sturchio, A.; Catanzaro, F. Natural Esters Distribution Transformers: A Solution for Environmental and Fire Risk Prevention. In Proceedings of the IEEE International Conference, Annual Conference (AEIT), Capri, Italy, 5 October 2016.
28. Perrier, C.; Ryadi, M.; Bertrand, Y.; Tran Du, C. Comparison between Mineral and Ester Oils. In Proceedings of the Cigre 10 Conference, Paris, France, 22 August 2010.
29. Hernandez-Herrera, H.; Silva-Ortega, J.I.; Mejia-Taboada, M.; Jacome, A.D.; Torregroza-Rosas, M. Natural Ester Fluids Applications in Transformers as a Sustainable Dielectric and Coolant. In Proceedings of the AIP Proceeding Conference 2123, Beirut, Lebanon, 10 April 2019.
30. Liu, Q.; Venkatasubramanian, R.; Matharage, S.; Wang, Z. Effect of oil regeneration on improving paper conditions in a distribution transformer. *Energy* **2019**, *12*, 1665. [CrossRef]
31. Choi, H.; Veriansyah, B.; Kim, J.; Duck, K.J.; Lee, Y. Recycling of transformer oil contaminated by polychlorinated biphenyls (PCBs) using catalytic hydrodechlorination. *J. Environ. Sci. Health Part A* **2009**, *44*, 494–501. [CrossRef]
32. Schulte, J.; Hallstedt, S. Workshop Method for Early Sustainable Product Development. In Proceedings of the International Design Conference-DESIGN, Dubrovnik, Croatia, 21 May 2018.
33. Robèrt, K.; Broman, G.; Waldron, D.; Ny, H.; Hallstedt, S.; Cook, D.; Johansson, L.; Oldmark, J.; Basile, G.; Haraldsson, H.; et al. Planning and acting strategically towards sustainability. In *Sustainability Handbook*, 2nd ed.; Studentlitteratur AB: Lund, Sweden, 2019.
34. Broman, G.; Robèrt, K. A framework for strategic sustainable development. *J. Clean. Prod.* **2017**, *140*, 17–31. [CrossRef]
35. Yin, R.K. *Case Study Research and Applications: Design and Methods*, 6th ed.; SAGE Publications: Thousand Oaks, CA, USA, 2018.
36. Filippis, C.; Maiber, I.; Vidler, H.; Wilbrink, T. Taking Care to Change Trajectory-Exploring an Integrated Process of Collective Narrative Practices and Strategic Sustainable Development. Master's Thesis, Blekinge Institute of Technology Karlskrona, Blekinge, Sweden, 2019.
37. Schulte, J.; Hallstedt, S. Company risk management in light of the sustainability transition. *Sustainability* **2018**, *10*, 4137. [CrossRef]
38. Robert, K. Tools and concepts for sustainable development, how do they relate to a general framework for sustainable development, and to each other? *J. Clean. Prod.* **2000**, *8*, 243–254. [CrossRef]
39. Schulte, J.; Villamil, C.; Hallstedt, S.I. Strategic sustainability risk management in product development companies: Key aspects and conceptual approach. *Sustainability* **2020**, *12*, 10531. [CrossRef]
40. IEC 60567. Oil-Filled Electrical Equipment—Sampling of Gases and of Oil for Analysis of Free and Dissolved Gases—Guidance. In Proceedings of the International Electrotechnical Commission, Geneva, Switzerland, 4 June 2013.
41. IEC 62021-3. Insulating Liquids—Determination of Acidity—Part 3: Test Methods for Non-Mineral Insulating Oils. In Proceedings of the International Electrotechnical Commission, Geneva, Switzerland, 19 March 2014.
42. ASTM D5504. Standard Test Method for Determination of Sulfur Compounds in Natural Gas and Gaseous Fuels by Gas Chromatography and Chemiluminescence. In Proceedings of the American Society for Testing and Materials, West Conshohocken, PA, USA, 1 November 2020.
43. ASTM D5580. Standard Test Method for Determination of Benzene, Toluene, Ethylbenzene, p/m-Xylene, o-Xylene, C9 and Heavier Aromatics, and Total Aromatics in Finished Gasoline by Gas Chromatography. In Proceedings of the American Society for Testing and Materials, West Conshohocken, PA, USA, 1 April 2021.
44. C57.106. *IEEE Guide for Acceptance and Maintenance of Insulating Mineral Oil in Electrical Equipment*; The Institute of Electrical and Electronics Engineers: New York, NY, USA, 5 December 2015.
45. ASTM D3487. *Standard Specification for Mineral Insulating Oil Used in Electrical Apparatus*; American Society for Testing and Materials: West Conshohocken, PA, USA, 15 June 2016.
46. CIGRE TB 771. Advances in DGA Interpretation. Available online: https://e-cigre.org/publication/771-advances-in-dga-interpretation (accessed on 1 July 2019).
47. Arrhenius, K.; Fischer, A.; Buker, O.; Adrien, H.; El Masri, A.; Lestremaub, F.; Robinsonc, T. Analytical methods for the determination of oil carryover from CNG/biomethane refueling stations recovered in a solvent. *R. Soc. Chem.* **2020**, *10*, 11907–11917. [CrossRef]
48. Seo, J.; Ma, H.; Saha, T. A Joint vibration and arcing measurement system for online condition monitoring of onload tap changer of the power transformer. *IEEE Trans. Power Deliv.* **2017**, *32*, 1031–1038. [CrossRef]
49. Stone, C. Partial discharge diagnostics and electrical equipment insulation condition assessment. *IEEE J. Mag.* **2005**, *5*, 891–904. [CrossRef]
50. IEC 60270. Consolidated Version High-Voltage Test Strategy—Partial Discharge Measurements. In Proceedings of the International Electrotechnical Commission, Geneva, Switzerland, 27 November 2015.
51. Qingsong, C.; Yongxiang, L.; Zhixiang, L.; Zhenyu, Z.; Shen, Z.; Zhiyuan, W. Analysis of Transformer Abnormal Heating Based on Infrared Thermal Imaging Technology. In Proceedings of the 2nd IEEE Conference on Energy Internet and Energy System Integration, Beijing, China, 20 October 2018.

52. Yehye, W.; Abdul Rahman, N.; Ariffin, A.; Abd Hamid, S.B.; Alhadi, A.; Kadir, F.; Yaeghoobi, M. Understanding the chemistry behind the antioxidant activities of butylated hydroxytoluene (BHT): A review. *Eur. J. Med. Chem.* **2015**, *101*, 295–312. [CrossRef] [PubMed]
53. Rehman, S.; Alhems, L.; Jadim, R.; Al Faraj, B.; Balasubramanian, K.; Al Mutairi, K.; Al-Yemni, A.; Shinde, D.; Al-Hsaien, S. Maximum acceptable concentrations of dbds, sulphur mercaptan and optimal concentration of metal deactivators for safe and prolonged operation of power transformers. *IEEE Trans. Dielectr. Electr. Insul.* **2016**, *4*, 2438–2442. [CrossRef]

 *energies*

Article

Case Study of Pollution with Particulate Matter in Selected Locations of Polish Cities

Remigiusz Jasiński, Marta Galant-Gołębiewska, Mateusz Nowak, Monika Ginter, Paula Kurzawska, Karolina Kurtyka * and Marta Maciejewska

Faculty of Civil Engineering, Institute of Combustion Engines and Powertrains, Poznan University of Technology, Piotrowo 3 Str., 60965 Poznan, Poland; remigiusz.jasinski@put.poznan.pl (R.J.); marta.galant@put.poznan.pl (M.G.-G.); Mateusz.s.nowak@put.poznan.pl (M.N.); monika.t.kardach@doctorate.put.poznan.pl (M.G.); paula.kurzawska@put.poznan.pl (P.K.); marta.maciejewska@put.poznan.pl (M.M.)
* Correspondence: karolina.t.kurtyka@doctorate.put.poznan.pl

Abstract: Despite the introduction of increasingly restrictive regulations, the air quality in Poland is still considered one of the worst in Europe. Two cities (Wroclaw and Cracow) were selected for this study, so they represent a pair of Polish cities with poor air quality, and at the same time are academic cities, popular with tourists. The article focuses on the emission of particulate matter, which is one of the most dangerous components of air pollution. The focus was on particles less than 10 μm in diameter which are most often neglected at measuring stations. We have identified the sources of particulate emissions in selected locations in Wroclaw and Cracow, and then measured particles in terms of their mass and number distribution. It was noted that the PM_{10} emission values obtained as a result of the measurements were different from the value specified by the Inspectorate of the Environmental Protection in Poland.

Keywords: urban air quality; $PM_{2.5}$; PM_{10}; particle emission sources

 check for **updates**

Citation: Jasiński, R.; Galant-Gołębiewska, M.; Nowak, M.; Ginter, M.; Kurzawska, P.; Kurtyka, K.; Maciejewska, M. Case Study of Pollution with Particulate Matter in Selected Locations of Polish Cities. *Energies* **2021**, *14*, 2529. https://doi.org/10.3390/en14092529

Academic Editor: T. M. Indra Mahlia

Received: 31 March 2021
Accepted: 25 April 2021
Published: 28 April 2021

Publisher's Note: MDPI stays neutral with regard to jurisdictional claims in published maps and institutional affiliations.

1. Introduction

The monitoring and evaluation of ambient air quality is the first important step in controlling air pollution. Monitoring pollutant concentrations in air is pivotal to properly manage air quality [1,2]. Air pollution can be defined as a phenomenon harmful to the ecological system and the normal conditions of human existence and development when some substances in the atmosphere exceed a certain concentration [3]. Air pollution comes from a wide variety of sources, which discharge of harmful substances into the atmosphere, causing adverse effects to humans and the environment. They can be natural or anthropogenic [4]. One of harmful pollutants are particulate matters. Fine particulate matters of aerodynamic diameter lower than 2.5 μm ($PM_{2.5}$) are considered to be one of the most important environmental factors contributing to the global human disease burden [5]. As half of the world's population (55%) lives in urban areas, the environmental degradation produced by cities threatens the health and quality of life of a fair share of the world's population [6]. Many articles focus on explaining the connections between pollutants and their toxicological effects on the environment [7]. Outdoor air pollution led to an estimated 4.2 million premature deaths worldwide and half a million in the European Union (EU) in 2016 [8,9]. Exposure to ambient $PM_{2.5}$ particulate matter has been found to be associated with different negative health endpoints, from minor respiratory symptoms to premature mortality. This worldwide burden of disease includes 7–10 million premature deaths, mostly in developing countries [10]. Acute lower respiratory infections (ALRI), including pneumonia and bronchiolitis of bacterial and viral origin, are the largest single cause of mortality among young children worldwide and thus account for a significant burden of disease worldwide [11]. Outdoor air pollution has been associated with increased symptoms and increased ALRI mortality.

169

Toxicologic evidence suggests that exposure to particulate air pollution can cause pulmonary inflammation and affect host defenses against infection [12]. The World Health Organization (WHO) estimated, that in 2016, 91% of the world population was living in places where the WHO air quality guidelines levels were not met. According to WHO, about 440,000 people die every year due to air pollution in Europe. In Poland only, which is at the forefront of European countries with the most polluted air, it is about 44,000 people a year [8]. Based on the specificity of formation and the influence of particles on the human body it can be stated that one of the basic issues in assessing air quality is the concentration of particles. Particulate matter is a term generally used to describe a type of air pollutant consisting of a complex of various mixtures of suspended particles that vary in size, composition and place of formation [13]. The main sources of this type of pollution are factories, power plants, incinerators, motor vehicles and many others. The basic division of particles is based on their aerodynamic diameter, which allows for the determination of two main groups: $PM_{2.5}$ and PM_{10}, i.e., particulate matter with diameters smaller than 2.5 μm and 10 μm, respectively [14].

Air quality monitoring is carried out in Poland by the Environmental Protection Inspectorate. On the basis of the reports and their analyses, it was concluded that the air in Poland is one of the most polluted in Europe. In 2018, the WHO published a list [15] of the 50 most polluted cities in Europe. Unfortunately, Poland is at the top of this list. In terms of $PM_{2.5}$ as many as 36 of those cities are located in Poland. One of them is Cracow, which is the subject of this article's analysis. As mentioned, the assessment of air quality in Poland is carried out by the Inspectorate of Environmental Protection. Measurement stations and measurement methods are established in accordance with the relevant Regulations of the Environment Ministry.

On problem is the insufficient number of measuring stations in Poland and the fact that only a small part of them measures $PM_{2.5}$ particles and none of them measures particles up to 1 micrometer in diameter (e.g., in the city of Poznan there are five measuring stations, and only three of them allow the measurement of both PM_{10} and $PM_{2.5}$). The ambient air pollution is particularly noticeable in urban agglomerations, but unfortunately it is not limited to them. The deterioration of air quality is already observable in much smaller towns and even villages. In the case of the developing countries, the problem is more serious due to overpopulation and uncontrolled urbanization along with the development of industrialization [16]. The author of [17] pointed out that in Poland the particularly negative impact on air quality is related to households and transport. The transport sector is being developed in the direction of electromobility, however, Polish roads are still dominated by vehicles equipped with conventional engines. Recent scientific studies [18–20], carried out in Poland, analyse the impact of particulate emissions from both road and air transport. The authors of [19] have made an analysis to classify the vehicles and also to identify the advantages of the latest developments in conventional, hybrid and electric vehicles.

The aim of the article was the assessment of particulate air pollution in two selected cities in Poland. Firstly, the particle emission sources were identified in selected locations in two Polish cities (Wroclaw and Cracow), and then measurement of particles in terms of their mass distribution and numbers was performed.

2. Materials and Methods

2.1. Research Object

The choice of Wroclaw and Cracow as the study object was mainly based on their excessive emissions of air pollutants, but not only that. Cracow was chosen because it is the most polluted city in Poland, which is mainly influenced by its geographical location. The unfavorable location of the city in the Vistula valley, surrounded (on three sides) by terrain elevations, means the polluted air hangs over the city instead of being removed. In addition, the buildings of air corridors, which so far allowed to remove some pollutants from the city and introduce fresh air into it, and low emissions significantly affect the pollution distribution of this city. Wroclaw, on the other hand, due to its characteristic dense buildings (mainly old tenement houses and estates of old single-family houses), also struggles with high air pollution, especially in the autumn and winter period. In this case, the first step (identification of sources) is understood as the determination of all factors that create the possibility of an increase in the concentration of particles in the air. The distribution of emission sources should be adopted according to the modeling needs. For this reason, the emission can be divided into:

- point (emission from single sources, most often from high chimneys),
- household (emission from many different sources, e.g., from inhabited areas),
- transport (emissions from means of transport, mainly roads).

The research area was chosen based on data from documents such as the National Air Quality Program or Annual Air Quality Assessments for the Poland Voivodeships, performed by the Inspectorate of Environmental Protection in Poland. Based on the analysis of the maps prepared in those documents, the areas of increased emission were indicated, and its source was identified. The maps contained in the above-mentioned documents concern the distribution of the average annual concentration of PM_{10} in Wroclaw and Cracow [21,22]. In Wroclaw, the highest average annual concentration of PM_{10} occurs in the south of the agglomeration and in the eastern part of the city. In case of Cracow, the highest average annual concentration is found in the central part of the agglomeration. The air quality is worst in the following districts: Kazimierz, Old Town, Prądnik Biały, Prądnik Czerwony.

The second step concerns air quality measurements. Air quality measurements were taken in two popular tourist and academic cities in Poland—Wroclaw and Cracow—respectively. Both cities are known to their poor air quality. In 2018, Cracow appeared in 8th place on the list of European cities with the worst air quality. The poor air quality in Cracow is mainly due to the city's location. According to [23], the inflow of air from the surrounding areas results in a deterioration of the air quality in Cracow. Unfortunately, the city is surrounded by communities where low-quality coal stoves are still the dominant method of heating houses [23]. Measurement locations were selected in the first step and they represent each type of air pollution source.

Measurements were carried out in the following locations of Wroclaw:

- Przemkowska Street (point sources),
- Slezna Street (transportation sources),
- Ustronie Street (household sources).

Measurements were carried out in the following locations of Cracow:

- Ujastek Street (point sources),
- Debnicki Bridge (transportation sources),
- Zapolskiej Street (household sources).

It is worth noting that the locations of the measurement points do not coincide with those used by Environmental Protection Inspectorate. Measurement points have been selected to reflect the real problems of residents suffering from poor air quality. The measurements took place in two sessions: morning and afternoon. Each measurement lasted 30 min. During this time the measuring device collected 186 samples (one sample/10 s). With those 12 measurements (six in Wroclaw and six in Cracow) a total of 2232 samples were collected. Measurements were taken in early autumn, when the heating season was slowly beginning. The ambient temperature was in the 14–20 °C range. The weather conditions during the measurements were not the same. Rainfall accompanied all morning measurements only in Wroclaw. Moreover, it intensified with each subsequent measurement.

2.2. Measuring Equipment

The study used an outdoor environmental air quality monitoring device in selected cities (Wroclaw and Cracow). The number and mass distribution, and immissions of particulate matter were used as criteria for urban air quality. A TSI Optical Particle Sizer (OPS) 3330 was used to measure the number and mass of particulate matter (PM) in

Wroclaw and Cracow. At this point, it is important to emphasize the difference between emissions and immissions. Emission is the mass of substances released directly into the environment, both from natural and anthropogenic sources. The amount of these substances is determined in units of weight over time, e.g., g/h, kg/year. Immission, also called concentration of pollutants, is the amount of a dust or gas pollutants in a given volume of air unit. Immission (the object of our interest) is determined in units of weight of the substance per unit volume, e.g., μ/m^3, g/m^3 [24].

The analyzer used in our research enables the measurement of particles in the range from 0.3 to 10 μm for concentrations from 0 to $3000/cm^3$. TSI's 3330 Optical Particle Sizer (OPS) is a lightweight, portable device that provides fast and accurate measurement of particle concentration and size distribution using single particle counting technology. Sixteen measurement channels were used during the tests [25]. The quality of the measurement is ensured by the manufacturer's calibration standards.

3. Results

3.1. Analysis of Particle Number (PN) Distribution in Wroclaw and Cracow

The obtained results were analysed in terms of the type of air pollution immission source. They were divided into transportation, household, and point sources of immission. The analysis has shown that for all measurements performed (irrespective of the source of immission), particulate matter occurred in the whole measurement range, i.e., 0.34–10.00 μm (Table 1). The highest results (irrespective of immission source) were obtained during morning measurement in Wroclaw. The graphs show the results obtained during two measurements (a morning one and an afternoon one) from each type of immission source. Figure 1 shows the characteristics of the particles number concentration as a function of their size (dN/dlogDp).

Table 1. Detailed results of PN distribution in Wroclaw and Cracow.

Pollutant	Point-Source		Transportation Source		Household Source	
	Wroclaw	Cracow	Wroclaw	Cracow	Wroclaw	Cracow
range of the diameter of the largest particles number [μm]	0.3–0.5	0.3–1	0.3–0.5	0.3–1	0.3–0.5	0.3–1
share of particles with a diameter smaller than 1 μm	<99%	<96%	<99%	<98%	<99%	<98%

A more detailed description is provided in Figure 2 which is showing the percentage share of each particle diameter in the test in each type of immission source. The green color indicates the shares up to 5%, yellow the range between 5% and 50%, and red above 50%. Additionally, the range of the diameter of the largest particles number, and the share of particles with a diameter smaller than 1 μm, are shown in Table 1.

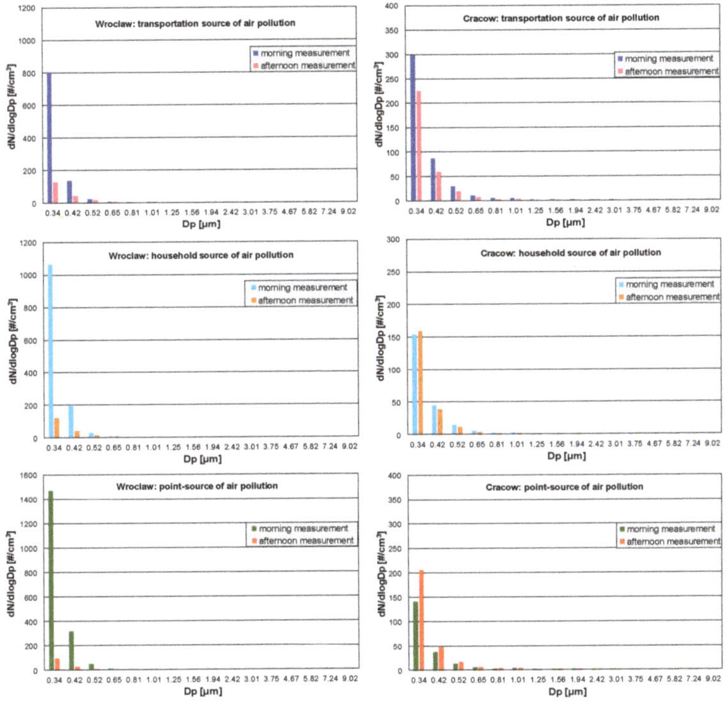

Figure 1. Particle number (PN) distributions in Wroclaw and Cracow.

Diameter	Wroclaw						Cracow					
	morning measurement			afternoon measurement			morning measurement			afternoon measurement		
	transportation	household	point	transportation	household	point	transportation	household	point	transportation	household	point
0.34	81.60%	81.55%	79.14%	46.79%	63.48%	89.99%	67.03%	67.09%	65.82%	63.29%	72.16%	77.13%
0.42	13.87%	14.83%	7.37%	12.84%	21.98%	31.52%	19.46%	19.42%	16.92%	14.74%	17.58%	20.35%
0.52	2.50%	2.36%	1.33%	4.52%	7.83%	11.60%	6.70%	6.46%	6.23%	4.90%	5.18%	6.60%
0.65	0.65%	0.51%	0.35%	1.68%	2.57%	4.11%	2.27%	2.35%	2.63%	1.87%	1.79%	2.37%
0.81	0.32%	0.22%	0.17%	0.92%	1.28%	1.56%	1.11%	1.15%	1.35%	1.04%	0.80%	1.06%
1.01	0.29%	0.17%	0.15%	0.77%	0.94%	2.04%	1.13%	1.03%	1.90%	1.12%	0.73%	1.25%
1.25	0.13%	0.07%	0.07%	0.34%	0.37%	0.68%	0.44%	0.42%	0.88%	0.49%	0.30%	0.52%
1.56	0.15%	0.07%	0.08%	0.40%	0.42%	0.66%	0.43%	0.45%	0.95%	0.53%	0.31%	0.52%
1.94	0.16%	0.08%	0.08%	0.41%	0.44%	0.76%	0.44%	0.44%	1.02%	0.55%	0.32%	0.51%
2.42	0.11%	0.05%	0.06%	0.22%	0.24%	0.44%	0.32%	0.34%	0.75%	0.39%	0.24%	0.36%
3.01	0.07%	0.03%	0.03%	0.13%	0.13%	0.23%	0.21%	0.24%	0.49%	0.25%	0.16%	0.22%
3.75	0.05%	0.02%	0.03%	0.09%	0.10%	0.16%	0.17%	0.19%	0.37%	0.20%	0.13%	0.17%
4.67	0.04%	0.01%	0.02%	0.07%	0.07%	0.14%	0.12%	0.15%	0.26%	0.14%	0.10%	0.13%
5.82	0.03%	0.01%	0.02%	0.07%	0.06%	0.11%	0.08%	0.10%	0.20%	0.12%	0.08%	0.09%
7.24	0.02%	0.01%	0.01%	0.06%	0.05%	0.09%	0.06%	0.10%	0.14%	0.09%	0.07%	0.06%
9.02	0.02%	0.01%	0.01%	0.05%	0.04%	0.07%	0.04%	0.08%	0.11%	0.07%	0.05%	0.05%

Figure 2. Percentage of particles number by their size in tests (all sectors).

3.2. Analysis of Particulate Mass (PM) Distribution in Wroclaw and Cracow

The next stage of the analysis was particulate mass (PM) distribution in Wroclaw and Cracow. Figure 3 presents mass distribution of particulate matter in measurement points in the Wroclaw and Cracow areas. The results obtained are presented in the form of characteristics of the particles mass concentration as a function of their diameter (dM/dlogDp). This analysis showed that particulate matter occurred throughout the whole measurement range, i.e., 0.34–10.00 μm.

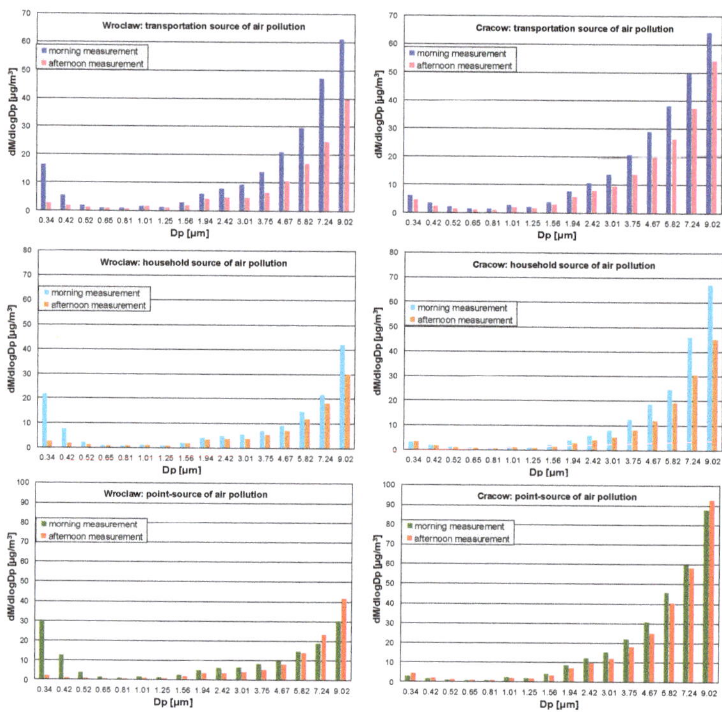

Figure 3. Particulate mass (PM) distribution in Wroclaw and Cracow.

A more detailed description is provided in Figure 4 which shows the share percentage of each particle diameter in the test in each type of immission source. The green color indicates the shares up to 5%, yellow the range between 5% and 50%. Additionally, the range of the diameter of the largest particles number, and the share of particles with a diameter smaller than 1 μm are shown in Table 2.

| | Wroclaw | | | | | | Cracow | | | | | |
| | morning measurement | | | afternoon measurement | | | morning measurement | | | afternoon measurement | | |
Diameter	transporatation	household	point	transporatation	household	point	transporatation	household	point	transporatation	household	point
0.34	7.20%	14.74%	19.67%	2.09%	2.67%	1.76%	2.36%	1.58%	0.97%	2.39%	2.38%	1.50%
0.42	2.36%	5.17%	8.23%	1.41%	1.78%	0.93%	1.32%	0.88%	0.48%	1.22%	1.12%	0.67%
0.52	0.82%	1.59%	2.37%	1.00%	1.22%	0.63%	0.88%	0.56%	0.34%	0.76%	0.63%	0.43%
0.65	0.41%	0.66%	0.85%	0.69%	0.77%	0.45%	0.57%	0.40%	0.28%	0.53%	0.42%	0.32%
0.81	0.40%	0.56%	0.64%	0.50%	0.75%	0.48%	0.54%	0.38%	0.27%	0.45%	0.36%	0.34%
1.01	0.69%	0.82%	1.04%	1.26%	1.06%	0.77%	1.06%	0.64%	0.74%	1.03%	0.64%	0.71%
1.25	0.59%	0.64%	0.72%	0.81%	0.80%	0.67%	0.81%	0.51%	0.66%	0.83%	0.51%	0.59%
1.56	1.32%	1.31%	1.62%	1.53%	1.75%	1.50%	1.51%	1.05%	1.38%	1.61%	1.00%	1.24%
1.94	2.64%	2.75%	3.19%	3.40%	3.58%	2.95%	2.98%	1.98%	2.86%	3.03%	2.03%	2.52%
2.42	3.49%	3.26%	3.95%	3.83%	3.69%	3.11%	4.15%	2.96%	4.06%	4.15%	2.94%	3.42%
3.01	4.14%	3.67%	4.25%	3.75%	3.97%	3.42%	5.34%	4.01%	5.12%	4.94%	3.83%	4.26%
3.75	6.07%	4.76%	5.25%	5.26%	5.62%	4.67%	8.05%	6.22%	7.43%	7.16%	5.95%	6.49%
4.67	9.21%	6.29%	6.41%	8.53%	7.55%	7.15%	11.29%	9.41%	10.35%	10.41%	8.77%	8.98%
5.82	12.98%	10.14%	9.68%	13.66%	12.66%	12.65%	14.83%	12.44%	15.41%	13.78%	14.00%	14.47%
7.24	20.76%	14.88%	12.39%	19.92%	19.62%	21.00%	19.31%	23.16%	20.15%	19.45%	22.38%	20.86%
9.02	26.94%	28.76%	19.74%	32.36%	32.49%	37.88%	25.01%	33.81%	29.50%	28.26%	33.04%	33.20%

Figure 4. Percentage of particulate matter by their size in tests (all sectors).

Table 2. Detailed results of PM distribution in Wroclaw and Cracow.

Pollutant	Point-Source		Transportation Source		Household Source	
	Wroclaw	Cracow	Wroclaw	Cracow	Wroclaw	Cracow
range of the diameter of the largest particulate matter [μm]	0.3–0.5 1.5–10	2–10	0.3–0.4 2–10	2–10	0.3–0.4 2–10	2–10
share of particulate matter with a diameter smaller than 1 μm	<19%	<4%	<10%	<7%	<16%	<5%

3.3. Total Average Number and Mass Concentrations of Particulate Matter in Wroclaw and Cracow

The final stage of the analysis of the results of the studies carried out in Wroclaw and Cracow consisted in creating the characteristics presenting the total average concentration of particulate matter in terms of number and mass with a division into air pollution sources. The results obtained are shown in Figure 5a,b.

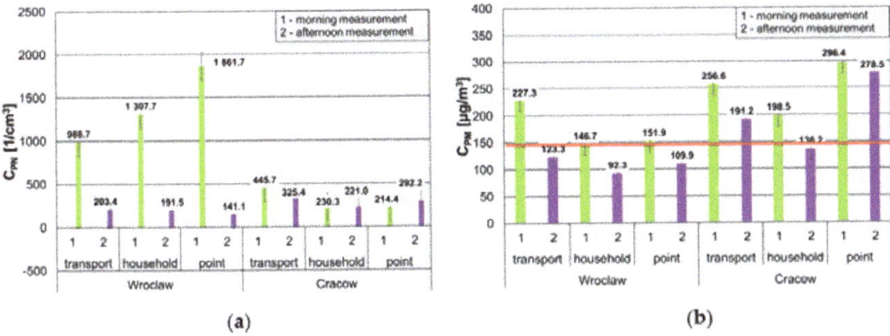

Figure 5. Total average concentration of particle number (**a**) and particulate mass (**b**) in each city (in red line marked alarm level for PM$_{10}$ emission settled by Inspectorate of Environmental Protection).

Figure 5a presents an analysis of the total average number concentration (CPN) of particulate matter obtained during air quality measurements in Wroclaw and Cracow. Data were divided according to the type of source of measured immission (transport, household and point). The results obtained for individual sources were as follows:

- point immission sources 1000 cm^{-3} (in Wroclaw), 253 cm^{-3} (in Cracow),
- household immission sources 750 cm^{-3} (in Wroclaw), 226 cm^{-3} (in Cracow),
- transportation immission sources 596 cm^{-3} (Wroclaw), 386 cm^{-3} (Cracow).

Figure 5b presents an analysis of the total average mass concentration (C$_{PM}$) of particulate matter obtained during air quality measurements in Wroclaw and Cracow. Data were divided according to the type of source of measured immission (point, household and transport). The results obtained for individual sources were as follows:

- point immission sources 131 μg/m^3 (in Wroclaw), 287 μg/m^3 (in Cracow),
- household immission sources 120 μg/m^3 (in Wroclaw), 167 μg/m^3 (in Cracow),
- transportation immission sources 175 μg/m^3 (Wroclaw), 224 μg/m^3 (Cracow).

Figure 5b presents a total average mass concentration of particulate matter obtained during measurements at 1 September 2020 in Wroclaw and 3 September 2020 in Cracow. Equipment, which was used to measurements has range to 10.00 μm. This mean that prepared graph shows concentration of particulate matter to 10.00 μm (PM$_{10}$). The red line in Figure 5b indicates the limit above which the smog alarm is announced.

In order to compare, the results of measurements taken on the same day by the measuring station of the Environmental Protection Inspectorate were checked. The differences between the results from the Environmental Protection Inspectorate and those obtained during the research result from the location of the research stations. The measurements carried out by the authors were focused on points with increased emissions. Therefore, the measurements showed a large exceedance of PM_{10} standards and indicate the regionalization of the city's pollution. During the measurements hours, the PM_{10} results obtained by the Inspectorate of Environmental Protection did not exceed 40 $\mu g/m^3$ for Wroclaw and 50 $\mu g/m^3$ for Cracow. The data was obtained from the measurement data bank of the Environmental Protection Inspectorate for the selected station located in the vicinity of the measurement points identified in the research. Figure 6a,b show the averaged results of the performed measurements. The value of the immission obtained from transport was set as the baseline. Thanks to this, the percentage differences obtained in other sectors were determined. The obtained results were also presented in terms of low and high (point) imission. Low imission is the sum of the values obtained for the transport and household sectors (Figure 7a,b).

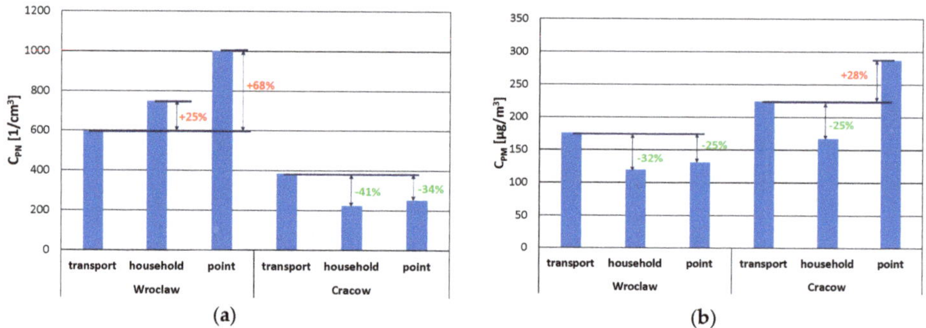

Figure 6. Total average concentration of particles distribution in terms of: (**a**) number, (**b**) mass.

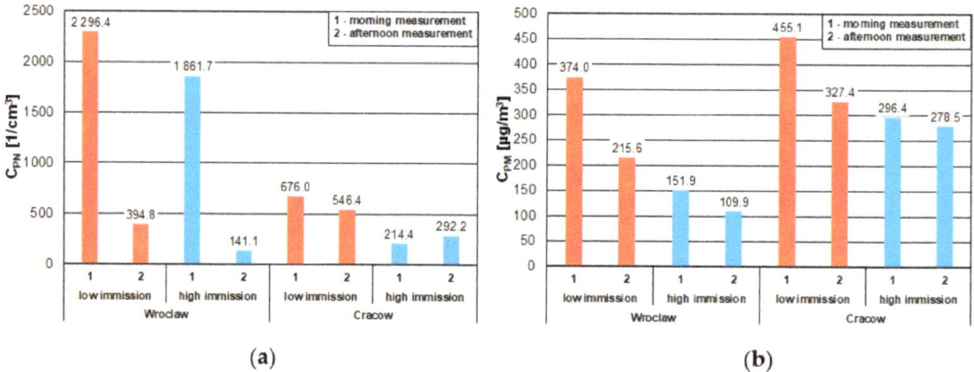

Figure 7. Low and high particle emission results in terms of: (**a**) number, (**b**) mass.

4. Discussion

The area of interest during testing was particles up to 10.00 μm in aerodynamic diameter. At each of the measurement points, both morning and afternoon particulate matter occurred over the entire measurement range. Due to the number distribution, small diameter particles (up to 0.3 μm) predominated during the tests. The largest number of particles was in the size range up to 1 μm. In the case of Wroclaw, there was a significant difference between the number of particles obtained during the morning measurement compared to the afternoon one. The weather conditions were considered the cause. Rainfall accompanied all morning measurements in Wroclaw. Moreover, it intensified with each subsequent measurement. A relation between the degree of rainfall intensity and the accompanying wind with increased PM immission was noticed. During the rainfall, a very large number of solid particles were observed, in terms of mass and number. The authors of the study [26,27] mention the influence of humidity on air quality. In effect, that morning rain was effective in reducing the particulate immission, i.e., cleaning the air. In Cracow, on the other hand, the values obtained during the morning and afternoon measurements were similar.

The predominant share of particles with a diameter smaller than 1 μm (almost 100%) was observed during the measurements. This means that it is these smallest particles that are most abundant in the air. This effect is visible in both the Wroclaw and Cracow agglomerations. It is worth noting that the point-source measurement is also nearby a highway as well as a nearby railway station where the emissions from the train and diesel engine locomotive are present not only household exhaust emissions. Wind and wind turbulences which mix the particulate matter around the city and country also exist. The conducted assessment of pollution with particles pays particular attention to the share of the smallest particle size (up to 1 micrometer) in the total emissions. It is these particles that are not filtered by the human body and go directly into the respiratory system.

In terms of mass distribution, particles with a diameter greater than 1.5 μm represented the largest shares in the mass of particles. According to the obtained mass distributions of particulate matter within Wroclaw and Cracow agglomerations, the particles with a larger diameter, i.e., about 5 μm, have the greatest impact on the particulate matter immissions. The share of particulate matter with a diameter smaller than 1 μm is significantly decreased compared to the numerical distribution. This means that both in Cracow and Wroclaw (from the perspective of mass distribution of particulate matter) particles with larger diameters dominate.

The results in Figure 7 show that the transport sector is not at all the largest source of the measured particulate pollution immission in all cases. Due to the number of particles, it would be better to take a closer look at transport in Cracow. The obtained data show that the imission from households and the point imission are about 40% lower. On the other hand, in Wroclaw the biggest problem (during the research) was the point imissions. Due to the mass of particles, transport is the main source of air pollution with particles. It follows from the results obtained by us that the main source of PM pollution is low emission (surface and transport). Industrial chimneys, responsible mainly for point emission, are high, therefore the pollution is dispersed over a much larger area, far beyond the city limits.

Figure 5 presents the results of the particulate matters up to 10 μm. According to the legislation, these are PM_{10}. The results obtained during the measurements were compared with the data from official sources of Environmental Protection Inspectorate. As expected, the obtained results not only significantly exceeded the values presented by the Environmental Protection Inspectorate, but also the daily PM_{10} standards (in some cases, even by twice). This means that despite meeting the legal requirements, it would be worth considering a more precise location of the measuring stations.

The results obtained indicate that it is necessary to improve measurement methods, especially for the smallest particles, smaller than $PM_{2.5}$. As mentioned in the Introduction the number of measuring stations, not only in Wroclaw and Cracow but also across Poland, is insufficient. This also applies to the measurement of particles smaller than 2.5 μm. Our measurements were not made in accordance with the guidelines of Environmental Protection Inspectorate, however indicated that it is necessary to extend the measurements performed by the smallest particles. The assessment thus carried out highlights the shortcomings of air quality monitoring stations measuring these smallest particles. Perhaps the application of the method [28], but extended by the measurement of particulate matter, would help in determining better measuring station locations. The choice of the location of the measurement points in our research was not accidental. Similarly to [6], we were guided not only by the data from official reports, but also by taking into account subjective factors that are not taken into account in the analyses of the Environmental Protection Inspectorate. Unfortunately, both of these cities struggle with the problem of poor air quality and nuisance smog. It is related not only to the location of both cities, but also to the habits of the inhabitants. Coal stoves are still used as the main source of heating in many homes. Fortunately, the situation is improving year by year. City authorities introduce new restrictions and activate programs aimed at achieving clean air such as replacement of stoves, development of bicycle infrastructure, restrictions on vehicle entry to city centers. This applies both to subsidies encouraging residents to use green energy, and to changing the city's infrastructure.

5. Conclusions

Nowadays, air pollution is considered to be one of the biggest problems affecting society worldwide [29–32]. Specific attention is paid to one component of pollution, i.e., excessive particulate emissions. This is a crucial contaminant since it contains other pollutants adsorbed on particle surfaces, such as heavy metals. Unfortunately, Poland is at the top of the European countries with the worst air quality. Particle immission tests were carried out in two Polish cities (Wroclaw and Cracow). A total of 12 tests were carried out during the measurements (two for each type of immission source). Each measurement location represented one type of emission: transport, household and point. The conducted research shows that it is the low immission, i.e., from transport and households, that is the largest source of air pollution in the locations selected by us. The results obtained during the measurements clearly indicate that the data are not consistent with those presented by the Environmental Protection Inspectorate. The data analyzed in this article is much larger, and additionally significantly exceeds the daily PM_{10} standard. Despite the fact that the Environmental Protection Inspectorate's measuring stations are arranged in accordance with legal requirements, they are still insufficiently precise. There are still places, popular on the map of Wroclaw and Cracow, where excessive PM_{10} emissions occur, despite satisfactory data from official sources.

Moreover, the measurements performed by the Inspectorate of Environmental Protection are not sufficient. It should be necessary to expand research to include monitoring of particulate matter PM_1 and even smaller ones. The next step should be to perform further, more detailed research. On that basis, it will be possible to propose a new method for selecting the location of measurement points, especially in those places where excessive particulate emissions are a problem. Obviously, this will not directly improve air quality, but it will certainly enable better monitoring. This will enable local improvements to be made, which in the long run will have a positive impact on air quality throughout the city. Additionally, in order to reduce the level of pollution in the studied cities, it is suggested to develop social programs aimed at actively acting against air pollution. In addition, the city authorities in their environmental protection programs should focus on replacing furnaces, changing road infrastructure, expanding bicycle paths, or introducing various benefits for people using electric motor vehicles.

Author Contributions: Conceptualization, R.J. and M.G.-G.; methodology, R.J.; software, M.N.; validation, P.K., M.M., M.G. and K.K.; formal analysis, K.K.; investigation, P.K.; resources, M.M.; data curation, M.N.; writing—original draft preparation, K.K. and M.M.; writing—review and editing, K.K. and M.M.; visualization, M.M., K.K. and P.K.; supervision, M.G.-G.; project administration, R.J.; funding acquisition, R.J. All authors have read and agreed to the published version of the manuscript.

Funding: This research and APC was funded by Interdisciplinary Rector's Grant, grant number 559 ERP/33/32/SIGR/0004.

Institutional Review Board Statement: Not applicable.

Informed Consent Statement: Not applicable.

Data Availability Statement: Data available on request due to restriction e.g., privacy or ethical. The data presented in this study are available on request from the corresponding author. The data are not publicly available due to ongoing project.

Conflicts of Interest: The authors declare no conflict of interest.

References

1. Suman. *Air Quality Indices: A Review of Methods to Interpret Air Quality Status, Materials Today: Proceedings, 2214-7853/2020*; Elsevier Ltd: Amsterdam, The Netherlands, 2020. [CrossRef]
2. Doval Minarro, M.; Banon, D.; Egea, J.A.; Costa-Gomez, I.; Baeza Caracena, A. *A Multi-Pollutant Methodology to Locate a Single Air Quality Monitoring Station in Small and Medium-Size Urban Areas, Environmental Pollution, 0269-7491/2020*; Elsevier Ltd: Amsterdam, The Netherlands, 2000. [CrossRef]
3. Bai, L.; Wang, J.Z.; Ma, X.J.; Lu, H.Y. Air Pollution Forecasts: An Overview. *Int. J. Environ. Res. Public Health* **2018**, *15*, 780. [CrossRef] [PubMed]
4. Khelfi, A. *Sources of Air Pollution, Handbook of Research on Microbial Tools for Environmental Waste Management, Advances in Environmental Engineering and Green Technologies*; IGI Global: Hershey, PA, USA, 2018. [CrossRef]
5. Fantke, P.; Jolliet, O.; Evans, J.S.; Apte, J.S.; Cohen, A.J.; Hänninen, O.O.; Hurley, F.; Jantunen, M.J.; Jerrett, M.; Levy, J.I.; et al. Health effects of fine particulate matter in life cycle impact assessment: Findings from the Basel Guidance Workshop. *Int. J. Life Cycle Assess.* **2015**, *20*, 276–288. [CrossRef]
6. Chiarini, B.; D'Agostino, A.; Marzano, E.; Regoli, A. Air quality in urban areas: Comparing objective and subjective indicators in European countries. *Ecol. Indic.* **2021**. [CrossRef]
7. Koolen, C.D.; Rothenberg, G. Air Pollution in Europe. *ChemSusChem* **2019**, *12*. [CrossRef] [PubMed]
8. World Health Organization. Ambient (Outdoor) Air Pollution. 2018. Available online: https://www.who.int/news-room/fact-sheets/detail/ambient-(outdoor)-air-quality-and-health (accessed on 22 February 2021).
9. Sicard, P.; Agathokleous, E.; De Marco, A.; Paoletti, E.; Calatayud, V. Urban population exposure to air pollution in Europe over the last decades. *Environ. Sci. Eur.* **2021**, *33*, 1–12. [CrossRef] [PubMed]
10. Burnett, R.; Chen, H.; Szyszkowicz, M.; Fann, N.; Hubbell, B.; Pope, C.A.; Apte, J.S.; Brauer, M.; Cohen, A.; Weichenthal, S.; et al. Global estimates of mortality associated with long-term exposure to outdoor fine particulate matter. *Proc. Natl. Acad. Sci. USA* **2018**, *115*, 201803222. [CrossRef] [PubMed]
11. Cohen, A.J.; Anderson, H.R.; Ostro, B.; Pandey, K.D.; Krzyzanowski, M.; Künzli, N.; Gutschmidt, K.; Pope, C.A., III; Romieu, I.; Samet, J.M.; et al. Urban air pollution. In *Comparative Quantification of Health Risks: Global and Regional Burden of Disease Attributable to Selected Major Risk Factors*; LA Ezzati, M., Rodgers, A., Murray, C.J.L., Eds.; World Health Organization: Geneva, Switzerland, 2004; Volume 2, pp. 1153–1433.
12. Mehta, S.; Shin, H.; Burnett, R.; North, T.; Cohen, A.J. Ambient particulate air pollution and acute lower respiratory infections: A systematic review and implications for estimating the global burden of disease. *Air Qual. Atmos. Health* **2013**, *6*, 69–83. [CrossRef] [PubMed]
13. Wielgosiński, G.; Czerwińska, J. Smog episodes in Poland. *Atmosphere* **2020**, *11*, 277. [CrossRef]
14. Jasiński, R. Estimation of particles emissions from a jet engine in real flight. In *E3S Web Conf, Proceedings of the 11th Conference on Interdisciplinary Problems in Environmental Protection and Engineering EKO-DOK, Polanica Zdrój, Poland, 8–10 April 2019*; EDP Sciences: Les Ulis, France, 2019; Volume 100, p. 00029.
15. EPA. What Is Particulate Matter? Available online: https://www3.epa.gov/region1/eco/uep/particulatematter.html (accessed on 21 February 2021).
16. Manisalidis, I.; Stavropoulou, E.; Stavropoulos, A.; Bezirtzoglou, E. Environmental and health Impacts of air pollution: A review. *Front. Public Health* **2020**, *8*. [CrossRef] [PubMed]
17. Burchard-Dziubinska, M. Air pollution and health in Poland: Anti-smog movement in the most polluted Polish cities. *Ekon. Śr.* **2019**, *2*, 76–90. [CrossRef]
18. Nowak, M.; Jasinski, R. Ecological Comparison of Domestic Travel by Air and Road Transport. *SAE Tech. Paper* **2020**, *1*, 2137. [CrossRef]
19. Pielecha, J.; Skobiej, K.; Kurtyka, K. Exhaust Emissions and Energy Consumption Analysis of Conventional, Hybrid, and Electric Vehicles in Real Driving Cycles. *Energies* **2020**, *13*, 6423. [CrossRef]
20. Pielecha, J.; Gis, M.; Skobiej, K.; Kurtyka, K. Measurements of particulate emissions from Euro 5/6 passenger cars in different drive settings. *IOP Conf. Ser. Earth Environ. Sci.* **2021**. [CrossRef]
21. *Annual Air Quality Assessment in the Dolnośląskie Voivodeship*; Voivodship Report for 2019; Inspectorate of Environmental Protection: Warshawa, Poland, 2019.
22. Inspectorate of Environmental Protection. Annual Air Quality Assessment in the Malopolskie Voivodeship. Voivodship Report for 2017; 2018. Available online: http://powietrze.gios.gov.pl/pjp/rwms/publications/card/1163 (accessed on 26 April 2021).
23. Traczyk, P.; Gruszewska-Kosowska, A. The Condition of Air Pollution in Kraków, Poland, in 2005–2020, with Health Risk Assessment. *Int. J Environ. Res. Public Health.* **2020**, *17*, 6063. [CrossRef] [PubMed]
24. LIFE-MAPPINGAIR/PL. What's the Difference between Emission and Immission? Available online: https://mappingair.meteo.uni.wroc.pl/en/2020/04/whats-the-difference-between-emission-and-immission/ (accessed on 24 January 2021).

25. Optical Particle Sizer 3330. Available online: https://tsi.com/products/particle-sizers/particle-size-spectrometers/optical-particle-sizer-3330/ (accessed on 12 March 2021).
26. Ding, J.; Dai, Q.; Li, Y.; Han, S.; Zhang, Y.; Feng, Y. Impact of meteorological condition changes on air quality and particulate chemical composition during the COVID-19 lockdown. *J. Environ. Sci.* **2021**, *109*, 45–56. [CrossRef]
27. Kwak, H.Y.; Ko, J.; Lee, S.; Joh, C.H. Identifying the correlation between rainfall, traffic flow performance and air pollution concentration in Seoul using a path analysis. *Transp. Res. Proc.* **2017**, *25*, 3552–3563. [CrossRef]
28. Minarro, M.D.; Banon, D.; Egea, J.A.; Costa-Gomez, I.; Baeza Caracena, A. A multi-pollutant methodology to locate a single air quality monitoring station in small and medium-size urban areas. *Environ. Pollut.* **2020**, *266*, 115279. [CrossRef] [PubMed]
29. Nowak, M.; Galant, M.; Kardach, M.; Maciejewska, M. Using the simulation technique to improve efficiency in general aviation. *AIP Conf. Proc.* **2019**, *2078*, 020097.
30. Galant, M.; Nowak, M.; Jasinski, R. Implementation of the LTO cycle in flight conditions using FNTP II MCC simulator. *IOP Conf. Ser. Mater. Sci. Eng.* **2018**, *421*, 042060.
31. Jasiński, R. Mass and number analysis of particles emitted during aircraft landing. In Proceedings of the E3S Web Conference, 10th Conference on Interdisciplinary Problems in Environmental Protection and Engineering EKO-DOK, Polanica-Zdrój, Poland, 16–18 April 2018; Volume 44, p. 00057. [CrossRef]
32. Jasiński, R. Number and mass analysis of particles emitted by aircraft engine. In Proceedings of the MATEC Web Conference, VII International Congress on Combustion Engines, Poznań, Poland, 27–29 June 2017; Volume 118, p. 0023.

 *energies*

Article

Mathematical Modelling and Operational Analysis of Combined Vertical–Horizontal Heat Exchanger for Shallow Geothermal Energy Application in Cooling Mode

Sarwo Edhy Sofyan [1], Eric Hu [2], Andrei Kotousov [2,*], Teuku Meurah Indra Riayatsyah [1] and Razali Thaib [1]

[1]   Department of Mechanical Engineering, Universitas Syiah Kuala, Banda Aceh 23111, Indonesia;
    sarwo.edhy@unsyiah.ac.id (S.E.S.); indraayat@unsyiah.ac.id (T.M.I.R.); razalithaib@unsyiah.ac.id (R.T.)

[2]   School of Mechanical Engineering, The University of Adelaide, Adelaide, SA 5005, Australia;
    eric.hu@adelaide.edu.au

*   Correspondence: andrei.kotousov@adelaide.edu.au

Received: 27 October 2020; Accepted: 11 December 2020; Published: 14 December 2020

Abstract: Geothermal heat exchangers (GHEs) represent a buried pipe system, which can be utilised to harness renewable thermal energy stored in the ground to improve the efficiency of heating and cooling systems. Two basic arrangements of GHEs have been widely used: vertical and horizontal. Vertical GHEs generally have a better performance in comparison with the horizontal arrangement, and these systems are particularly suitable for confined spaces. Nevertheless, the main technical challenge associated with GHEs, for either the vertical or the horizontal arrangement, is the performance deterioration associated with an increase in the operation times during summer or winter seasons. In this paper, a combined horizontal-vertical GHE arrangement is proposed to address the current challenges. The combined GHE arrangement can be operated in five different modes, corresponding to different thermal loading conditions. These five operation modes of the combined GHE are analysed based on the transient finite difference models previously developed for the horizontal and vertical arrangements. The simulation results reveal that for the single operation mode (horizontal or vertical only), the vertical GHE performs better than the horizontal GHE due to relatively stable ground temperature deep down. While, for the combined operation mode, the series operations (horizontal to vertical or vertical to horizontal) of the GHE are superior to the split mode. It is found that the effect of the fluid mass flow rate ratio is trivial on the heat dissipation of the split mode GHE. The highest heat transfer rate in the split flow operational mode is rendered by the ratio of the mass flow rate of 40% horizontal and 60% vertical. In addition, the climate condition has more effect on GHE's performance and the increase of the fluid flow rate it can enhance the amount of energy released by the GHE.

Keywords: geothermal heat exchangers; combined arrangement; operation analysis

1. Introduction

Increasing energy demands and crude oil depletion have forced the world to explore new energy resources, those that are not only sustainable but also have a minor environmental footprint. Shallow geothermal energy is one of the sustainable energy resources that could be applied to improve the efficiency of current and future heating and cooling systems. This renewable thermal energy is usually exploited in a geothermal heat exchanger (GHE); a system of pipes buried in the ground, in which the heat carrier fluid is circulated. GHEs are normally integrated with heat pumps and air conditioning systems. These systems use the ground as a heat source/sink during heating/cooling to provide the thermal comfort condition of buildings' interiors in both winter and summer. Compared with the traditional air source heat pumps and air conditioning systems, the application of GHEs as a thermal energy rejecter/absorber can reduce energy consumption by 30%–60%, since the ground has a relatively stable temperature [1].

There have been two basic arrangements of GHEs: vertical and horizontal. Vertical GHEs have been widely installed in confined areas such as an urban area or an area where the earth is rocky close to the surface. The typical depth of vertical GHEs varies from 20 to 300 m [2]. Vertical GHEs have a better efficiency since the ground temperature at the deeper depth below the surface remains relatively largely constant all year around. The drawback of vertical GHEs, however, is their installation cost, which is higher than for the horizontal arrangement. In addition, the degradation in their thermal performance is relatively hard to recover because of the poor soil thermal conductivity and relatively deep depth of burring, which hinder the heat transfer rate to and from the atmosphere. Horizontal GHEs are mostly installed where a land area is available at low cost. They are buried in a horizontal trench, with a typical depth up to 2 m below the surface. The continuous thermal interaction between the ground and the atmosphere could improve the GHE's performance through an appropriate operation mode. The diverse mechanisms of the heat transfer occurring on the ground surface could also facilitate the thermal recovery in the case of a horizontal arrangement.

The technical challenges associated with GHEs are their performance degradation with an increase in the operation time, which often happens in heating or cooling seasons. This performance degradation is slow to restore, especially for vertical GHEs. Meanwhile, it can be minimised if the GHEs are run intermittently or with a smaller thermal load, which can be achieved with the proposed GHE system, which combines both vertical and horizontal arrangements. To address the degradation performance problem caused by an imbalance in the heating and cooling loads, a number of researchers have proposed hybrid GHE systems and different operation strategies [3]. For example, hybrid GHE, solar energy systems, and heat pump have been suggested by a number of researchers [4–8], to cope with performance deterioration. Dai et al. [4] carried out an experimental study on a solar assisted ground source heat pump system. In the parallel mode, the performance of the heat pump, under three different flow rate ratios of working fluid inside both solar and shallow geothermal systems, is investigated. The outcomes of this study demonstrated that the hybrid systems could recover the degradation of soil temperature much faster than happens during natural recovery. In addition, the effect of the flow rate ratio in mode 4 has a significant impact on the electricity consumption. The electricity consumption decreases with the increase in the flow rate ratio of the solar heat storage water tank. Furthermore, mode 3 is recommended for use in the coldest months.

Kjellsson et al. [5] analysed different systems with a combination of a ground source heat pump and a solar collector to provide heating for houses and domestic hot water systems. The outcomes demonstrated that the system works in an optimal regime when the solar energy is utilised to recharge the borehole during winter time and to produce the domestic hot water during summer time. The performance of a solar assisted ground source heat pump, which is used for green house heating, was investigated by Ozgener et al. [6]. The archived exergetic efficiency of the overall system was reported to be at 67.7%. Yang et al. [7] presented a theoretical and experimental study on a solar-ground source heat pump system. The experimental coefficient of performances (COPs) of modes 2, 3, and 4 were obtained 2.69, 2.65, and 2.59, respectively. While the theoretical COPs were found to be 3.67, 3.64, and 3.52, which are quite different from the theoretical values. In addition, research on small ammonia heat pumps for space and hot tap water heater with a capacity of 8.4 kW to provide space heating and hot water has been done by Aleksandrs Zajacs et al. [8]. In this study, equation calculations have been carried out using the engineering equation solver (EES) to estimate the heat pump performance. The results of the calculations are able to provide electrical energy savings of up to 75% compared to using electric heating. The volume of the tank also affects the efficiency of the heat pump and compressor usage. The optimal tank volume is 1000 L to cover 2–3 h of high electricity price peaks.

Studies on hybrid ground source heat pump systems with a cooling tower as a supplemental heat rejecter were presented by a number of researchers [9–13]. Park et al. [9] proposed a new parallel system comprising of a hybrid ground source heat pump-cooling tower. The new parallel system enables the GHE to be switched off during the recovery period of the soil thermal condition. The performance of the heat pump was investigated at different flow rates of the fluid in the primary flow loop (the heat pump), GHE, and cooling tower. The results indicated that the parallel system of the ground source heat pump-cooling tower generates 21% COP more than that produced by the conventional ground source heat pump system. Man et al. [10] provided the technical and economic analysis of a hybrid cooling tower-ground source heat pump, based on the hourly load of a two storey residential building located in Hong Kong. The hybrid system in this study not only solves the thermal degradation problem, but also reduces the operation and capital costs of the air conditioning system. A study of the operation strategy of a hybrid cooling tower-ground source heat pump system was presented by Wang et al. [11]. The operation strategy consists of a fixed cooling set point, outside air reset, wet bulb reset, and load reset. Fan et al. [13] presented a theoretical design of a hybrid cooling tower-ground source heat pump, which takes into account the effect of borehole distance, borehole depth, and thermal properties of the grout. In this study, a combined strategy of operation was introduced. The results showed that the lowest energy consumption was obtained when the control strategies of the entering water temperature and wet-bulb temperature differences are combined.

Canelli et al. [14], presented an analysis of the energy, economic, and environmental performances of three different hybrid ground source heat pump systems including (1) a hybrid boiler-chiller-ground source heat pump, (2) a hybrid boiler-chiller-ground source heat pump and fuel cell, and (3) a hybrid boiler-chiller-ground source heat pump and photovoltaic thermal system. The system was optimised to meet the heating and cooling conditions of both residential and commercial buildings, which are in a sharing load. The results indicated that the hybrid system with the fuel cells and photovoltaic thermal system has a definite advantage in terms of energy savings, operational costs, and carbon emission reductions.

A detailed numerical study was conducted by Zhu et al. [15] to investigate the performance and economic characteristics of a combined vertical ground source heat pump and phase change material cooling storage system. The optimal performance of the hybrid system was achieved by varying the ratio of the phase change material cooling storage system to the total cooling load of the system. The obtained optimal cooling storage ratio was close to 40%.

A hybrid system of the ground source electrical heat pump and ground source absorption heat pump was proposed by Wu et al. [16]. The motivation for this study was to combine the features of both heat pumps, as the ground source electrical heat pump has higher energy efficiency in the cooling mode. Conversely, the ground source absorption heat pump has higher energy efficiency in the heating mode. This study also presented the effects of supply ratios on thermal imbalance ratios, annual primary energy efficiency, and cost-efficiency characteristics.

A theoretical study on a hybrid air source heat compensator-vertical ground source heat pump system was conducted by You et al. [17], using a simulation tool, TRNSYS. Four operation strategies were analysed including an air source heat compensator for direct heat compensation, a combined air source heat compensator-ground source heat pump for heat compensation, a combined air source heat compensator-ground source heat pump for space heating, and an air source heat compensator-ground source heat pump for domestic hot water. The results showed that the hybrid air source heat compensator-ground source heat pump reduces energy consumption by 23.86% compared with the boiler-split air conditioner system. Additionally, the operational costs are reduced by 50%. Another study has done by Xianting Li et al. [18] by developing an energy-efficient heat pump system using the TRNSYS simulation tool in a combination of a hybrid source and three types of hybrid source heat pumps to calculate the efficiency of year-round operation. The results show that the hybrid heat pump system can maintain heat reliability during year-round operation including winter. This hybrid source heat pump system saves energy up to 15% and the payback period is approximately five years.

In addition, Gaoyang Hou et al. [19] also conducted a simulation using the TRNSYS tool to analyse the work system of heat pumps sourced from hybrid soil with optimal control strategies. His research combines a horizontal ground loop and a liquid dry cooler with a short and long term simulation process on TRNSYS to analyse soil thermal conditions and energy variations. Gaoyang Hou et al. also performed a simulation using the TRNSYS tool to analyse the heat pump working system sourced from hybrid soil with optimal control strategies. Their research combines a horizontal ground loop and a liquid dry cooler with a short and long term simulation process on TRNSYS to analyse soil thermal conditions and energy variations. The simulation results show that the overall performance in the short-term simulation is influenced mainly by the temperature of the diverter heating set rather than by cooling. Diverter heating set was recommended about 8–10 °C in climatic zones similar to the Birmingham area by combining long-term coefficient of performance (COP) values and soil thermal variation.

A hybrid active air source regeneration-ground source heat pump system was studied by Allaerts et al. [20]. In this study, the borehole area is divided into two different regions, namely warm and cold regions. These two different regions were proposed to balance the extraction/rejection of heat during heating and cooling periods. In addition, a supplementary dry cooler was used to capture heat/cold during summer/winter to recover the degradation of the soil thermal condition. According to Allaerts et al. the proposed hybrid system can significantly reduce the size of the borehole area by up to 47% in the cost-optimal configuration.

The brief review of the literature [1–22] above shows that most researchers proposed a hybrid cooling tower-ground source heat pump for a cooling load dominated regime and a hybrid solar system-ground source heat pump for heating load dominated conditions. Some other studies proposed a combined system of a ground source heat pump with different additional heat rejecter/absorber systems including a boiler, chiller, fuel cell, photovoltaic, and phase change material cooling storage systems. Additionally, a ground source heat pump system with two different regions of BHEs including warm and cold was recently suggested and analysed. However, it seems a comprehensive study of a combined GHE arrangement with different operation modes has yet to be presented.

This paper is focused on the operational analysis of a combined horizontal-vertical GHE to address various demands and thermal loading conditions. The performance of the combined GHE is studied based on the results of a transient finite difference model developed in this paper. The effects of continuous and intermittent operation conditions, climate condition and fluid mass flow rate on the GHE's performance are investigated with this new model. Based on the analysis of the outcomes of numerical simulations, the recommendations for the optimum operation of combined GHEs are summarised in the conclusion.

2. Development of a Combined Horizontal–Vertical Geothermal Heat Exchanger Model

2.1. Physical Model

The physical model is described first including a number of simplifications, which are usually needed for quantitative analysis of complex systems or phenomena. A horizontal GHE, as shown in Figure 1a, represents a multi-U-shaped pipe buried in a trench at a specific depth h close to the ground surface. A horizontal GHE has a total length L, diameter d, wall thickness t, and pipe spacing L_s. The heat transfer process at the horizontal GHE occurs when the fluid at a temperature T_{fi} enters the GHE and exchanges the heat energy with the pipe's inner surface by convection. The heat then conducts through the pipe wall to the surrounding soil. The ground exchanges the heat at the ground surface with the atmosphere through diverse heat transfer mechanisms including reflection, convection, radiation, and evaporation. A vertical GHE, as shown in Figure 1b, could be modelled as a single U-shaped pipe, which is buried inside a borehole having a diameter D_b and depth z. The pipe is described by diameter d, length L, thickness t, and shank spacing L_s. The borehole is usually filled with a material, which has a relatively higher thermal conductivity than the surrounding soil, in order to enhance the GHE's thermal performance. The borehole could be confined by different types of soil/rock layers that have different thermal conductivities. The heat transfer at GHE occurs when the fluid is operating at a temperature of T_{fi} and the mass flow rate m is circulated through a U-shaped pipe, exchanges its heat with the pipe's wall by convection. The heat then flows through the pipe's wall, grout, and surrounding soil by conduction sequentially. The convection occurs when the soil surface exchanges the heat with the atmosphere. As mentioned above, a vertical GHE is appropriate for installation in areas with limited land access. It may benefit from the relatively stable ground temperature at the deeper ground depth, and it leads to an improvement in performance. However, this performance decreases with the increase in the operation time, and it may take a season to recover fully. The poor soil conductivity and relatively deep depth of burring are the main reasons for the slow recovery of the soil temperature.

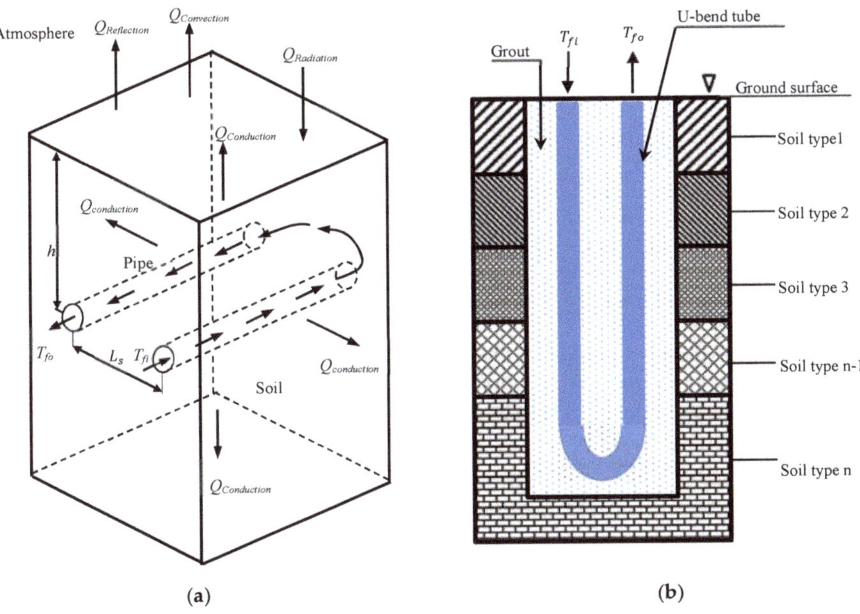

Figure 1. (a) Schematic of the horizontal geothermal heat exchanger (GHE). (b) Schematic of the vertical GHE.

To take advantage of different arrangements and negate their disadvantages, a combined, structured, horizontal–vertical GHE was proposed, see Figure 2. This figure represents a schematic diagram of the proposed combined horizontal–vertical GHE, coupled with a heat pump. The hybrid horizontal vertical GHE can be operated in five different modes through the adjustment of the valves, namely (1) the horizontal GHE only, (2) the vertical GHE only, (3) the horizontal to the vertical GHE, (4) the vertical to the horizontal GHE, and (5) both the horizontal and the vertical GHEs concurrently, by splitting the fluid flow.

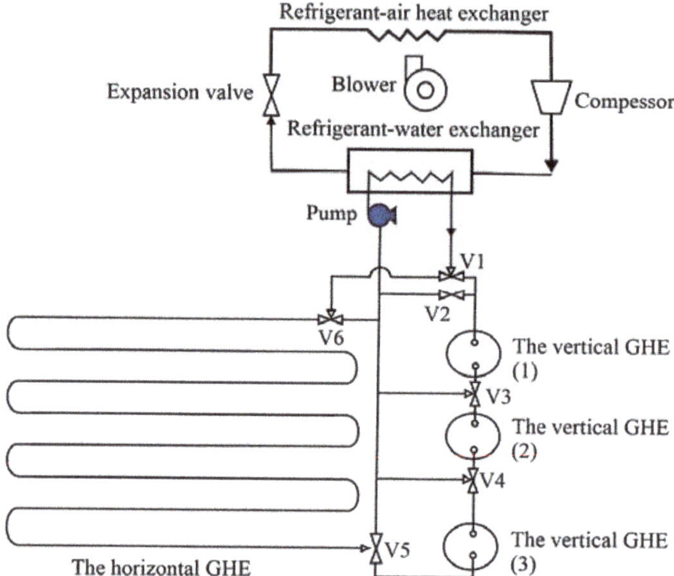

Figure 2. Schematic of the hybrid horizontal vertical GHE.

2.2. Mathematical Model

The performance of the combined GHE was studied based on the previously developed mathematical models for the horizontal [21] and vertical [22] GHE, respectively. The key equations in these models are briefly summarised below.

2.2.1. Governing Equation of the Soil

The thermal model of the soil domain around the horizontal GHE is developed based on the Cartesian coordinates as it is easy to handle when applying the soil's internal source term, in which the value varies with the increase in the soil depth. The governing equation of the soil domain around the horizontal GHE is given as:

$$\frac{1}{\alpha_s}\frac{\partial T_s}{\partial t} = \frac{\partial^2 T_s}{\partial x^2} + \frac{\partial^2 T_s}{\partial y^2} + \frac{H_s}{k_s} \tag{1}$$

The thermal model of the soil domain around the vertical GHE is developed by considering the cylindrical coordinates. The cylindrical coordinates are selected because these represent the shape of the borehole, thus the boundary conditions can be easily defined and prescribed. As a result, an accurate result can be obtained. The governing equation of the soil domain around the vertical GHE is given as

$$\frac{1}{\alpha_s}\frac{\partial T_s}{\partial t} = \frac{\partial^2 T_s}{\partial r^2} + \frac{1}{r}\frac{\partial T_s}{\partial r} + \frac{\partial^2 T_s}{\partial z^2} + \frac{H_s}{k_s} \tag{2}$$

An internal heat source term was used to take into account the effect of seasonal changes in soil temperature [21,23], expressed by the following equation:

$$H_s = \rho_s c_s \frac{\Delta T_s}{\Delta t} \tag{3}$$

The direct measurement or analytical approach can be used to determine deviations in soil temperature during the process of seasonal change. The experimental results are preferred as they represent the actual soil temperature. However, it is sometimes hard to obtain the measurement data, especially for a specific location and a certain depth. Thus, the analytical equation presented by Baggs [24] was used in this work:

$$T(x,t) = (T_m \pm \Delta T_m) + 1.07 k_v A_s e^{\left(-0.00316x\left(\frac{1}{\alpha}\right)^{0.5}\right)} \cos\left[\frac{2\pi}{365}\left(t - t_0 - 0.1834x\left(\frac{1}{\alpha}\right)^{0.5}\right)\right] \tag{4}$$

2.2.2. Governing Equation of the Temperature Exchange in the Grout for the Vertical GHE

The governing equation of the grout is given by a similar equation to the one above:

$$\frac{1}{\alpha_g}\frac{\partial T_g}{\partial t} = \frac{\partial^2 T_g}{\partial r^2} + \frac{1}{r}\frac{\partial T_g}{\partial r} + \frac{\partial^2 T_g}{\partial z^2} \tag{5}$$

2.2.3. Governing Equation of the Pipe

The energy balance equation for the pipe is

$$c_p \rho_p V_p \frac{\partial T_p}{\partial t} = A h_f \left(T_f - T_p\right) + \frac{k_g A}{0.5\Delta r}\left(T_s - T_p\right) \tag{6}$$

The area of the pipe for the vertical GHE was determined based on an equivalent pipe diameter, which is given later by Equation (8).

2.2.4. Governing Equation of the Fluid

For the working fluid, the energy balance equation states

$$c_f \rho_f A \frac{\partial T_f}{\partial t} = \pi d_{in} h_f \left(T_p - T_f\right) - \dot{m}_f c_f \frac{\partial T_f}{\partial z} \tag{7}$$

Equation (6) can be applied directly to the evaluation of performance of horizontal GHE. While for the vertical GHE, the internal pipe diameter d_{in} was replaced with an equivalent pipe diameter D_{eq} since the single U shaped pipe is represented as a pipe, which has an equivalent diameter (refer to Figure 3). The equation to calculate the equivalent diameter has been developed and validated by Gu and Oneal [25] as:

$$D_{eq} = \sqrt{2 D_p L_s} \tag{8}$$

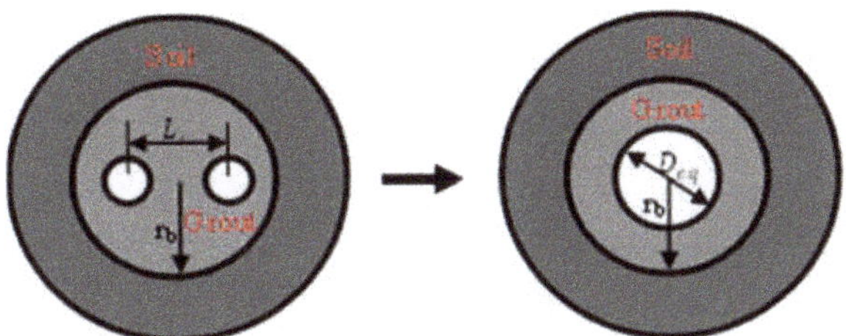

Figure 3. Schematic of the vertical GHE with an equivalent diameter.

The equation is valid if $D_p \leq L_s \leq r_b$

2.2.5. The Initial and Boundary Conditions

- The initial conditions are:

$$T_s = T\,(x,t),\ t = 0 \tag{9}$$

$$T_g = T_s, \; t = 0 \tag{10}$$

- The boundary conditions of the computational domain of the horizontal GHE are:

$$\frac{\partial T_s}{\partial x}\Big|_{x=0} = 0 \tag{11}$$

$$\frac{\partial T_s}{\partial x}\Big|_{x=x_{max}} = 0 \tag{12}$$

$$Q_{ca}, y = 0 \tag{13}$$

$$\frac{\partial T_s}{\partial y}\Big|_{y=y_{max}} = 0 \tag{14}$$

- The boundary conditions of the computational domain of the vertical GHE are:

$$Q_{cf}, r = r_{in} \tag{15}$$

$$\frac{\partial T_s}{\partial r}\Big|_{r=r_{max}} = 0 \tag{16}$$

$$Q_{ca}, z = 0 \tag{17}$$

$$\frac{\partial T_s}{\partial z}\Big|_{z=z_{max}} = 0 \tag{18}$$

The system of differential Equations (1), (2), and (4)–(6) were solved by using an explicit finite difference scheme, which provided the accurate solution for the temperature distribution in the soil, grout, pipe, and fluid. As an example, Figure 4 shows the computational domain of the vertical GHE. The working fluid exchanges the heat energy with the pipe's surface, which then flows through the pipe wall in a 1D way. The heat is then transferred through the grout and surrounding soil in a 2D manner. Two different boundary conditions are applied at the boundary of the soil domain. The convection heat transfer is considered at the ground surface, while the adiabatic conditions are applied at the bottom and lateral edges of soil domain boundaries.

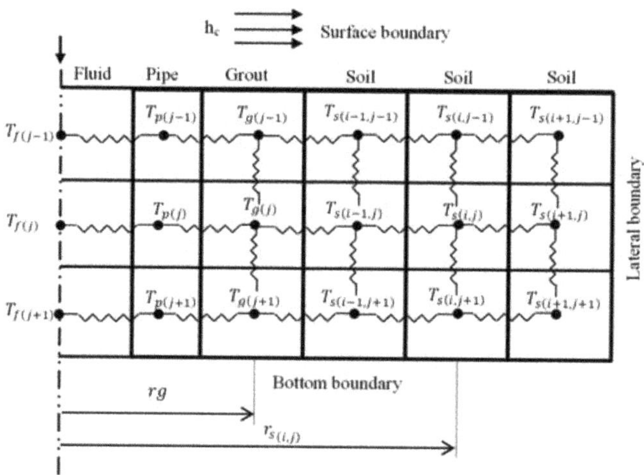

Figure 4. Computational domain of the vertical GHE.

To achieve the stability of the explicit numerical calculations, the time step Δt must be within the Courant–Friedrichs–Lewy stability range [26], which is given by:

$$|\psi| \leq 1, \ \Delta t \leq \frac{\Delta z}{v_f} \tag{19}$$

Total of heat exchange by the GHE was calculated in accordance with the following equation:

$$Q = \dot{m}c_p\left(T_{fi} - T_{fo}\right)\Delta t \tag{20}$$

For the split flow mode, the final temperature of mixed fluid from both the horizontal and the vertical GHE was found as:

$$T_{fmix} = \frac{\left(\dot{m}_{fh}\, c_f\, T_{foh} + \dot{m}_{fv}\, c_f\, T_{fov}\right)}{\left(\dot{m}_{fh}\, c_f + \dot{m}_{fv}\, c_f\right)} \tag{21}$$

The described GHE's models were validated against the experimental data [21–23] and will be used to simulate the performance of a combined GHE.

3. A Hypothetical Case Study

In this study, each arrangement of the GHEs (horizontal or vertical) was set to have the same parameters including the pipe length, pipe diameter, and fluid flow rates. The soil domain around the GHEs was assumed to be a single soil layer, which was homogeneous and isotropic. Table 1 presents the parameters of the reference case used in the simulation to be described next.

Table 1. The parameters of the reference case.

Parameters	Value	Unit	Parameters	Value	Unit
Horizontal GHE	-	-	*Circulation fluid (water)*	-	-
Total pipe length (L)	200	m	Inlet water temperature (T_{fi})	50	°C
Burial depth (h)	0.25	m	Flow rate ($\dot{m}_f$)	0.6	kg/s
Pipe internal diameter (d_{in})	0.04	m	Specific heat (c_f)	4188	J/kg K
Pipe outer diameter (d_{out})	0.044	m	Density (ρ_f)	980	kg/m^3
Centre distance between pipe (L_s)	0.28	m	*Soil type in Adelaide: layered old dune sands [27]*	-	-
Distance of soil domain in x direction (z)	0.14	m	Thermal conductivity (k_s)	1.3	W/m K
Distance of soil domain in y direction (z)	1.5	m	Specific heat (c_s)	1140	J/kg K
Vertical GHE	-	-	Density (ρ_s)	1500	kg/m^3
Total pipe length (L)	200	m	*Soil type in Brisbane: clay [28]*	-	-
Borehole depth (h)	100	m	Thermal conductivity (k_s)	1.1	W/m K
Borehole diameter (D_b)	0.15	m	Specific heat (c_s)	1500	J/kg K
Soil domain diameter (D_s)	4	m	Density (ρ_s)	1300	kg/m^3
Pipe internal diameter (d_{in})	0.04	m	*Grout (vertical GHE)*		
Pipe outer diameter (d_{out})	0.044	m	Thermal conductivity (k_g)	2	W/m K
Centre distance between pipes (L_s)	0.07	m	Specific heat (c_g)	1140	J/kg K
			Density (ρ_g)	1500	kg/m^3
			Wind speed (Adelaide)	4.9	m/s
			Wind speed (Brisbane)	3.8	m/s

The initial temperature of the ground was estimated using the analytical equation (4) suggested by Baggs [24]. The Baggs' equation was determined based on the input parameters of the climate conditions data. The amplitude data for annual air temperature and the average air temperature was obtained from the Australian Bureau of Meteorology [29]. The variable ground temperature for the local site was based on the data given by Baggs [24]. Table 2 summarises the parameters of the reference used to estimate the soil temperature conditions. Figure 5 shows the typical changes of soil temperatures in Adelaide and Brisbane at the end of winter (in August) and summer (in February). As observed from the figure that the soil temperature in Adelaide was lower than in the Brisbane. This tendency could be affected by mild, and generally a warm and temperate climate of Adelaide compared to humid subtropical climate of Brisbane as represented by the parameter of average annual air temperature (T_m) in Equation (11). Besides the local site variable of the ground temperature (ΔT_m), which varied with geographic location, it also affects the soil temperature. As an example, Figure 6 shows the ambient temperature in Adelaide and Brisbane on three consecutive summer days. These were used in the case study. In general, it can be seen from the figure that the pattern of air temperature was almost the same. The profile of air temperature fluctuated during day and night. For these three randomly selected summer days it was found that the air temperature in Adelaide was slightly higher than that in Brisbane.

Table 2. The parameters reference used to estimate the soil temperature.

Parameter	Adelaide	Unit	Brisbane	Unit
Average annual air temperature (T_m)	16.45	°C	25.4	°C
Amplitude of the annual air temperature (A_s)	11.9	°C	11.2	°C
The local site variable for the ground temperature (ΔT_m)	2.5	°C	3	°C
Vegetation coefficient (k_v)	1	-	1	-
Soil thermal diffusivity (α)	0.76	$10^{-2}\ \mathrm{cm^{-2}\ s^{-1}}$	0.55	$10^{-2}\ \mathrm{cm^{-2}\ s^{-1}}$
Phase of air temperature wave (t_0)	10 (10 January)	day	19 (19 January)	Day

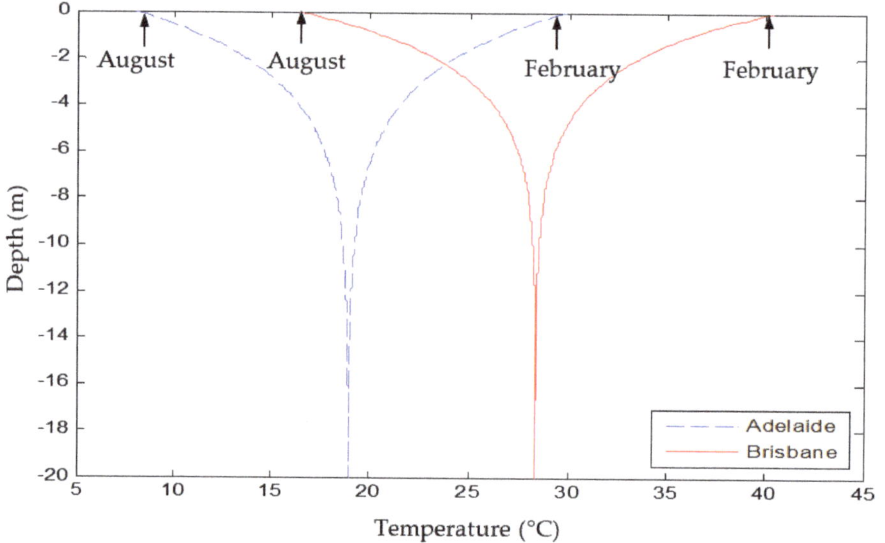

Figure 5. Typical soil temperature in Adelaide and Brisbane.

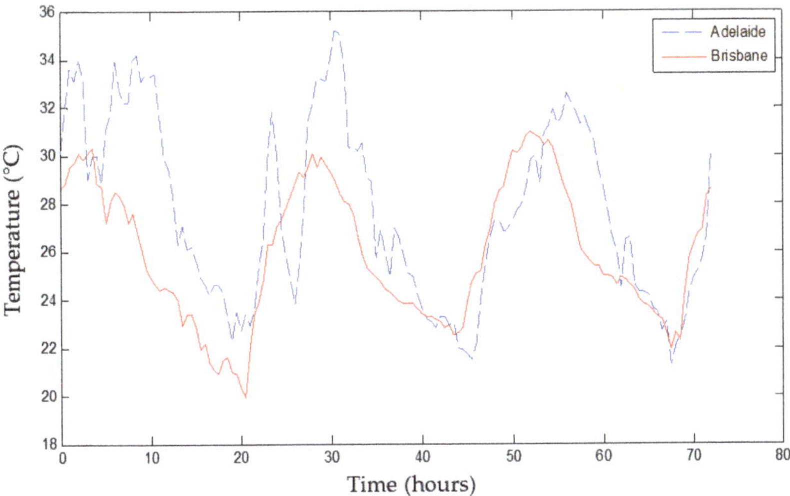

Figure 6. An example of ambient temperature in Adelaide and Brisbane on 3 consecutive summer days.

The effect of seasonal changes on the soil temperature is modelled by incorporating the internal source term concept into the GHE model [21]. The value of the internal source term varied with the soil depths. It was higher in a shallow region and lower in a deeper zone. At a depth of 12 m below the ground surface, the value of the internal source term was assumed to be zero (this corresponded to the depth at which the effect of the ambient temperature on the soil temperature was negligible). Figure 7 shows the absolute value of the internal source term at various depths for two reference locations, namely Adelaide and Brisbane. As is shown in Figure 7, at the upper layer (0–3.5 m depth), the internal source term of the soil in Brisbane was relatively higher than in Adelaide. However, at a deeper layer, the condition was in the opposite. This phenomenon could be affected by meteorological, terrain, and subsurface conditions. The value of the internal heat source was positive when the ground temperature was warm (August to February) and conversely it was negative when the ground temperature was cool (February to August), as is shown in Figure 5.

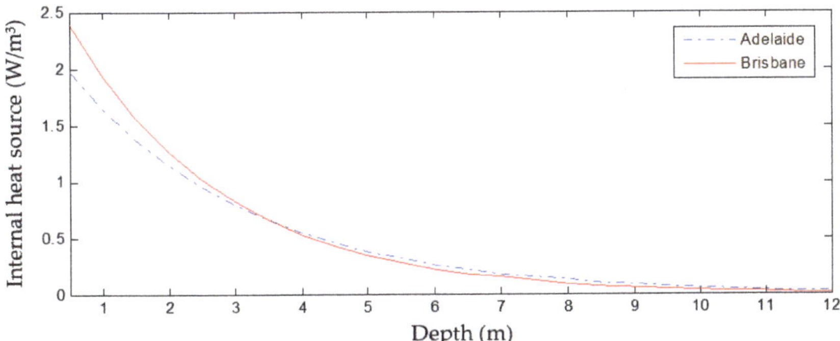

Figure 7. The absolute value of the internal heat source at various depths.

4. Results and Discussion

This section shows the simulation outcomes of the GHE subjected to five operational modes, as described above. The analysis and discussion on continuous operation, intermittent operation, split flow operation, climate condition, and variations of the fluid mass flow rate will be discussed in the following section.

4.1. Continuous Operation

Figure 8 shows the simulation results of the GHE, which was operated on three consecutive summer days, with respect to the geological and climate conditions corresponding to Adelaide (South Australia). The amount of energy released by the GHE was calculated using Equation (21). It is observed that the lowest value of the outlet fluid temperature could be attained so the highest amount of energy can be released. As an example, Figure 9 provides the profile of the fluid temperature generated by operation of the horizontal, vertical, and the horizontal to the vertical GHE. This figure demonstrated that the outlet fluid temperature increased with the increase in the operation period. The accumulation of heat in the surrounding soil during the operation of the GHE led to a reduction in the heat transfer rate. As is shown in the Figure 9, the profile of the outlet fluid temperature of the vertical GHE was relatively stable when compared with those generated by the horizontal and the series operation modes. This phenomenon occurred because the ambient temperature, which fluctuated diurnally, did not have a significant effect on the performance of the vertical GHE. From Figure 8, it followed that the lowest energy was released when operating the horizontal GHE only. The horizontal GHE released 1002 kW less energy than that released in 3 days by the vertical GHE. The relatively stable temperature of the ground at a deeper layer might enhance the heat transfer capacity of the vertical GHE and led to better thermal performance. The amount of energy released could be increased by using the combined operation mode of the GHE, including the split flow and series operation modes. It can be seen that the series operation mode, from horizontal to vertical, could release slightly higher energy, namely 22 MJ more than the opposite operation mode. While the split flow mode, with a ratio of fluid flow at 50% in the horizontal GHE and 50% in the vertical GHE, released less energy than the series operation mode. The average heat transfer rate was calculated by dividing the amount of energy released with the operation period, namely 72 h. The single arrangement of the horizontal or the vertical GHE generated an average heat transfer rate of 6.3 kW and 10.2 kW, respectively. The combined arrangements produced a relatively higher heat transfer rate namely: 15.7 kW, 15.6 kW, and 15.4 kW for operation modes from the horizontal to the vertical, from the vertical to the horizontal GHE, and by splitting the fluid flow, respectively. Therefore, the horizontal or vertical GHE may be operated when the loading load was relatively low. At peak loads or when heating/cooling demands were relatively high, the combined GHE's operation could have a significant advantage, see Figure 8.

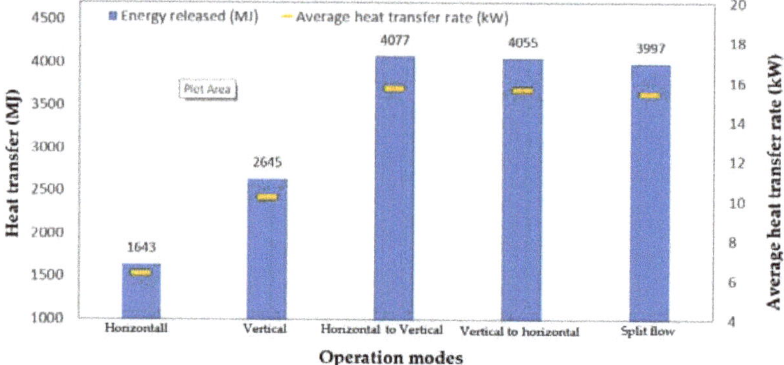

Figure 8. Energy released in 3 days and average heat transfer rate of the GHE under continuous operation condition in Adelaide (where the inlet fluid temperature = 50 °C, fluid mass flow rate = 0.6 kg/s, length of horizontal GHE = 200 m, length of vertical GHE = 200 m).

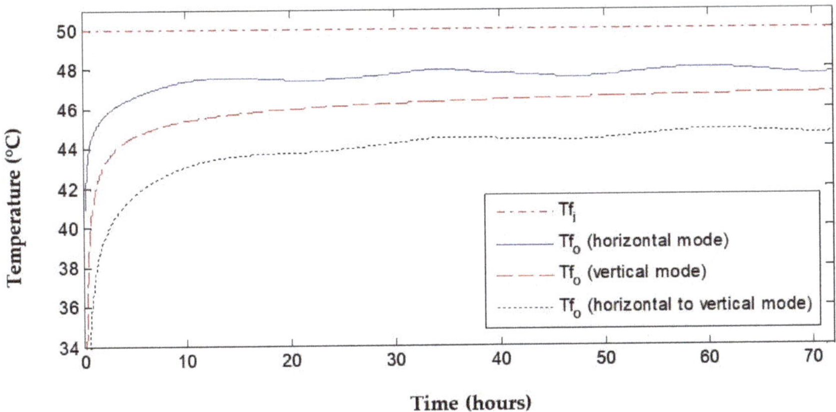

Time (hours)

Figure 9. Profile of fluid temperature of the horizontal, vertical, and horizontal to the vertical mode in Adelaide.

4.2. Intermittent Operation

The GHE can also be operated in the intermittent condition to cope with cyclic load conditions. Figure 10 shows the amount of energy released by the GHE operating intermittently in three summer days, see Figures 5 and 6. The GHE ran for 8 h (during the working hours) and was off for 16 h daily. The results display that during 24 h of operation, the horizontal GHE released 41% less energy than the vertical GHE. The combined operation modes of the GHE could increase the amount of energy released, as the contact area, where heat was exchanged with the surrounding soil, increased. It is seen that the series operations could release 40.8% and 39.6% more energy than the vertical mode, for the horizontal to vertical and vice versa modes, respectively. In the split flow mode, the fluid is split to flow with a ratio of 50% in the horizontal GHE and 50% in the vertical GHE. The results demonstrate that the split flow mode released 2.8% less energy than the horizontal to vertical mode. Even though, the amount of energy released in the intermittent operation regime was less than that generated by the continuous operation due to the total operation period in the intermittent condition was less compared with the continuous condition. However, the average heat transfer rates increased. They were 60.1% for the horizontal mode, 68.5% for the vertical mode, 54% for the horizontal to vertical mode, 53.5% for the vertical to horizontal mode, and 52.6% for the split flow mode. In the intermittent operation condition, the deterioration of the ground temperature during the operation hours is possible to recover during the time when the system is switched off. As a result, it increases the heat transfer rate of the GHE and produces lower outlet fluid temperature over the next day's operation as reflected by an example of fluid temperature generated by the vertical to horizontal mode (see Figure 11).

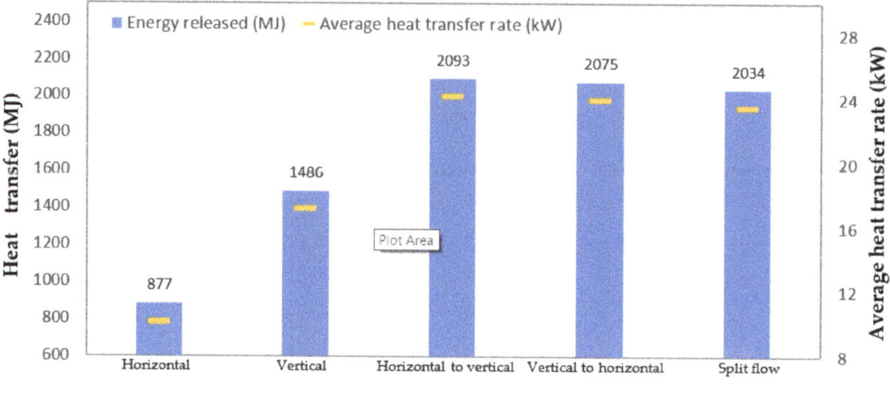

Figure 10. Energy released in 3 days and average heat transfer rate of the GHE under intermittent operation in Adelaide (where the inlet fluid temperature = 50 °C, fluid mass flow rate = 0.6 kg/s, length of horizontal GHE = 200 m, and length of vertical GHE = 200 m).

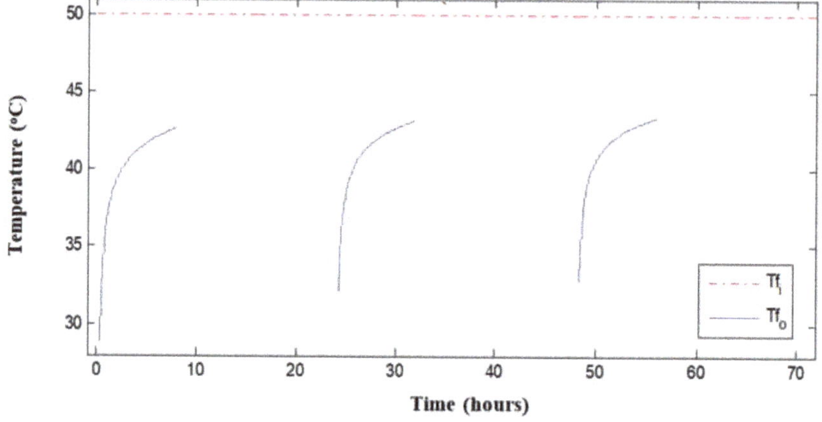

Figure 11. Profile of the fluid temperature of the horizontal to the vertical mode in Adelaide.

4.3. Split Flow Operation

In this section, the effect of the fluid mass flow rate ratio in the split flow operation mode was investigated. The ratio of the fluid mass flow rate varied as follows: 30%:70%; 40%:60%; 50%:50%; 60%:40%; and 70%:30% for the horizontal and the vertical GHE, respectively. From the current numerical simulations, it was found that the ratio of fluid mass flow rate in the split flow mode did not significantly affect the amount of energy released by the GHE, as shown in Figure 12. As an example, Figure 13, presents the outlet fluid temperature of the GHE at three different flow rate ratios, namely: 30%:70%; 50%:50%; and 70%:30% for the horizontal and the vertical GHE, respectively. It is found that the difference in the outlet fluid temperature was relatively small. The highest fluid outlet temperature was yielded by the GHE with a flow rate ratio of 70%:30%. It can be seen that the GHE operated with a flow rate ratio of 70% horizontal and 30% vertical released the lowest amount of energy across three consecutive days of operation, namely, 3% less than the one with a flow rate ratio of 60% horizontal and 40% vertical. The amount of energy released gradually increased with the increase of the ratio of the fluid mass flow rate in the vertical GHE and declined at the ratio of 30% horizontal and 70% vertical

GHE namely, 0.7% less than that with a ratio of 40% horizontal and 60% vertical GHE. This tendency was due to a significant reduction in the fluid mass flow rate in the horizontal GHE (only 30% of the total fluid flow rate). At a lower mass flow rate, the thermal resistance of the horizontal GHE increased as a reduction in fluid mass flow rate was directly proportional to the decrease in the convective heat transfer coefficient between the working fluid and the inner pipe surface. As a result, it decreased the heat transfer rate of the horizontal GHE. In addition, with 70% of mass flow rate in the vertical GHE, the surrounding soil temperature deteriorated quickly due to heat accumulation, leading to degradation in the heat transfer capacity of the vertical GHE. From the Figure 12, it can be seen that the highest heat transfer rate was produced by the GHE with a ratio of flow rate of 40% horizontal and 60% vertical namely, 15,490 kW.

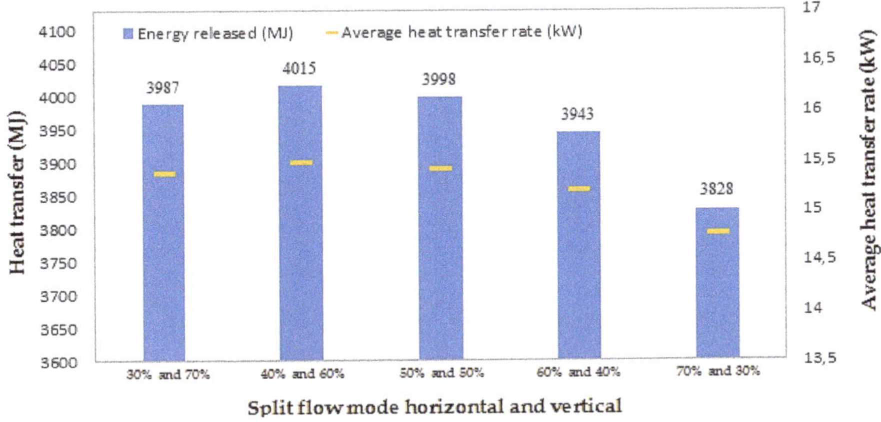

Figure 12. Energy released in 3 days by the GHE with different ratio of mass flow rate in Adelaide (where the inlet fluid temperature = 50 °C, fluid mass flow rate = 0.6 kg/s, length of horizontal GHE = 200 m, and length of vertical GHE = 200 m).

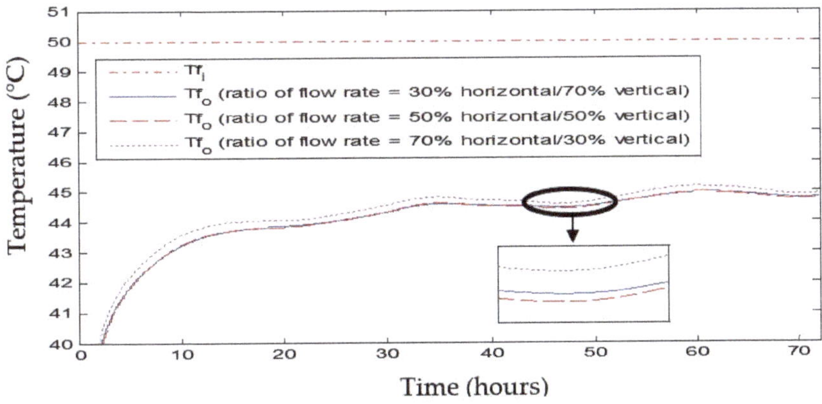

Figure 13. Profile of the outlet fluid temperature of the GHE (with a split flow operation mode) under different flow rate ratios (Adelaide case).

4.4. Climate Condition

The performance of the GHE installed in a temperate climate corresponding to Adelaide was compared with that installed in a subtropical climate, such as near Brisbane. Figure 14 summarises the performance of the GHE, which was installed in two different regions, with different climate conditions. It was found that during 72 h (3 days: 10–12 January) of the continuous operation, the GHE installed in Adelaide's temperate climate could release 34.3% and 26.9% more energy compared with that installed in Brisbane, for the horizontal and vertical mode, respectively. While for the combined modes including horizontal to vertical, vertical to horizontal, and split flow were 31.7%, 31.8%, and 31% higher, respectively. It is observed that operating the vertical GHE only, in the temperate climate, Adelaide, could release an amount of energy that was almost the same as that released by the combined operation mode, in the subtropical climate. This tendency occurred because the climate condition influenced the initial soil temperature in both regions. The initial soil temperature in Adelaide was lower than that in Brisbane, as seen in Figure 5. In addition, the difference in soil types in both locations contributed to the heat transfer capacity of the GHE. As a result, the GHE installed in a temperate climate, Adelaide, produced a lower outlet fluid temperature as demonstrated in Figure 15.

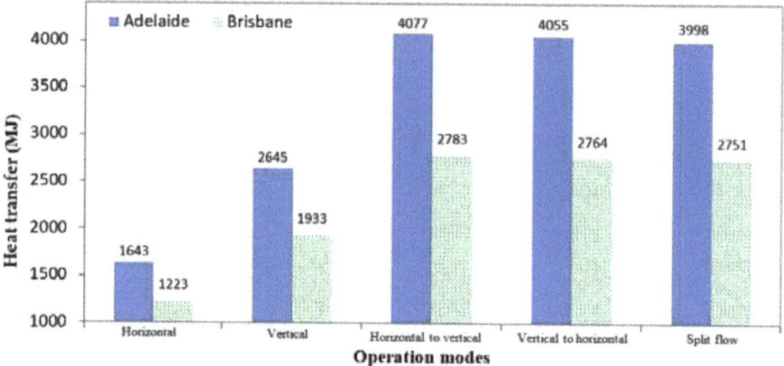

Figure 14. Energy released in 3 days by the GHE under different climate conditions (where the inlet fluid temperature = 50 °C, fluid mass flow rate = 0.6 kg/s, length of horizontal GHE = 200 m, and length of vertical GHE = 200 m).

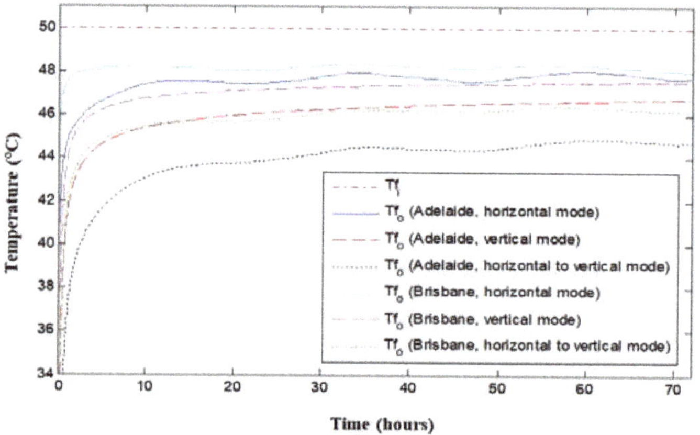

Figure 15. Profile of fluid temperatures of the horizontal, vertical, and the horizontal to vertical modes under different climate conditions.

4.5. Variations of the Fluid Mass Flow Rate

Figure 16 shows the effect of the fluid mass flow rate on the amount of energy released by the GHE in 3 summer days in Adelaide. The performance of the GHE under two different fluid mass flow rates namely, 0.6 kg/s and 1 kg/s, was compared. The results of the simulation indicate that the amount of energy released by the GHE increased as the fluid flow rate increased. This tendency occurred because the mass flow rate affected the convective heat transfer coefficient of the fluid inside the GHE. The higher the mass flow rate, the higher the coefficient convective heat transfer was attained. As a result, it increased the heat transfer capacity and the amount of energy released by the GHE. As an example, Figure 17 shows the profile of the fluid temperature generated from the horizontal, vertical, and horizontal to vertical modes at two different mass flow rates namely, 0.6 kg/s and 1 kg/s. This figure shows that the outlet fluid temperature increased with the increase of fluid mass flow rate. This tendency occurred because at a higher mass flow rate, the time period during which the fluid makes contact with the pipe was shorter when compared to a relatively low flow rate. In addition, at a higher mass flow rate, the GHE released more heat into the surrounding soil and led to a quick increase in soil temperature, resulting in increased outlet fluid temperature. Varying the mass flow rate from 0.6 to 1 kg/s increased the amount of energy released by 2.4% for the horizontal mode and 3.3% for the vertical mode. The rate increases were 4.9%, 5.1%, and 5.6% for the horizontal to vertical, vertical to horizontal, and split flow mode, respectively.

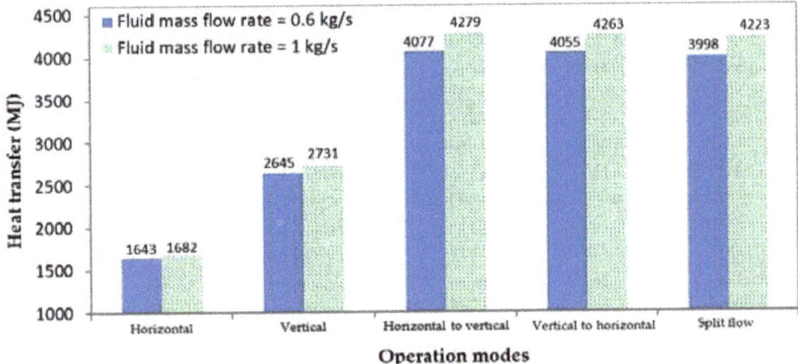

Figure 16. Energy released in 3 days by the GHE with different mass flow rate in Adelaide (where the inlet fluid temperature = 50 °C, length of horizontal GHE = 200 m, and length of vertical GHE = 200 m).

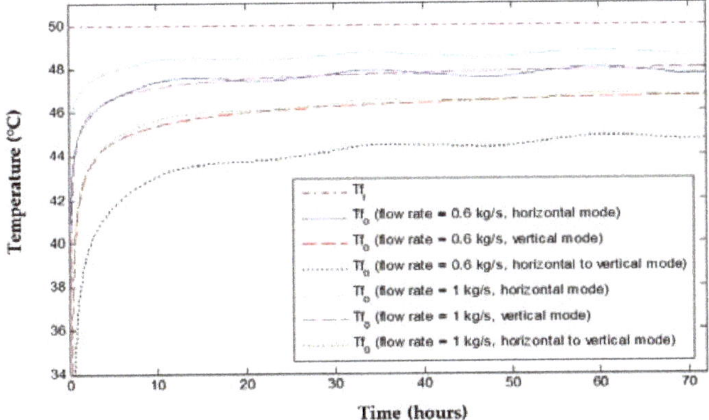

Figure 17. Profile of fluid temperature of the horizontal, vertical, and the horizontal to vertical modes with variations in the fluid mass flow rates, in Adelaide.

5. Conclusions

In this paper, a combined GHE was proposed and its performance in five operational modes was analysed by using the validated models previously developed for horizontal and vertical arrangements of GHEs. The main conclusions, which followed from the calculations, are summarised as follows:

1. With the same length of the pipe system, the vertical GHE could release more energy than the horizontal GHE, as the initial soil temperature (in summer) at a deeper layer was lower than that at a shallow region;

2. When the GHE operated in combined modes, the amount of energy released by the GHE was increased, as the contact area, where heat was exchanged with the surrounding soil, was increased;

3. The series operations (horizontal to vertical or vertical to horizontal) of the GHE could release more energy than could be done in the split mode. The difference in the fluid velocity in the split flow and series modes contributed to the amount of energy released by the GHE;

4. The intermittent operation of the GHE might be conducted to cope with cyclic thermal loading. The intermittent operation could benefit the performance of the GHE as the degradation of the ground temperature during the operation of the GHE was largely recovered during the system shut down.

5. In the split flow mode, the ratio of the fluid mass flow rate did not significantly affect the amount of energy released by the GHE. The GHE with a ratio of mass flow rate of 40% horizontal and 60% vertical releases the highest amount of energy in the split flow operational mode.

6. Climate conditions had a significant effect on the GHE's performance. The GHE installed in a temperate climate, corresponding to Adelaide's conditions, could release more energy than the same installation located in a subtropical climate, such as Brisbane. This was due to the difference in the initial soil temperature.

7. Increasing the fluid mass flow rate could enhance the amount of energy released by the GHE as the fluid mass flow rate affected the coefficient of the convective heat transfer of the working fluid.

Author Contributions: Conceptualization, methodology, and results and were initiated and wrote by S.E.S., E.H. and A.K.; S.E.S. contributed to the experiment and writing—original draft preparation; Writing—review and editing, T.M.I.R., R.T.; Supervision by E.H. and A.K. All authors have read and agreed to the published version of the manuscript.

Funding: This research received no external funding.

Acknowledgments: The authors graciously acknowledge the support provided by the School of Mechanical Engineering, the University of Adelaide.

Nomenclature

A	area of the pipe (m^2)		T_g	grout temperature (K)
A_s	amplitude of the annual air temperature (K)		T_m	average annual air temperature (°C)
c_f	fluid specific heat (J/kg K)		T_p	pipe temperature (K)
c_s	soil specific heat (J/kg K)		$T(x,t)$	ground temperature at a given depth x on calendar day t (°C)
c_p	pipe specific heat (J/kg K)		T_{fi}	inlet fluid temperature (K)
d_{in}	internal pipe diameter (m)		T_{fo}	outlet fluid temperature
D_{eq}	equivalent diameter of the pipe (m)		T_{fmix}	final temperature of the mixed fluids (K)
h_f	convective heat transfer coefficient of fluid (W/(m^2 K))		T_s	soil temperature (K)
H_s	soil heat source (W/m^3)		T_{foh}	outlet temperature of the horizontal GHE (K)
k_s	soil thermal conductivity (W/(m K))		T_{fov}	outlet temperature of the vertical GHE (K)
k_g	grout thermal conductivity (W/(m K))		v_f	velocity of the working fluid (m/s)

k_v	vegetation coefficient ($k_v = 1$ for bare ground, $k_v = 0.22$ for year round full vegetation cover)	V_p	volume of the pipe's wall (m^3)
L_s	centre to centre distance between two legs of the U pipe	x	soil depth (cm)
$\dot{m}$	fluid mass flow rate (kg/s)	z	axial distance of grout domain (m)
$\dot{m}_f$	fluid mass flow rate (kg/s)	α_s	soil diffusivity (m^2/s)
$\dot{m}_{fh}$	fluid mass flow rate in the horizontal GHE (kg/s)	α_g	grout diffusivity (m^2/s)
$\dot{m}_{fv}$	fluid mass flow rate in the vertical GHE (kg/s)	ρ_f	fluid density (kg/m^3)
Q	energy released by the GHE (J)	ρ_p	pipe density (kg/m^3)
$\dot{Q}_{ca}$	Convective heat transfer on the ground surface (W)	ρ_s	soil density (kg/m^3)
$\dot{Q}_{cf}$	Convective heat transfer on the inner pipe surface (W)	Δt	time period (s)
r	radius of soil domain (m)	ΔTm	local site variable for the ground temperature (K)
r_b	radius of the borehole	ρ_f	fluid density (kg/m^3)
t	time period (s)	ΔT_s	soil temperature difference in summer and winter (K)
t_0	phase of air temperature wave (day)	Δr	radial increment (m)
t_c	calendar day, where 1 January = 1 and so forth	Δz	distance in the direction parallel to the pipe (m)
T_f	fluid temperature (K)	ψ	Courant number

References

1. Sliwa, T.; Gonet, A. Theoretical Model of Borehole Heat Exchanger. *J. Energy Resour. Technol.* **2004**, *127*, 142–148. [CrossRef]

2. Florides, G.; Christodoulides, P.; Pouloupatis, P. Single and double U-tube ground heat exchangers in multiple-layer substrates. *Appl. Energy* **2013**, *102*, 364–373. [CrossRef]

3. Zhai, X.; Qu, M.; Yu, X.; Yang, Y.; Wang, R. A review for the applications and integrated approaches of ground-coupled heat pump systems. *Renew. Sustain. Energy Rev.* **2011**, *15*, 3133–3140. [CrossRef]

4. Dai, L.; Li, S.; Duanmu, L.; Li, X.; Shang, Y.; Dong, M. Experimental performance analysis of a solar assisted ground source heat pump system under different heating operation modes. *Appl. Therm. Eng.* **2015**, *75*, 325–333. [CrossRef]

5. Kjellsson, E.; Hellström, G.; Perers, B. Optimization of systems with the combination of ground-source heat pump and solar collectors in dwellings. *Energy* **2010**, *35*, 2667–2673. [CrossRef]

6. Ozgener, O.; Hepbasli, A. Experimental performance analysis of a solar assisted ground-source heat pump greenhouse heating system. *Energy Build.* **2005**, *37*, 101–110. [CrossRef]

7. Yang, W.; Sun, L.; Chen, Y. Experimental investigations of the performance of a solar-ground source heat pump system operated in heating modes. *Energy Build.* **2015**, *89*, 97–111. [CrossRef]

8. Zajacs, A.; Lalovs, A.; Borodinecs, A.; Bogdanovics, R. Small ammonia heat pumps for space and hot tap water heating. *Energy Procedia* **2017**, *122*, 74–79. [CrossRef]

9. Park, H.; Lee, J.S.; Kim, W.; Kim, Y. Performance optimization of a hybrid ground source heat pump with the parallel configuration of a ground heat exchanger and a supplemental heat rejecter in the cooling mode. *Int. J. Refrig.* **2012**, *35*, 1537–1546. [CrossRef]

10. Man, Y.; Yang, H.; Wang, J. Study on hybrid ground-coupled heat pump system for air-conditioning in hot-weather areas like Hong Kong. *Appl. Energy* **2010**, *87*, 2826–2833. [CrossRef]

11. Wang, S.; Liu, X.; Gates, S. Comparative study of control strategies for hybrid GSHP system in the cooling dominated climate. *Energy Build.* **2015**, *89*, 222–230. [CrossRef]

12. Sagia, Z.; Rakopoulos, C.D.; Kakaras, E. Cooling dominated Hybrid Ground Source Heat Pump System application. *Appl. Energy* **2012**, *94*, 41–47. [CrossRef]

13. Fan, R.; Gao, Y.; Hua, L.; Deng, X.; Shi, J. Thermal performance and operation strategy optimization for a practical hybrid ground-source heat-pump system. *Energy Build.* **2014**, *78*, 238–247. [CrossRef]

14. Canelli, M.; Entchev, E.; Sasso, M.; Yang, L.; Ghorab, M. Dynamic simulations of hybrid energy systems in load sharing application. *Appl. Therm. Eng.* **2015**, *78*, 315–325. [CrossRef]

15. Zhu, N.; Hu, P.; Lei, Y.; Jiang, Z.; Lei, F. Numerical study on ground source heat pump integrated with phase change material cooling storage system in office building. *Appl. Therm. Eng.* **2015**, *87*, 615–623. [CrossRef]

16. Wu, W.; Li, X.; You, T.; Wang, B.; Shi, W. Combining ground source absorption heat pump with ground source electrical heat pump for thermal balance, higher efficiency and better economy in cold regions. *Renew. Energy* **2015**, *84*, 74–88. [CrossRef]

17. You, T.; Shi, W.; Wang, B.; Wu, W.; Li, X. A new ground-coupled heat pump system integrated with a multi-mode air-source heat compensator to eliminate thermal imbalance in cold regions. *Energy Build.* **2015**, *107*, 103–112. [CrossRef]

18. Li, X.; Lyu, W.; Ran, S.; Wang, B.; Wu, W.; Yang, Z.; Jiang, S.; Cui, M.; Song, P.; You, T.; et al. Combination principle of hybrid sources and three typical types of hybrid source heat pumps for year-round efficient operation. *Energy* **2020**, *193*, 116772. [CrossRef]

19. Hou, G.; Taherian, H.; Li, L.; Fuse, J.; Moradi, L. System performance analysis of a hybrid ground source heat pump with optimal control strategies based on numerical simulations. *Geothermics* **2020**, *86*, 101849. [CrossRef]

20. Allaerts, K.; Coomans, M.; Salenbien, R. Hybrid ground-source heat pump system with active air source regeneration. *Energy Convers. Manag.* **2015**, *90*, 230–237. [CrossRef]

21. Sofyan, S.E.; Hu, E.; Kotousov, A.; Kotousov, A. A new approach to modelling of a horizontal geo-heat exchanger with an internal source term. *Appl. Energy* **2016**, *164*, 963–971. [CrossRef]

22. Sofyan, S.E.; Hu, E.; Kotousov, A.; Riayatsyah, T.M.I.; Khairil; Hamdani. A new approach to modelling of seasonal soil temperature fluctuations and their impact on the performance of a shallow borehole heat exchanger. *Case Stud. Therm. Eng.* **2020**, *22*, 100781. [CrossRef]

23. Sofyan, S.E.; Hu, E.; Kotousov, A. Modelling of a Horizontal Geo Heat Exchanger with an Internal Source Term Approach. *Energy Procedia* **2014**, *61*, 104–108. [CrossRef]

24. Baggs, S.A. Remote prediction of ground temperature in Australian soils and mapping its distribution. *Sol. Energy* **1983**, *30*, 351–366. [CrossRef]

25. Gu, Y.; O'Neal, D.L. Development of an equivalent diameter expression for vertical U-tube used in ground-coupled heat pumps. *ASHRAE Trans.* **1998**, *104*, 347–355.

26. Zima, W.; Dziewa, P. Modelling of liquid flat-plate solar collector operation in transient states. *Proc. Inst. Mech. Eng. Part A J. Power Energy* **2011**, *225*, 53–62. [CrossRef]

27. Geoserver, S.A.R.I. Soil Association Map. Available online: https://sarigbasis.pir.sa.gov.au/WebtopEw/ws/plans/sarig1/image/DDD/200471-235 (accessed on 4 October 2020).

28. Prangnell, J.; McGowan, G. Soil temperature calculation for burial site analysis. *Forensic Sci. Int.* **2009**, *191*, 104–109. [CrossRef]

29. Australian Government Bureau of Meteorology. Climate Data Online. Available online: http://www.bom.gov.au/climate/data/?ref=ftr (accessed on 4 October 2020).

Publisher's Note: MDPI stays neutral with regard to jurisdictional claims in published maps and institutional affiliations.

Article

Stakeholders' Recount on the Dynamics of Indonesia's Renewable Energy Sector

Satya Widya Yudha [1], Benny Tjahjono [2,*] and Philip Longhurst [1]

[1] School of Water, Energy and Environment, Cranfield University, Cranfield, Bedford MK43 0AL, UK; s.widya-yudha@cranfield.ac.uk (S.W.Y.); p.j.longhurst@cranfield.ac.uk (P.L.)

[2] Centre for Business in Society, Coventry University, Coventry CV1 5FB, UK

* Correspondence: benny.tjahjono@coventry.ac.uk

Abstract: The study described in this paper uses direct evidence from processes applied for the developing economy of Indonesia, as it defines the trajectory for its future energy policy and energy research agenda. The paper addresses the research gap to make explicit the process undertaken by key stakeholders in assessing and determining the suitability, feasibility, and dynamics of the renewable energy sector. Barriers and enablers that are key in selecting the most suitable renewable energy sources for developing economies for the renewable energy development have been identified from extensive analyses of research documents alongside qualitative data from the Focus Group Discussions (FGD). The selected FGD participants encompass the collective views that cut across the political, economic, social, technological, legal, and environmental aspects of renewable energy development in Indonesia. The information gained from the FGD gives insights into the outlook and challenges that are central to energy transition within the country, alongside the perceptions of renewable energy development from the influential stakeholders contributing to the process. It is notable that the biggest barriers to transition are centred on planning and implementation aspects, as it is also evident that many in the community do not adhere to the same vision.

Keywords: focus group discussion; sustainability; renewable energy development; Indonesia; geothermal

Citation: Yudha, S.W.; Tjahjono, B.; Longhurst, P. Stakeholders' Recount on the Dynamics of Indonesia's Renewable Energy Sector. *Energies* **2021**, *14*, 2762. https://doi.org/10.3390/en14102762

Academic Editors: David Borge-Diez and Almas Heshmati

Received: 29 March 2021
Accepted: 8 May 2021
Published: 12 May 2021

Publisher's Note: MDPI stays neutral with regard to jurisdictional claims in published maps and institutional affiliations.

1. Introduction

The development of alternative renewable energy sources in Indonesia is of paramount importance not only to fulfil the ever-increasing energy demand in the country but also to contribute to reducing the carbon emission, as well as combating the devastating effects of climate change. During the Conference of Parties (COP) 21 in 2015, known as Paris Agreement, countries around the world committed to reducing carbon emissions by ratifying the agreement. All the countries who participated set the targets in regard to carbon reduction, according to their respective capabilities, known as Nationally Determined Contributions (NDCs).

As one of the countries who participated in COP, Indonesia plans to reduce carbon emission by 29% with its own effort or 41% with international aid by 2030 [1]. Renewable energy sector plays an important role to reduce the carbon emissions, and Indonesia is currently aiming to increase the share of renewable energy to become 23% by 2025 within the National Energy Mix [1]. Due to its unique geographical contour features, Indonesia hosts an enormous potential for renewable energy from various sources, such as geothermal, hydropower, solar energy, bioenergy, wind energy, and ocean energy. The country is undergoing a journey to seek the most suitable renewable energy sources to be developed.

By 2020, Indonesia has only reached halfway towards the 2025 renewable energy target. The development of renewable energy in Indonesia is currently suffering from many obstacles ranging from technical to policy aspects that have significantly hindered its progress. Exploiting renewable energy sources requires a careful appraisal of the potential key predicaments and enablers of its development. Identifying and focusing on a specific type of renewable energy source is therefore deemed essential to mitigate the risks of failure. Analysing the progress of existing development of renewable energy can be done using various ways, for example, by pulling together insights from all relevant stakeholders, which in the case of Indonesia, broadly encompass both state and market players.

In order to acquire a more comprehensive and dynamic outlook, it is necessary to move from the preliminary stakeholder analyses onto probing some crucial information directly from the sources, i.e., the stakeholders. This paper therefore aims to obtain the stakeholder's recount on the assessment and of the suitability, feasibility, and dynamics of the renewable energy sector in Indonesia, that will ultimately pave a trajectory of our future agenda of research. To ascertain a transparent, repeatable, and credible research execution, the following research questions have subsequently been set.

RQ1 According to the stakeholders' recount and outlook, what are the main key challenges, barriers or problems associated with renewable energy development in Indonesia?

RQ2 What are the stakeholders' views on the potential key enablers for the renewable energy development in Indonesia?

RQ3 What is the most suitable renewable energy type to be developed in Indonesia?

RQ4 Depending on the type of the renewable energy selected, what can be proposed to support the development of that particular renewable energy in Indonesia?

Such information gathering can be done via interviews, surveys or focus group discussions involving key stakeholders of the renewable energy in Indonesia, with an ultimate goal to collate the previously disparate information, experience and decision-making processes.

This structure of this paper is as follows. Section 2 set out the foundation of this work by elaborating the various work not only pertinent but also relevant to this paper. In this way, gaps in the existing literature that this work will address can be clearly identified. Section 3 details how the research will be conducted, including the data collection method and analysis. Sections 4 and 5 discuss the findings and their implications to the body of knowledge. Section 6 concludes the paper by showing the process of how a set of proposals are defined that lay down a pathway for the development of renewable energy in Indonesia, a research gap and process that had not previously been evident.

2. Related Work

A successful transition of energy sources from fossil fuel to renewables requires careful evaluation in terms of the planning system and renewable energy selection [1]. Evaluating the renewable energy system can be a complicated process. Such an evaluation process requires appropriate tools that support the data analysis of the availability of the renewable energy sources, selection criteria and methods used in the selection process [2].

2.1. Renewable Energy Selection and Decision Making

Many researchers have carried out the evaluation and selection of the most suitable renewable energy in many countries and different scenarios. There are many decision-making methods that can be applied for renewable energy selection. In this section, we will have a look at some of the previous research on renewable energy selection in different countries, using different research approaches.

One of the many popular methods that can be used for assessing the most suitable renewable energy to develop is a mathematical modelling method. Gonçalves da Silva (2010) [3] used a conceptual framework and a set of mathematical models to evaluate the energy balance of energy conversion technologies for renewable energy development in Brazil. The result showed that wind energy was the most favourable renewable energy source to develop in Brazil, while solar power was the least suitable for development. Another method that can be applied for evaluating and selecting renewable energy is Multi-Criteria Decision Making (MCDM) method. Emir (2014) [4] performed the selection of renewable energy for small islands using the MCDM method, applicable for Malta,

Cyprus, Cuba, Jamaica, Dominican Republic, and Singapore. They considered cost analysis, technical issues, social issues, locations, and environmental issues as the criteria for evaluation. Solar energy was deemed the most suitable renewable energy to invest and develop in small islands.

Analytical Hierarchy Process remains the most popular method to use for selecting the most suitable renewable energy in many different countries. The research conducted with this type of method typically employed a variety of criteria, such as technical performance and efficiency, ecological integrity, economic expedience, sustainable development, socially responsible operation, and technological innovativeness. Based on these criteria, different countries have different results in regard to the most suitable renewable energy [5–8].

2.2. Renewable Energy Selection in Indonesia

Indonesia is one of the countries with abundant potential for many different types of renewable energy development; therefore, selecting the most suitable type of renewable energy to develop is very vital for the energy transition. Despite having quite a few choices for renewable energy with abundant potential, research that is specific to the selection of the most suitable renewable energy in Indonesia is still lacking.

Rumbayan and Nagasaka (2012) [9] used the Analytical Hierarchy Process (AHP) method to identify and rank the most suitable renewable energy in Indonesia, using the level of availability of renewable energies as the primary consideration. Three types of renewable energy were analysed for this study, including solar energy, wind energy, and geothermal energy. The result shows from this study that geothermal is the best criteria, followed by solar and wind alternatives. Tasri and Susilawati (2014) [10] used the Fuzzy Analytical Hierarchy Process (F-AHP) to determine the most appropriate type of renewable energy to develop in Indonesia. This research used several selection criteria, such as sustainability, economic, social, and technological point of view. They evaluated renewable energies for this research include solar energy, hydropower, geothermal energy, wind energy and biomass energy.

Based on the previous research, it is evident that the quantitative approach, especially the Analytical Hierarchy Process (AHP), remains the most popular method to use for renewable energy selection. While the quantitative approach can provide tangible information by generating numerical data, which is beneficial for statistics, it also has certain limitations. A quantitative approach may not be able to provide a deep understanding and insights that explain the underlying reasons behind those numbers. Quantitative approaches are most suitable for countries with a more mature system of renewable energy development, the abundance of data, and clear parameters. However, this method does not provide an explanation that gives insights suitable for countries developing with limited resources. Use of a qualitative approach here provides an extended explanation and reasoning behind each dataset. Collating a deep understanding and explanation of the decision-making process from qualitative analysis is thus important. In the case of renewable energy selection, this information can be used not only as an input for evaluating and selecting the best renewable energy for an area, but it can also provide the thoughts, opinions, and essential information that can be useful for potential follow-up research, for example, research on policy development, which involves many parties and stakeholders. In addition to that, policymaking is often performed through a qualitative approach or science diplomacy. Therefore, for this research, the qualitative approach is preferable over the quantitative approach, and it is chosen to evaluate and to select the most appropriate renewable energy technology in Indonesia. The research could be a reference for performing renewable energy selection in other countries with limited resources in terms of data abundance, parameters, a less mature system of renewable energy development. Moreover, Indonesia is dealing with the power balance issue, where the stakeholders are mostly taken into account in terms of regulatory aspects. Because of that, stakeholders' recount plays an important role in the early stage of policymaking. This paper can be applied to the countries or locations that have similar characteristics to Indonesia.

3. Methodology

In terms of renewable energy evaluation and selection, the quantitative approach appears to be the most popular method used [2–10] as it offers a number of benefits, notably providing a more tangible data analysis thus preventing perceived biases [11]. The qualitative approach, on the other hand, takes the benefits of the flexibility of qualitative data and the level of feedback that is capable of explaining phenomena that are difficult to be quantified [11]. In the context of renewable energy selection, many unquantifiable parameters need to be considered, for instance, appropriate technology, political impacts, capacity building, stakeholder engagement, community acceptance, etc. [12]. Obtaining a deeper understanding of these unquantifiable parameters in renewable energy development can be done by incorporating the roles of stakeholders within this industry. Yudha and Tjahjono (2019) [13] performed a stakeholder analysis to map out the actors in the renewable and sustainable energy sector in Indonesia using PESTLE (Political, Economic, Social, Technological, Legal and Environmental) analysis. Each stakeholder encompasses specific areas, for example, the Ministry of Energy and Mineral Resources covers the political and economic aspect, while the public covers the social aspect. According to this study, there are numerous stakeholders in the renewable energy sectors that can provide thoughts and opinion in regard to renewable energy development in Indonesia [13]. However, incorporating all this input using a quantitative approach would be less effective than using a qualitative approach.

This research employed a qualitative approach based on a Focus Group Discussion (FGD) as a primary research method, complemented by the document analysis (Figure 1).

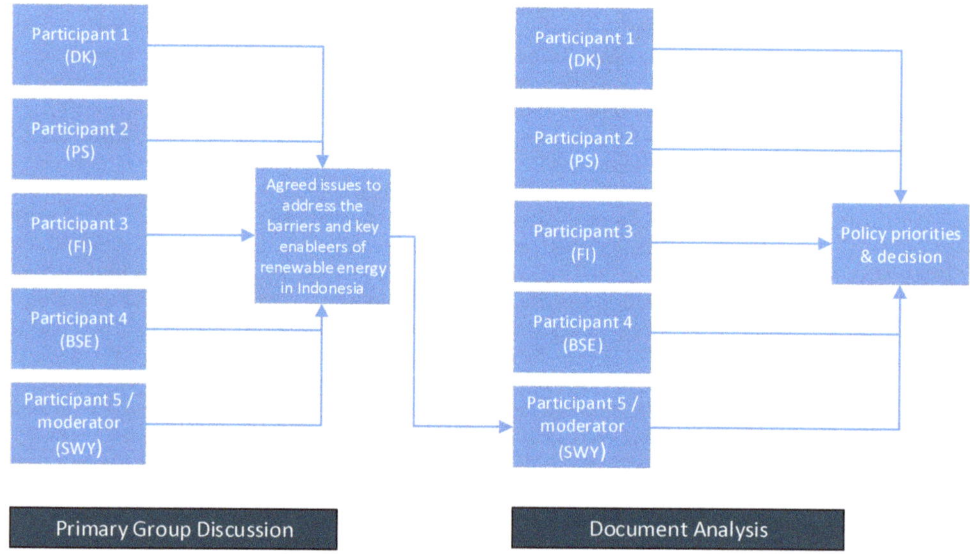

Figure 1. Flowchart of primary group discussion and document analysis.

This method is selected since it allows the researcher/interviewer to question several individuals systematically and simultaneously [14] or in this case, the stakeholders in sustainable and renewable energy. FGD, or also known as the group interviewing method can be based on structured, semi-structured, or unstructured interviews [15,16] and can generate data [17–19] which can be both descriptive and explanatory [20]. This method is frequently used as a qualitative approach to gain an in-depth understanding of complex issues [21]. Krueger (1994) [11] warned that there are advantages and disadvantages to conducting an FGD. The clear advantage of FGD is that it can capture real-life data within a social environment, and it has high flexibility, high face validity, and a speedy result, in addition to its low cost. However, when conducting FGDs, care must be taken when moderating it, as it can potentially be a problem when there are differences within the group, in which case it could lead to a great deal of difficulty in analysing the outcomes. Following the primary group discussion, document analysis was performed to corroborate the points raised during the FGD and to formulate the policy priorities on Indonesia's renewable energy. The process of policy development and confirmation in this FGD is shown in Figure 2.

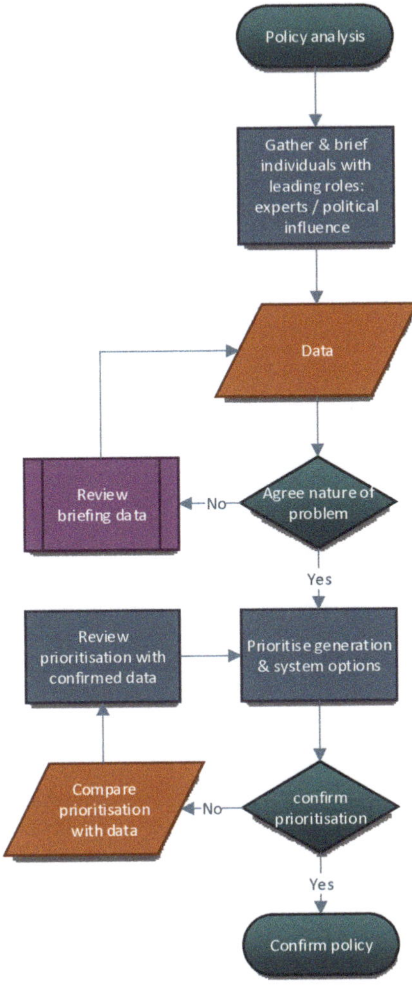

Figure 2. FGD process of policy development and confirmation.

The group size can range from as few as four to as many as 12 people within a conducive environment to engage in a guided discussion of a certain topic or issue [19], in this case, prospects and challenges in Indonesia's renewable energy development. The participating subjects are selected on the basis of relevance to the topic under study. In addition to this, special consideration is given to the role of the researcher/interviewer, as the moderator in the focus group process. As Babbie (2010) [22] comments: *"In a focus group interview, much more than in any other type of interview, the interviewer has to develop the skills of a moderator"*, thus there is a need to control the dynamics within the group.

The FGD was conducted on 15 January 2020, incorporating four participants as the sample population, plus one of the researchers, who acted as the moderator. The participants were chosen due to their expertise and experience in renewable energy development in Indonesia. Their representation encompasses the collective view of stakeholders identified by Yudha and Tjahjono [13], cutting across the political, economic, social, technological, legal, and environmental (PESTLE) aspects of the renewable energy development in Indonesia (Table 1). The participants represent the government of the Republic of Indonesia, comprised of the Ministry of Energy and Mineral Resources (executory) and the Special Task Force for Upstream Oil and Gas Business Activities (SKK Migas; regulatory); industry actors, comprised of the renewable energy division of the Indonesian House of Commerce, as well as a business actor and observer (member of public).

Table 1. Summary of FGD participants and their representation in the renewable energy sector in Indonesia (modified from Yudha and Tjahjono (2019) [13]).

Participant	Representing	Political	Economic	Social	Technology	Legal	Environment
DK	Ministry of Energy and Mineral Resources	✓	✓				
PS	Special Task Force for Upstream Oil and Gas Business Activities		✓		✓		✓
FI	Indonesian House of Commerce	✓	✓			✓	
BSE	Member of public			✓			
SWY	House of Representative, the Republic of Indonesia	✓	✓				✓

Without necessarily reducing the essence of the perspective of PESTLE analysis, the numbers of the participants were purposely kept to a small number to keep the forum conducive. This purposive sampling of participants was ensured to cater for the range of expertise at hand, to enable the document analysis that was also used as a basis of the analysis during the FGD.

The unit of analysis chosen, consistent to the previously applied PESTLE method of analysis, is the stakeholders' outlooks and responses with regards to renewable energy development in Indonesia. In this research, the FGD also served other purposes, such as developing specific insight and practical knowledge, as well as obtaining the feedbacks and propositions for renewable energy development, based on each stakeholder's perspectives.

In addition, the FGD was also open to members of the press, including the House of Representative's official press, covering the political, environment and legal aspect, to inquire and provide input to the participants during the questions and answer session.

4. Findings

The FGD began with an introductory opening by the moderator, who introduced the participants and laid out the overall theme of the discussion. Each of the participants was then given the time and the opportunity to share their recounts and outlooks on the renewable energy in Indonesia, including the challenges associated with renewable energy development in Indonesia, and the propositions for moving forwards and overcoming these challenges.

Following the discussion, all the participants proceeded to analyse each type of renewable energy in Indonesia, specifically wind, solar, ocean, biomass, hydropower, and geothermal. Using multiple secondary information and research documents as a basis of the analysis, the group then appraise these energy sources in terms of their potentials, current development, limitations, and opportunities for development. The main objective was to map out the progress of each type of renewable energy development. Unlike the previous session, in this session, every participant was encouraged to voluntarily give their opinion and constructively rebut each other in an open discussion. The outcome of both sessions will be used to pave the way forward to deciding the most suitable and feasible renewable energy type for further development in Indonesia.

4.1. A Snapshot of Renewable Energy Development in Indonesia

Following the introduction, the moderator described the precarious situation of Indonesia's inevitably declining fossil energy supply and the urgent need for a transition from fossil to renewable energy. Using this opening statement, the moderator then invited the FGD participants to voice out their views.

First to speak was DK, Secretary to the Director-General of New and Renewable Energy of the Ministry of Energy and Mineral Resources, representing the Director-General of New and Renewable Energy, RM. In general, DK outlined the Indonesian government's readiness in developing Indonesia's renewable energy sector, as well as provided the government perspectives as to the current situation and challenges of the industry. For example, DK highlighted the imperative of developing renewable energy in Indonesia, not only from the aspect of promoting environmental consciousness but also as a crucial element in the realisation of Indonesia's sorely needed and ambitious national electrification goal.

> *"Renewable energy is driven by its environmental aspect, given its environmentally friendly and clean nature. For us, aside from the environmental aspect (there is a presidential regulation already in effect concerning emissions), renewable energy contributes to reducing greenhouse gases".* (DK)

According to DK, what was deemed important from the point of view of energy and mineral resources was the ultimate goal of developing renewable energy is to help accelerate energy access for the large population of the nation who live in far-flung areas from the capital.

> *"In Java, Madura, or Bali, electricity is sufficiently supplied by PLN [State Electricity Company], but if we travel to the eastern regions and islands, there are still many of our brothers and sisters who have not yet benefited from electricity."* (DK)

Recent data indicated there are 12,500 villages in the eastern Indonesia regions have been electrified, but this figure is far from ideal as there are at least 2500 villages are still without any access to electricity. Responding to the queries from the audience, he further stated:

> *"We will carry out our village electrification program until 2019. The Director-General of New and Renewable Energy has been tasked by the Minister of Energy and Mineral Resources to assist in the provision of access to electricity sources".* (DK)

Furthermore, he mentioned that it has been promulgated in Government Regulation No. 79/2014, also known as the National Energy Policy, that renewable energy is targeted to comprise 23% of the primary energy mix by 2025. However, Indonesia currently has only 7% of renewable energy in its energy mix. To make up for the relatively significant difference in renewable energy composition within the energy mix, DK highlighted key renewable energy potentials as well as several ongoing government strategies for renewable energy development. These include, among others, Indonesia's 11,000 hectares of oil palm plantation which can be used for biodiesel.

The second speaker was PS, Deputy of Finance and Monetisation at the Special Task Force for Upstream Oil and Gas Business Activities (SKK Migas). PS made several key observations, firstly concerning present obstacles in Indonesia's oil and gas industry stemming from the global decline of oil prices. As a representative of the government regulatory body for oil and gas, PS also interestingly noted that

"... [the oil and gas] business has become over-regulated. Our oil and gas management practices are currently under scrutiny. Are our current regulations capable of providing incentives to bring results to our oil and gas resources?" (PS)

Concerning renewables, PS posed important statements from his observation:

" ... considering the fact that major corporate players in fossil-fuel energy have been uniformly and consistently diversifying their portfolios into the renewable energy sector, is our capability in managing the fossil energy business transferrable to the renewable energy business?" (PS)

This was subsequently responded by other participants proposing differing views. Nonetheless, they in the end reached a collective view, acknowledging that although the technicalities differ, long-time corporate players have arguably brought along their managerial and economic know-how of the fossil energy industry to leverage their business activities in renewables, particularly geothermal. It can therefore be concluded that learning from Indonesia's experience in managing fossil fuel, the country is hopeful to use its wealth of experience and know-how to manage the renewable energy sector. Exactly how these are going to be managed indeed needs further elaboration and thoughts.

Following PS, the third speaker was FI from the Renewable Energy Division of the Indonesian Chamber of Commerce (Kamar Dagang Indonesia or Kadin). FI expressed his disappointment with the present state of Indonesia's renewable energy sector. He then proceeded to identify the primary barrier of uptake from the private sector:

"Kadin is pushing forward in the renewable energy sector, but what is the obstacle? Regulation!" (FI)

Referring to tenurial disputes over several renewable energy projects, particularly geothermal, FI also mentioned that the development of renewable energy in Indonesia is often *"... hampered by NGOs, indigenous communities, and others."*

FI also hit on the barriers to renewable energy development that cause the slow uptake by investors, mentioning that

" ... feed-in-tariffs must also be fairer and involve stakeholders, not suddenly prescribed. This is indeed a problem in the renewable energy sector; as initial technologies are exorbitant, investors choose to wait and see." (FI)

Lastly, FI sees the need for a strong local manufacturing and supply chain, so that components for renewable energy would be cheaper to produce domestically rather than that of an import.

The final speaker was BSE, an observer of the renewable energy industry. BSE opened by hypothesising that energy sustainability is linearly correlated with welfare and the wealth of a nation. BSE proceeded to outline his solutions:

"The question that follows is how to satisfy the large amount of energy needed by low-cost, clean energy sources? We cannot do business as usual. We must push for breakthroughs." (BSE)

He asserted that electrification consists of three large components: power generation, transmission, and distribution. Therefore, it would be sensible to clearly split the responsibilities between those components. BSE argued that this was necessary to stimulate a healthy competition and to foster the core competence.

"PLN [State Electricity Company] should only focus on transmission and distribution of electricity, giving an opportunity for other parties, including private sectors to 'play' in the renewable energy generation arena, especially clean, large-scale power generation. There are only three options: hydro, geothermal, and nuclear." (BSE)

The moderator concluded this first session with a summary of key findings and lessons learnt.

Finding 1: There remain problems in both planning and implementation stages of renewable energy, mainly due to the [lack of] regulations, but this does not necessarily mean that both stages do not adhere to the same vision.

Finding 2: Lessons learnt from the oil and gas sector should later be transferred over for the future development of renewable energy, so as not to fall into the same pitfalls that impede and create inefficiency in the oil and gas sector.

4.2. Renewable Energy Types in Indonesia

The second session of the FGD analysed in detail several documents, mainly government policy analyses of various renewable energy sources. The moderator led the discussion (following the method illustrated in Figure 1) and asked the participants of FGD to comment on the suitability of the six sources of renewable energy and come up with a collective decision on the most suitable renewable energy source that Indonesia should develop going forward.

In order to hit 23% of renewables in the primary energy mix by 2025 and 31% by 2050, Indonesia has been attempting to achieve the targets [23]. Renewables accounted for just 15.7% of the country's primary energy mix by 2019, while fossil fuels accounted for 87.6% by 2019 [23,24]. Indonesia is a host to a variety of renewable energy sources, namely, wind energy, solar energy, ocean energy, biomass energy, hydropower, and lastly geothermal energy [25]. The development of each type of energy sources varied, and the FGD looked at each type of renewable energy, how they have been developed in Indonesia, and the challenges that each energy type encounters, with an expectation that the group came up with a collective view on the preferred renewable energy type.

4.2.1. Wind Energy

Wind energy is the type of energy that uses the conversion of wind speed into a useful source of power, and it can be used for multiple purposes, such as electricity generation, mechanical power wind turbines, water-driven wind turbines, or ship propellers [26]. Wind energy is one of the renewable energy innovations, as it does not contribute to air pollution or greenhouse gases and has a slight impact on the climate.

The use of wind power as an energy source in Indonesia has great potential for further growth, especially in coastal areas where the wind is quite abundant. According to the previous research, Indonesia has an estimated total potential for onshore wind energy of 9.3 GW. With the range of wind speeds between 2 and 6 m per second, Indonesia is suitable for installing small-scale (10 kW) and medium-scale (10–100 kW) wind-driven generators [27]. Indonesia has installed five units of windmill generators across the country each with a capacity of 80 kW and seven other units with the same capacity have been established in four places, North Sulawesi, the Pacific Islands, Selayar Island, and Nusa Penida in Bali [28]. Several wind-based power plants, namely, Sidrap, Tanah Laut, and Jenepoto, have currently been under construction, while the ones in Sukabumi, Banten, and Bantul are currently being considered to be placed under construction planning [28].

Price is one of the biggest obstacles that Indonesia faces in installing wind energy. The initial cost of developing wind energy is very high, particularly with the use of offshore wind turbines. IRENA suggests that it costs about US$3–US$4 million per megawatt (MW) to install offshore wind turbines compared to geothermal power plants that cost about US$2–US$3 million [29].

However, wind energy has some issues associated with the geographical locations the jeopardises the consistency of supply, as pointed out by a participant,

> " ... *indeed, we know that it [wind] is a promising form of energy in our country [Indonesia]. But its intermittent nature, makes it hard to provide electricity 24/7. It generates unstable and fluctuating electricity, and good source of wind power is only available in certain parts of our country [Indonesia]".* (FI)

Another difficulty is caused by the availability of other renewable energy sources available in Indonesia, such as geothermal, hydropower, biomass, and solar, which makes the cost analysis method challenging to carry out [29].

4.2.2. Solar Energy

Solar energy is a source of renewable energy that uses the power of the sun to produce electricity. Globally, solar energy has the fastest and highest growth compared to the rest of renewable energy types. It is currently considered as one of the most promising sources of clean, renewable energy and has greater potential than any other energy source to solve the world's energy problems.

As Indonesia is a tropical country situated on the equator line, the country has an abundant capacity for solar energy. Many areas of Indonesia have very strong solar radiation with average daily radiation of approximately 4 kWh/m^2 [27]. According to the report from Directorate of New and Renewable Energy (Ditjen EBTKE) in 2018, Indonesia has the potential to harness solar energy of up to 207.8 gigawatts peak (GWp) [30]. However, as of 2019, Indonesia has only installed 25.19 megawatts (MW), which is mainly used to meet energy demand in rural areas, including lighting for public service areas and places of worship [28]. The Institute for Energy Economics and Financial Analysis (IEEFA), an energy research institute, has reported that about 48 MW of solar power is currently under construction and about 326 MW is under construction planning [31].

Indonesia is still far behind other ASEAN countries in solar power utilisation. Thailand has 2.6 GW of installed solar capacity and the Philippines 868 MW of installed solar capacity [32]. Vietnam is also working on an expansion of more than 3000 MW in solar and wind power capacity by 2019 and 2020, and Malaysia is targeting an additional 3000 MW in 2020 [33].

There are many obstacles to the implementation of solar energy in Indonesia. One of them being the high prices of the solar cell, including the solar panels, inverters, batteries, wiring, installation, and battery storage [33]. This situation not only impacts financial institutions to provide resources to implement this program but also limits local communities' confidence and interests in using this system due to its exorbitant costs. Because of that, the implementation of this program is highly dependent on government funding. Despite its huge potential, a solar panel is very dependent on sunlight to effectively capture solar energy. Even though it can capture the energy during cloudy and rainy days, it can also give measurable technical effects on the energy system. Due to its intermittent nature, solar energy might be more suitable for a household scale, but it might not be the best option for a larger scale energy system. This is emphasised by one of the participants:

> *"It's fortunate that being in the equator, we [Indonesia] are blessed with warm climate all year round. We have the potential to develop solar energy. However, much like wind, it is intermittent and it [solar] will require storage systems to generate electricity after daylight".* (DK)

4.2.3. Ocean Energy

Ocean energy, or sometimes referred to as marine energy, is a type of energy that is carried by ocean's elements, such as ocean's tide, wave, salinity, and temperature. Each one of these elements can be exploited as different types of energy, namely, tidal energy, wave energy, salinity gradient energy, and ocean thermal energy conversion (OTEC). Wave energy converts the ocean waves to produce electricity, while tidal energy harvests the power that was produced during high and low tides [34]. Salinity gradient energy can generate electricity due to the difference in salt concentration between freshwater and seawater [35]. OTEC converts the temperature difference between cold seawater and warm surface seawater, typically at around 800 to 1000 m of depth, to produce electricity [36].

Indonesia is the world's largest archipelago with 70% of its area is covered by ocean, thus it has the largest potential of ocean energy. According to the research from the Indonesian Ocean Energy Association (INOCEAN), the potential of ocean energy resource that can be exploited is around 92.2 GW. The majority of its potential is coming from OTEC, with 43 GW of resource potential, followed by tidal and wave energy with 4.8 and 1.2 GW of resource potential, respectively. Despite its considerable amount of potential for harnessing ocean energy, the installed capacity for ocean energy is only 0.3 MW or 0.002% of the total energy use [24]. As of 2020, ocean energy is still under the stage of Research and Development, and this type of renewable energy has yet to be commercially developed in Indonesia [28].

> " . . . *unlike solar and wind, ocean energy can provide electricity throughout the entire day since it does not require the sun or the wind to harness electricity. Nevertheless, ocean energy is still rather far from the full commercialisation in Indonesia*". (DK)

4.2.4. Biomass

Biomass energy is one of the types of natural renewable energy that are generated from organisms, and it mostly comes from farm crops and residues, forest waste, farm foods, and animal waste [37]. Biomass is the only renewable energy that can be used to generate three types of fuels: liquid, solid, and gas. A proper biomass energy development could also reduce not only energy issues but also waste management issues.

As an agricultural country, the potential of biomass resources in Indonesia is relatively abundant. According to the report from Directorate of New and Renewable Energy in 2018, Indonesia has a potential of harnessing 32.6 GW of biomass energy [24]. However, only 167.54 MW of biomass energy in Indonesia has now been properly exploited. Currently, Indonesia's estimated total biomass production is around 146.7 million tons, which is equivalent to 470 gigajoules per year (GJ/y) [38]. Most of the biomass energy source comes from the rice residue and rubberwood, which contributes to 150 GJ/y and 120 GJ/y, respectively [38]. These are followed by sugar residues (78 GJ/y), palm oil residues (67 GJ/y), and other types of residues (20 GJ/y) [38]. Such biomass sources can help supply both heat and electricity to rural households and sometimes small-scale industries.

Bioenergy, in general, has faced similar problems as other renewable energy sources. One of the primary issues is the high investment cost for bioenergy installations, as claimed by one participant:

> " . . . *while Indonesia has an abundant amount of biomass energy that can be utilised to generate liquid, solid, and gas, it is still expensive to invest in and the lack of support from financial institutions has hindered its growth*". (DK)

Part of the reason is that bioenergy deployment feasibility studies are mostly not attractive for bank loans [39]. From the perspective of technology, the reliability and efficiency of the existing technology for biomass energy is still lower than those of fossil fuels [39], hence hindering bioenergy development.

4.2.5. Hydropower

Hydropower energy is a type of renewable energy source that extracts energy from flowing water, to produce electricity. As a potential future source of energy, hydropower has become an increasingly attractive choice for small capacity of the renewable energy. This type of energy, as well as other renewable energy sources, are the clean energy sources as they emit a negligible amount of greenhouse gas.

Hydropower is one of Indonesia's large-scale, commercially viable, renewable energy sources. According to the report from the Nippon Koei, the hydropower capacity in Indonesia is projected at around 26,321 MW [40]. Currently, the installed capacity of hydropower is 4938.64 MW from various hydropower plants all across the nation; the large-scale plants are operated by the state-owned electricity company (PLN), and many small-scale plants are owned by small enterprises. According to 2019–2028 Electricity Supply Business Plan issued by PLN, it has been reported that 5956 MW of hydropower capacity is currently under construction scattered in many places, while the new 16,027 MW of potential capacity has just been built as of 2018 and is considered for being placed under construction planning [24].

To date, the hydropower energy in Indonesia is still the most established and the most utilised small-scale renewable energy source, particularly for the rural areas. Hydropower systems provide unique operational versatility in that they can adapt to sudden fluctuating demand for electricity, which means that it can be tailored to satisfy market demand [41]. Hydropower is also able to provide the supports for the development of other renewable energy sources, for example, its storage capacity and flexibility can be the most cost-effective and efficient to support the utilisation of intermittent renewable energy sources such as solar and wind energy. Hydropower can generate vast quantities of energy, and the price is relatively stable, as it is less affected by market price fluctuations such as oil and gas, although the price advantage is proportional to the capacity of the plant, i.e., relatively small capacity. There is, however, a major drawback of hydropower development as it is heavily dependent on the geographical features (i.e., large rivers) to generate electricity. Therefore, large-scale utilisation of hydropower is only limited to certain places with specific geographical features, making micro-hydropower-with a lower price advantage-a more viable option, as asserted by FI:

> " … no, the majority of them [sources] are only suitable for small-scale power plants. Only certain areas of Indonesia have the full capability to generate large scale electricity". (FI)

4.2.6. Geothermal

Geothermal energy is the type of renewable energy that uses heat derived from the sub-surface of the Earth, which can be transmitted in the form of hot steam, hot water, or a mixture between both forms. Nowadays, it has been one of the most important alternatives for energy sources with significant growth potential. It not only provides alternative energy but also helps to reduce the effects of global warming and the risks to public health due to the use of conventional energy sources, as well as our dependence on fossil fuels. Geothermal energy may be used for district heating purposes or harnessed to produce renewable electricity, depending on its characteristics. Lower enthalpy type of geothermal is mostly suitable for direct use (e.g., room heater, tourism, agriculture/agro-industry, and fisheries), while medium to high enthalpy type of geothermal can be used for generating electricity, which is typical for the regions with active tectonics [42]. Indonesia has varying types of geothermal energy that can be utilised for both direct heating and generating electricity [43].

Indonesia is one of the countries in the world that falls on the "ring of fire", which traversed around the edges of the Pacific Ocean and is responsible for most active volcanoes and earthquakes. Due to its tectonic setting, Indonesia is a host to most of these active volcanoes, which accounted for 117 active volcanoes in total [44,45]. These active volcanoes are distributed in Sumatra, Java, Nusa Tenggara, Sulawesi, and Maluku. Consequently, Indonesia has a considerable amount of high heat flow, which makes it one of the countries with a large potential for geothermal energy.

According to studies, Indonesia has the world's largest geothermal energy potential, accounting for about 40% of the world's potential or approximately at 28,617 MW [42]. Most of these potential energy resources and reserve are distributed in several regions in Indonesia. Sumatra and Java have currently the highest total potential energy, which accounts for 12,760 and 9717 MW, respectively [42]. The rest of the potential are distributed in many other regions, namely Bali, Nusa Tenggara, Sulawesi, Maluku, Kalimantan, and Papua [42,43].

According to FI, geothermal, when compared to other sources of renewable energy except nuclear, can guarantee the provision of electricity at a stable rate throughout the entire year without being affected by weather patterns and conditions. He further added:

> "The cost of geothermal technology in the future will be increasingly competitive and is expected to continue to fall. Thus, the optimisation of geothermal energy in Indonesia is

vital in helping the government achieve its renewable energy target and reduce greenhouse gas emissions". (FI)

Despite having a considerable amount of potential, geothermal energy utilisation in Indonesia, especially for the electricity generation, is not quite optimal. Currently, the geothermal energy in Indonesia that has been utilised for generating electricity is 2130.6 MW [46], making it the second-largest country with installed geothermal capacity, putting Philippines in third place with 1868 MW of installed capacity and following United States with 3639 MW of installed capacity [47]. Most of the installed geothermal capacity in Indonesia comes from the geothermal power plant in Java, which accounts for a total of 1253.8 MW of installed capacity, followed by Sumatra with 744.3 MW, Sulawesi with 120 MW, and lastly, Nusa Tenggara with 12.5 MW of installed capacity [46,48]. The development of geothermal energy is still yet to be done in many other regions in Indonesia.

Up until 2019, the utilisation of geothermal in Indonesia was only 2130.6 MW out of 56,509.53 MW, or around 3.77% of the total energy utilisation [24]. This number is still very small compared to the Philippines that have already 44.5% of its energy use from geothermal energy [28]. There are a few factors that have been the reason for Indonesia's lagging development of geothermal energy utilisation. Exploration and resource commercialisation of geothermal utilisation is a costly process, in addition to a small market for the resource. Limited investment financing schemes for geothermal development has also contributed to the stagnation in this industry.

According to PS, Deputy of Finance and Monetisation at the Special Task Force for Upstream Oil and Gas Business Activities (SKK Migas),

" ... *geothermal development has a very unique characteristic, since it includes the upstream phase, similar to that of oil and gas sector".* (PS)

He further underlined the transferability of management know-how from the fossil fuel economy to the geothermal business. In essence, the pitfalls of Indonesia's oil and gas industry retrospectively provide lessons learnt for the management of geothermal. Although these two sectors quite different in terms of the nature of the commodities involved, PS believes that the technical management know-how from the oil and gas industry should be transferred and refined.

There is a difference between the outcome of the renewable energy selection provided by several authors in their previous works and the renewable energy selection as a result of the discussion between the stakeholders. For example, Tasri and Susilawati (2014) [10] evaluated that hydropower is more suitable than geothermal energy. On the other hand, earlier research on renewable energy, Rumbayan and Nagasaka (2012) [9], evaluated that geothermal, solar, and wind energy are the most suitable renewable energy. In this research, having considered many factors from the stakeholders' point of views, such as its availability; technology; and operational, financial, and market situation, geothermal energy is the most suitable renewable energy in Indonesia. Therefore, future research focusing on the geothermal energy development would certainly enhance the renewable energy development Indonesia. However, since the stakeholders contended that the lagging development of geothermal energy sources is mainly caused by the hefty initial cost, government intervention is needed, e.g., in the form of financing schemes.

Finding 3: Geothermal stood out during the FGD as the most promising renewable energy source that Indonesia should develop.

Finding 4: Exploration of geothermal utilisation needs an intervention from the government in the form of financing schemes, so as to alleviate the burden of upfront investment.

5. Discussion

Indonesia has a few choices when it comes to developing renewable energy sources. Having the capability to develop all types of renewable energy to a point where it cannot only fulfil the whole country's energy demand but also fully liberate the country out of fossil fuel dependency would be an ideal case. Although achieving this would require long years of huge effort and costly processes, having this ideal scenario would lead us to significantly contributing to reducing carbon emission and increasing the economic growth of the country. Reflecting on the participants' feedbacks reported during the FGD, we can glean from this dialogue that different stakeholders have alluded differing inherent interests and point of views based on their respective institutions. The differences are mainly related to the key enablers and barriers of renewable energy development in general leading to selection of the most suitable renewable energy to focus on future development.

5.1. Barriers and Key Enablers

Indonesia has tremendous renewable energy capacity which is still underutilised by the Indonesian power sector. The long dependency on fossil fuels, particularly coal, has proved difficult to break as the image of coal as cheap energy while renewable sources remain as expensive technologies. Although some steps to enhance renewable energy have been placed since many years ago, the Indonesian renewables sector has yet to take off.

The development of renewable energy sources in Indonesia has not been without any barriers, as it has encountered quite a few challenges from the operational, financial regulatory challenges. The operational challenge is mostly related to the nature of each type of renewable energy sources, which includes the availability and reliability issues. Financial challenge is mostly related to the exorbitant initial cost of installation, and it has been one of the major hurdles for the development of renewable energy, regardless of the type of renewable. Lastly, the regulatory challenge is viewed as the primary obstacle in the energy transition and renewable development, especially for the private sector, which hampers the development process, i.e., the regulation that define the conservation area makes it impossible for the exploration of renewable energy in the area with high potentials. Having analysed the barriers, the obstacles are quite evident in both the planning and implementation stage of renewable energy development. However, this does not necessarily mean that both stages of development do not adhere to the same vision. Therefore, regulation and policy refinement are indeed necessary, thus becoming the most important key enablers, as they allow us to tackle multiple present barriers effectively.

According to the stakeholders' points of view, some of the important key enablers are classified as follows, so identifying these enablers is of paramount importance for the transition to renewable and sustainable energy technologies.

1. The availability of an institutional framework of national targets and development plans that transcend organisational leadership is one of the most important initial key enablers, as it reflects the government's commitment to renewable energy endeavour. The imperative development of renewable energy in Indonesia needs to be viewed not only from the aspect of promoting environmental consciousness but also as a crucial element in the realisation of Indonesia's sorely needed and ambitious national electrification goal, as well as the national primary energy mix as a tangible target for now, which is achieving 23% of renewables in the primary energy mix by 2025 and 31% by 2050.

2. Focusing on the forward-thinking scheme of supply chain management for the manufacturing of renewable energy is viewed as another significant key enabler. Such a scheme needs to consider how local industries can effectively source the materials and technology needed for Indonesian-made renewable energy supply and market. Especially in the presently liberalised global economy, Indonesia needs to strengthen its local leverage in terms of competitiveness of goods and trade. The ultimate and obvious aim of this endeavour is to make local renewable energy supplies and technology cheaper to produce domestically than to import.

3. Refining the regulation and policy is one of the most vital key enablers for the energy transition. The courage of institutions and organisations is essential to break down

outdated and inadequate regulations, as well as to design new regulations and policy that can accommodate the interests of all relevant stakeholders. Therefore, having higher regulations covering the renewable energy sector, such as Renewable Energy Bill, Presidential Regulation, and Governmental Regulation, would have stronger impacts on renewable energy development.

4. Focusing on the clean, large scale types of renewable energy for power generation, for instance, hydropower and geothermal energy. Nuclear power as another new energy source has also been considered and is still categorised as a viable option, considering all the relevant safety and technical concerns being put in place.

5.2. Selection of Renewable Energy Type for Development

Focusing on developing the most suitable type of renewable energy would be an important first step towards better utilisation of renewable energy. Here, we will have a look at each type of renewable energy and use a comparative analysis to decide the most suitable renewable energy to focus on developing in Indonesia, considering the limitation and opportunity that each type of renewable energy has to offer, according to the expertise' inputs during the FGD.

Ocean energy is the least developed renewable energy in Indonesia. Despite having a huge potential for developing ocean energy, no commercial-scale ocean power plants now exist as it is now still under the research and development stage. According to 2019–2028 Electricity Supply Business Plan issued by PLN, there has been no technology manufacturer for ocean energy that has proven its reliability to operate commercially for at least 5 years. The development will be reconsidered once the technology is mature enough to generate electricity on a commercial scale.

Wind and solar energy can provide not only alternative sources for renewable energy but they can also give us a number of environmental benefits, as they produce a negligible amount of carbon footprints. However, the development of these renewable energy sources has major drawbacks. The initial deployment process of wind or solar energy can be quite costly, and it hinders the investment due to its exorbitant costs. Even though both solar and wind energy should be seen as low-risk investments with potentially major returns, they are hefty investments nonetheless. Moreover, both wind and solar energy are intermittent by nature, as they are heavily dependent on the weather. Therefore, the electricity generated by these renewables is most likely to be fluctuating, and it can potentially become a problem. Fluctuating supply of electricity is not a reliable power supply, as it is not best-suited for providing base-load. Furthermore, fluctuating electricity, particularly in the solar panel system, may have measurable effects on the power instrument. Therefore, relying solely only on solar and wind energy may not be the best option for now, especially for a massive commercial scale, but they may be suitable for smaller-scale development.

Biomass energy development is currently in a better state than previous sources of renewable energy now that the Minister of Energy and Mineral Resources Regulation (*Permen ESDM*) Number 50 of 2017 has been issued. Not only does it contribute to meeting the energy demands, but proper biomass energy development also allows us to cope with better waste management. However, the expensive upfront cost to get the power plants up and running has been the most common issue in developing the renewables, which also applies to biomass energy. In addition to the exorbitant upfront cost, biomass development would require additional costs associated with the extraction, transport, and storage of biomass prior to the generation of electricity. Biomass energy plants also require quite a bit of space with constant supply for biomass resources, which might not be suitable for big cities. Therefore, the development of biomass energy may be more suitable for underdeveloped, isolated region in the country.

Hydropower is by far the most established and the most utilised renewable source of energy. Hydropower has been operating as a commercial-viable, massive scale. Due to its operational flexibility, hydropower can adapt to sudden fluctuating demand, and it can be tailored to satisfy market demand, thus making it a very reliable source of energy.

Its development has been very steadily growing every year, with numerous ongoing projects. However, being heavily dependent on the geographical features (i.e., large rivers) to generate electricity has been a major drawback of hydropower development. Large-scale utilisation of hydropower is only limited to certain places with specific geographical features. Therefore, focusing more on a different type of renewable with huge potential for growth and suitability that needs further development might be the best option and more necessary in order to meet the 2025 national energy mix target.

Indonesia's abundant geothermal potential is not questionable. Despite having a considerable amount of potential, its utilisation, especially for the electricity generation, is not quite optimal. By 2020, only 2130.6 MW out of 28,617 MW had been properly utilised. Unlike wind and solar energy, geothermal energy is not an intermittent source of energy, and it has very high-capacity factors; thus, it can be a reliable source of energy.

Geothermal energy utilisation has very unique attributes, which can be viewed as a promising opportunity when it comes to massive development. In contrast to other renewable energy projects, geothermal power projects must include upstream activities to verify the resource and to determine the most favourable location for development. This upstream phase is very similar to the upstream process of an oil and gas field or that of a coal mine. Such a unique attribute may allow us to have a transferability of management know-how from the fossil fuel sector to the geothermal energy business. The knowledge transfer may be able to help us to reduce the hefty risk that can come during geothermal development, for instance, with the lesson from the oil and gas sector, the declining phase of production might be recognised earlier, giving us the time to work on the operational and managerial solution to avoid the pitfall. The nature of the geothermal operation and the availability of knowledge transfer can reflect on the managerial maturity that the geothermal sector can offer, compared to the rest of the renewable energy sources, thus becoming its greatest opportunity. It is also implied that geothermal energy, far from being a simple market competitor, is internalised within the oil and gas industry as the inevitable way forward, hence the emphasis on transferability of knowledge.

The geothermal energy sector has a fair share of obstacles when it comes to its development. Similar to the other renewables, the hefty cost has been one of the biggest obstacles that Indonesia has faced in the development of geothermal energy. The exploration and resource commercialisations are both costly processes, which makes it very reliance on heavy investments. Regulations have also become one of the major challenges for the geothermal sector, for instance, the land dispute caused by wavering regulations are most likely to hinder its development. There might have been many more underlying issues with regard to geothermal energy development; therefore, a further investigation on the geothermal supply chain trajectory is necessary to address the potential barriers in its development. Furthermore, developing a set of policies also can be done to bridge these potential barriers and to ultimately enhance the pace and magnitude of geothermal energy development [49]. Despite the obstacles, just as much as the other types of renewable energy, the high potential, reliability, and opportunity that geothermal energy can offer make it the most suitable renewable energy source to develop.

6. Conclusions

The FGD has provided unique inputs to this research via a combination of subjective and institutional leanings and experiences, particularly in identifying key enablers and barriers of renewable energy development. Throughout the course of FGD, the barriers were actually expected as there is an evident contradiction between policy and business, particularly in the planning and implementation stages. However, this does not necessarily mean that both stages do not adhere to the same vision. In particular, the private sector highlighted the lack of representative regulations truly needed to boost private participation in renewable energy development. The key enablers include constructing the national target as a framework and renewable development plans, building a forward-thinking scheme of supply chain management of renewable development, and regulation and policy refinement.

Regarding the renewable energy selection, geothermal energy has been considered as the most suitable and feasible renewable energy source to focus on further development in Indonesia. Not only does it have a considerable amount of potential for generating electricity, but it also has the most unique characteristics out of all the other options of renewable energy. The FGD has identified that the geothermal energy projects must include a set of supply chain trajectories, which include the upstream, midstream, and downstream. The upstream phase in this supply chain trajectory is in fact similar to that of the oil and gas sector. This allows us to perform the transferability of management know-how from the fossil fuel sector. The lessons learnt from the oil and gas sector should later be transferred over for the future development of geothermal energy, so as not to fall into the same pitfalls that impede and create inefficiency in the oil and gas sector. This implies a growing sense of corporate and institutional responsibility within the oil and gas sector, one that is visionary and should be capitalised on. Despite the tensions and disagreements between stakeholders, all parties agreed that the development of renewable energy, particularly in geothermal energy, should continue to be supported for the good of the public as well as the market. In order to do this, further investigation on geothermal supply chain trajectories needs to be done to identify the potential barriers and to design a set of policies that can bridge these barriers, thus enhancing the pace and magnitude of renewable energy development, or more specifically, geothermal energy.

Aside from directly absorbing the bold aspirations from each stakeholder, 'reading between the spoken lines' has provided plenty of room for abstraction and further inquiry. Most importantly, the FGD has succeeded in answering the four research questions posed at the beginning of this paper, meaning that we have acquired both (1) stakeholders' recount and outlook of institutional and market challenges associated with renewable energy development in Indonesia, as well as their (2) responses for overcoming the challenges, (3) their collective views on the most feasible renewable energy to develop in the near future, and, lastly, (4) the propositions to support the development of that particular renewable energy, i.e., geothermal, in Indonesia.

We recognise that, despite the benefits of FGDs, they have certain limitations, i.e., that it is possible that the participants may be hesitant to openly express their opinions when it is a sensitive topic or alternatively be dominant within the group about the topic being debated. Another limitation is that it may not be a true representation of the target group. Conscious of these limitations, we were diligent in selecting the participants plus followed specific sensitive questions individually.

Author Contributions: Conceptualization, S.W.Y. and B.T.; methodology, S.W.Y., B.T., and P.L.; validation, S.W.Y., B.T., and P.L.; formal analysis, S.W.Y., B.T., and P.L.; investigation, S.W.Y.; writing, S.W.Y. and B.T.; writing—review and editing, B.T. and P.L.; supervision, B.T. and P.L.; project administration, S.W.Y. All authors have read and agreed to the published version of the manuscript.

Funding: This research received no external funding.

Institutional Review Board Statement: Not applicable.

Informed Consent Statement: Not applicable.

Data Availability Statement: Not applicable.

Acknowledgments: The authors thank the reviewers for their constructive comments in improving the quality of this paper.

Conflicts of Interest: The authors declare no conflict of interest.

References

1. Gielena, D.; Boshella, F.; Saygin, D.; Bazilian, M.D.; Wagnera, N.; Gorinia, R. The role of renewable energy in the global energy transformation. *Energy Strategy Rev.* **2019**, *24*, 38–50. [CrossRef]

2. Al Irsyad, M.I.; Halog, A.B.; Nepal, R.; Koesrindartoto, D.P. Selecting Tools for Renewable Energy Analysis in Developing Countries: An Expanded Review. *Front. Energy Res.* **2017**, *5*, 1–13. [CrossRef]

3. Da Silva, C.G. Renewable energies: Choosing the best options. *Energy* **2010**, *35*, 3179–3193. [CrossRef]

4. Emir, F. Selecting Renewable Energy Sources for Small Islands Using Analytical Hierarchy Process (AHP). 2013. Available online: https://www.researchgate.net/publication/312630832_Selecting_Renewable_Energy_Sources_for_Small_Islands_Using_Analytical_Hierarchy_Process_AHP (accessed on 7 January 2021).

5. Sliogeriene, J.; Turskis, Z.; Streimikiene, D. Analysis and Choice of Energy Generation Technologies: The Multiple Criteria Assessment on the Case Study of Lithuania. *Energy Procedia* **2013**, *32*, 11–20. [CrossRef]

6. Ahmad, S.; Tahar, R.M. Selection of renewable energy sources for sustainable development of electricity generation system using analytic hierarchy process: A case of Malaysia. *Renew. Energy* **2014**, *63*, 458–466. [CrossRef]

7. Li-Bo, Z.; Tao, Y. The Evaluation and Selection of Renewable Energy Technologies in China. *Energy Procedia* **2014**, *61*, 2554–2557. [CrossRef]

8. Madhuri, S.; Yadav, A.; Hiwarkar, D. Selection of Appropriate Renewable Energy. Resources for Uttar Pradesh by using Analytical Hierarchy Process (AHP). *Int. J. Innov. Res. Sci. Eng. Technol.* **2017**, *6*, 2580–2587.

9. Rumbayan, M.; Nagasaka, K. Prioritization decision for renewable energy development using analytic hierarchy process and geographic information system. In Proceedings of the 2012 International Conference on Advanced Mechatronic Systems, Tokyo, Japan, 18–21 September 2012; pp. 36–41.

10. Tasri, A.; Susilawati, A. Selection among renewable energy alternatives based on a fuzzy analytic hierarchy process in Indonesia. *Sustain. Energy Technol. Assessments* **2014**, *7*, 34–44. [CrossRef]

11. Krueger, R.A.; Casey, M.A. *Focus Groups: A Practical Guide for Applied Research*; Sage Publications: Thousand Oaks, CA, USA, 1994.

12. Sovacool, B.K. A qualitative factor analysis of renewable energy and Sustainable Energy for All (SE4ALL) in the Asia-Pacific. *Energy Policy* **2013**, *59*, 393–403. [CrossRef]

13. Yudha, S.W.; Tjahjono, B. Stakeholder Mapping and Analysis of the Renewable. Energy Industry in Indonesia. *Energies* **2019**, *12*, 602. [CrossRef]

14. Glass, G.V. Primary, Secondary, and Meta-Analysis of Research. *Educ. Res.* **1976**, *5*, 3–8. [CrossRef]

15. Rahman, M.S. The Advantages and Disadvantages of Using Qualitative and Quantitative Approaches and Methods in Language "Testing and Assessment" Research: A Literature Review. *J. Educ. Learn.* **2017**, *6*, 102–112. [CrossRef]

16. Kvale, S. *Interviews: An Introduction to Qualitative Research Interviewing*; Sage: Thousand Oaks, CA, USA, 1996.

17. Carey, M.A. The group effect in focus groups: Planning, implementing, and interpreting focus group research. In *Critical Issues in Qualitative Research Methods*; Morse, J.M., Ed.; Sage: Thousand Oaks, CA, USA, 1994; pp. 225–241.

18. Stevens, P.E. Focus groups: Collecting aggregate-level data to understand community health phenomena. *Public Health Nurs.* **1996**, *13*, 170–176. [CrossRef] [PubMed]

19. McDaniel, R.W.; Bach, C.A. Focus group research: The question of scientific rigor. *Rehabil. Nurs. Res.* **1996**, *3*, 53–59.

20. Nyumba, T.O.; Wilson, K.; Derrick, C.J.; Mukher, N. The use of focus group discussion methodology: Insights from two decades of application in conservation. In *Methods in Ecology and Evolution*; John Wiley & Sons Ltd.: Hoboken, NJ, USA, 2017.

21. Miles, M.B.; Huberman, A.M. *Qualitative Data Analysis: An Expanded Sourcebook*; Sage: Thousand Oaks, CA, USA, 1994.

22. Babbie, E.R. *The Practice of Social Research*; Wadsworth Cengage: Belmont, CA, USA, 2010.

23. ESDM. *General Plan for National Energy*; Ministry of Energy and Mineral Resources (ESDM): Jakarta, Indonesia, 2017.

24. PLN. *Electricity Supply Business Plan from 2019 to 2028*; State-owned Electricity Company (PLN): Jakarta, Indonesia, 2019.

25. Simamora, P.; Mursanti, E.; Giwangkara, J.; Arinaldo, D.; Tampubolon, A.P.; Adiatma, J.C. Igniting A Rapid Deployment of Renewable Energy in Indonesia: Lessons Learned from Three Countries, Institute for Essential Services Reform (IESR). 2019. Available online: https://iesr.or.id/en/pustaka/igniting-a-rapid-deployment-of-renewables-in-indonesia (accessed on 3 January 2021).

26. Holttinen, H. Design and operation of power system with large amounts of wind power. In Proceedings of the Global Wind Power Conference, Adelaide, Australia, 18–21 September 2006.

27. *Indonesia Energy Outlook & Statistic*; Energy Reviewer; University of Indonesia: Depok, Indonesia, 2006.

28. Hasan, M.H.; Mahlia, T.M.I.; Nur, H. A review on energy scenario and sustainable energy in Indonesia. *Renew. Sustain. Energy Rev.* **2012**, *16*, 2316–2328. [CrossRef]

29. IRENA. Renewable Energy and Jobs Annual Review 2019, International Renewable Energy Agency IRENA. 2019. Available online: https://www.irena.org/publications/2019/Jun/Renewable-Energy-and-Jobs-Annual-Review-2019 (accessed on 8 March 2021).

30. DEN. *Indonesia Energy Outlook 2019*; Secretariat General National Energy Council (Dewan Energi Nasional/DEN): Jakarta, Indonesia, 2019.

31. IEEFA Asia Pacific. Indonesia's Solar Policies–designed to fail? Institute for Energy Economics and Financial Analysis (IEEFA), 2019. Available online: https://ieefa.org/ieefa-report-indonesias-solar-policies-designed-to-fail/ (accessed on 12 December 2020).

32. PwC. Renewable Energy Development: Large Potential of Solar Power Underoptimised, PricewaterhouseCoopers (PwC). 2019. Available online: https://www.pwc.com/id/en/media-centre/infrastructure-news/march-2019/large-potential-of-solar-power-underoptimised.html (accessed on 12 November 2020).

33. Dang, M.Q. Potential of Solar Energy in Indonesia. 2017. Available online: https://www.researchgate.net/publication/32484061 1_Potential_of_Solar_Energy_in_Indonesia (accessed on 15 January 2021).

34. Purba, N.P.; Kelvin, J.; Annisaa, M.; Teliandi, D.; Ghalib, K.G.; Ayu, I.P.R.; Damanik, F.S. Preliminary Research of Using Ocean Currents and Wind Energy to Support Lighthouse in Small Island, Indonesia. *Energy Procedia* **2014**, *47*, 204–210. [CrossRef]

35. IRENA. Salinity Gradient Energy Technology Brief, International Renewable Energy Agency IRENA. 2014. Available online: https://www.irena.org/publications/2014/Jun/Salinity-Gradient (accessed on 3 January 2021).

36. Siahaya, Y.; Salam, L. Ocean Thermal Energy Conversion (OTEC) Power Plant and its Byproducts Yield for Small Islands in Indonesia Sea Water. In Proceedings of the ICCHT-5th International Conference on Cooling and Heating Technologies, Bandung, Indonesia, 9–11 December 2010.

37. Fungenzi, T. Biomass as an Opportunity to Solve Indonesia's Energy Challenge. 2015. Available online: https://www.researchgate.net/publication/280075489_Biomass_as_an_opportunity_to_solve_Indonesia\T1\textquoterights_energy_challenge (accessed on 20 February 2021).

38. Abdullah, K. Biomass Energy Potentials and Utilization in Indonesia. 2006. Available online: https://www.researchgate.net/publication/228402035_Biomass_Energy_Potentials_And_Utilization_In_Indonesia (accessed on 26 January 2021).

39. Sapuan, D.; Aditya, W. Challenges and Policy for Biomass Energy in Indonesia. *Int. J. Bus. Econ. Law* **2018**, *15*, 41–47.

40. Nippon Koei. Project for the Master Plan Study of Hydropower Development in Indonesia. 2011. Available online: https://openjicareport.jica.go.jp/pdf/12037610.pdf (accessed on 10 December 2020).

41. Gokhale, P.; Date, A.; Akbarzadeh, A.; Bismantolo, P.; Suryono, A.F.; Mainil, A.K.; Nuramal, A. A review on micro hydropower in Indonesia. *Energy Procedia* **2017**, *110*, 316–321.

42. Nasruddin, N.; Alhamid, M.I.; Daud, Y.; Surachman, A.; Sugiyono, A.; Aditya, H.B.; Mahlia, T.M.I. Potential of geothermal energy for electricity generation in Indonesia: A review. *Renew. Sustain. Energy Rev.* **2016**, *53*, 733–740. [CrossRef]

43. Surya, D.; Harsoprayitno, S.; Setiawan, B.; Hadyanto; Sukhyar, R.; Soedibjo, A.W.; Ganefianto, N.; Stima, J. Geothermal Energy Update: Geothermal Energy Development and Utilization in Indonesia. In Proceedings of the World Geothermal Congress 2010, Bali, Indonesia, 25–29 April 2010.

44. Hamilton, W.B. Tectonics of the Indonesian Region. *US Govt.* **1979**. [CrossRef]

45. Manalu, P. Geothermal development in Indonesia. *Geothermics* **1988**, *17*, 415–420. [CrossRef]

46. PSDG. Kapasitas Terpasang Pembangkit Panas Bumi Indonesia, Pusat Sumber Daya Mineral Batubara dan Panas Bumi (PSDG). 2019. Available online: http://psdg.bgl.esdm.go.id/index.php?option=com_content&view=article&id=1306&Itemid=610 (accessed on 16 December 2020).

47. ThinkGeoEnergy. The Top 10 Geothermal Countries 2019—Based on Installed Generation Capacity (MWe). 2019. Available online: https://www.thinkgeoenergy.com/the-top-10-geothermal-countries-2019-based-on-installed-generation-capacity-mwe/ (accessed on 27 January 2021).

48. EBTKE ESDM. *Annual Performance Report 2019*; Direktorat Jenderal Energi Baru Terbarukan ESDM: Jakarta, Indonesia, 2019.

49. Halldorsson, A.; Svanberg, M. Energy resources: Trajectories for supply chain management. *Supply Chain Manag. Int. J.* **2013**, *18*, 66–73. [CrossRef]

MDPI
St. Alban-Anlage 66
4052 Basel
Switzerland
Tel. +41 61 683 77 34
Fax +41 61 302 89 18
www.mdpi.com

Energies Editorial Office
E-mail: energies@mdpi.com
www.mdpi.com/journal/energies

www.ingramcontent.com/pod-product-compliance
Lightning Source LLC
LaVergne TN
LVHW071301170726
843515LV00006BA/1454